The Lebesgue Integral for Undergraduates

© 2015 by
The Mathematical Association of America (Incorporated)
Library of Congress Control Number: 2015936109
Print ISBN: 978-1-93951-207-9
Electronic ISBN: 978-1-61444-620-0
Printed in the United States of America
Current Printing (last digit):
10 9 8 7 6 5 4 3 2

The Lebesgue Integral for Undergraduates

William Johnston
Butler University

 **MAA**

Published and distributed by
The Mathematical Association of America

Council on Publications and Communications
Jennifer J. Quinn, *Chair*

Committee on Books
Fernando Gouvêa, *Chair*

MAA Textbooks Editorial Board
Stanley E. Seltzer, *Editor*

Matthias Beck
Richard E. Bedient
Otto Bretscher
Heather Ann Dye
Charles R. Hampton
Suzanne Lynne Larson
John Lorch
Susan F. Pustejovsky

MAA TEXTBOOKS

Bridge to Abstract Mathematics, Ralph W. Oberste-Vorth, Aristides Mouzakitis, and Bonita A. Lawrence
Calculus Deconstructed: A Second Course in First-Year Calculus, Zbigniew H. Nitecki
Calculus for the Life Sciences: A Modeling Approach, James L. Cornette and Ralph A. Ackerman
Combinatorics: A Guided Tour, David R. Mazur
Combinatorics: A Problem Oriented Approach, Daniel A. Marcus
Complex Numbers and Geometry, Liang-shin Hahn
A Course in Mathematical Modeling, Douglas Mooney and Randall Swift
Cryptological Mathematics, Robert Edward Lewand
Differential Geometry and its Applications, John Oprea
Distilling Ideas: An Introduction to Mathematical Thinking, Brian P. Katz and Michael Starbird
Elementary Cryptanalysis, Abraham Sinkov
Elementary Mathematical Models, Dan Kalman
An Episodic History of Mathematics: Mathematical Culture Through Problem Solving, Steven G. Krantz
Essentials of Mathematics, Margie Hale
Field Theory and its Classical Problems, Charles Hadlock
Fourier Series, Rajendra Bhatia
Game Theory and Strategy, Philip D. Straffin
Geometry Illuminated: An Illustrated Introduction to Euclidean and Hyperbolic Plane Geometry, Matthew Harvey
Geometry Revisited, H. S. M. Coxeter and S. L. Greitzer
Graph Theory: A Problem Oriented Approach, Daniel Marcus
An Invitation to Real Analysis, Luis F. Moreno
Knot Theory, Charles Livingston
Learning Modern Algebra: From Early Attempts to Prove Fermat's Last Theorem, Al Cuoco and Joseph J. Rotman
The Lebesgue Integral for Undergraduates, William Johnston
Lie Groups: A Problem-Oriented Introduction via Matrix Groups, Harriet Pollatsek
Mathematical Connections: A Companion for Teachers and Others, Al Cuoco
Mathematical Interest Theory, Second Edition, Leslie Jane Federer Vaaler and James W. Daniel
Mathematical Modeling in the Environment, Charles Hadlock
Mathematics for Business Decisions Part 1: Probability and Simulation (electronic textbook), Richard B. Thompson and Christopher G. Lamoureux
Mathematics for Business Decisions Part 2: Calculus and Optimization (electronic textbook), Richard B. Thompson and Christopher G. Lamoureux
Mathematics for Secondary School Teachers, Elizabeth G. Bremigan, Ralph J. Bremigan, and John D. Lorch
The Mathematics of Choice, Ivan Niven
The Mathematics of Games and Gambling, Edward Packel
Math Through the Ages, William Berlinghoff and Fernando Gouvea
Noncommutative Rings, I. N. Herstein

Non-Euclidean Geometry, H. S. M. Coxeter
Number Theory Through Inquiry, David C. Marshall, Edward Odell, and Michael Starbird
Ordinary Differential Equations: from Calculus to Dynamical Systems, V. W. Noonburg
A Primer of Real Functions, Ralph P. Boas
A Radical Approach to Lebesgue's Theory of Integration, David M. Bressoud
A Radical Approach to Real Analysis, 2nd edition, David M. Bressoud
Real Infinite Series, Daniel D. Bonar and Michael Khoury, Jr.
Thinking Geometrically: A Survey of Geometries, Thomas Q. Sibley
Topology Now!, Robert Messer and Philip Straffin
Understanding our Quantitative World, Janet Andersen and Todd Swanson

MAA Service Center
P.O. Box 91112
Washington, DC 20090-1112
1-800-331-1MAA   FAX: 1-240-396-5647

Contents

Preface		ix
Introduction		1
1	**Lebesgue Integrable Functions**	**3**
	1.1 Two Infinities: Countable and Uncountable	6
	1.2 A Taste of Measure Theory	18
	1.3 Lebesgue's Integral for Step Functions	26
	1.4 Limits	36
	1.5 The Lebesgue Integral and L^1	49
	Notes for Chapter 1	61
2	**Lebesgue's Integral Compared to Riemann's**	**65**
	2.1 The Riemann Integral	66
	2.2 Properties of the Lebesgue Integral	85
	2.3 Dominated Convergence and Further Properties of the Integral	100
	2.4 Application: Fourier Series	108
	Notes for Chapter 2	121
3	**Function Spaces**	**125**
	3.1 The Spaces L^p	126
	3.2 The Hilbert Space Properties of L^2 and ℓ^2	140
	3.3 Orthonormal Basis for a Hilbert Space	147
	3.4 Application: Quantum Mechanics	165
	Notes for Chapter 3	178
4	**Measure Theory**	**183**
	4.1 Lebesgue Measure	184
	4.2 Lebesgue Integrals with Respect to Other Measures	191
	4.3 The Hilbert Space $L^2(\mu)$	200
	4.4 Application: Probability	205
	Notes for Chapter 4	211

5 Hilbert Space Operators 215
   5.1 Bounded Linear Operators on L^2 216
   5.2 Bounded Linear Operators on General Hilbert Spaces 224
   5.3 The Unilateral Shift Operator 237
   5.4 Application: A Spectral Theorem Example 248
   Notes for Chapter 5 256

Solutions to Selected Problems 259

Bibliography 275

Index 281

Preface

The Lebesgue Integral for Undergraduates presents the Lebesgue integral at an understandable level with almost no prerequisites. The text is accessible to anyone who has mastered the single-variable calculus concepts of limits, derivatives, and series. Its material is important to the development of mathematics majors in function theory. These ideas rely on many advanced topics, such as the definition of countable infinity (vs. uncountable infinity), but students should be able to work through the ideas successfully, as they are presented at an introductory level without need for prior exposure. In this way, the mathematics under discussion can be well understood.

This text provides a new ability to learn about Lebesgue integration as a standard course in the undergraduate curriculum. Mathematicians have long understood the course's benefits, since function theory depends on it, but until now have been able to offer it only at the advanced level. For example, it exists at a few institutions as Real Variables II (or a similar variation), and this text works well there. In fact, a student can now take Real Variables I and Real Variables II in either order whenever this text forms the basis of the latter course. Departments no longer need to worry if enough students are coming out of Real Variables I to populate the follow-up course. In addition, the book allows the mathematics undergraduate curriculum to do more. New options exist because the Lebesgue integral now does not depend on complicated measure theory nor a Real Variables I prerequisite. What are some of the results? A Lebesgue Integral course can serve as an elective, similar to Complex Analysis. It can enroll students immediately after Calculus II, or after a first course in mathematical proofs (a transition course), or as a required course in function theory. In fact, students can take the three courses The Lebesgue Integral, Complex Analysis, and Real Variables in any order, where each one enhances the other two. Along with Vector Calculus and Probability Theory, these courses allow students to experience a substantially complete undergraduate investigation into functions. *The Lebesgue Integral for Undergraduates* is the tool needed to provide such options.

The text also makes undergraduate research in modern function theory possible. In this way it rejuvenates function theory at a basic undergraduate level. After using this book, undergraduates should have better access to current research questions in the field. They should be able to perform collaborative research. They should be able to appreciate the developments in function theory in the 20th century and beyond, as the Lebesgue integral is the key. This new passage provides a way to lead undergraduates *en masse* to function theory research: institutions that currently do not teach the Lebesgue integral can now offer additional regular course opportunities or independent studies for students to learn it. The door to function theory research then swings open, and young mathematicians can enter.

The main reason for the textbook's success at the undergraduate level is its use of a method labeled the "Daniell-Riesz approach," which is briefly explained in the Introduction and then presented in Chapter 1. This text's main goal is to present Lebesgue's integral in an understandable way, including integrals defined in terms of general Borel measures. We want readers to get a real "feel" for what is involved with the integral, and so most of the exercises are calculational. Theoretical ideas and issues are also presented, but several concepts are intentionally omitted. For example, the equivalence of the L^1 spaces obtained by Lebesgue's method vs. the Daniell-Riesz approach is important and is verified in other texts (cf. [121, pp. 127–130] and [98, p. 96]), but it is not this book's focus. The text's expectation is thereby restricted but holistic: students gain a complete insight in *how* the integral is defined and works in fundamental ways, which is different from the traditional main goal of a graduate text on the subject (where students prove advanced theoretical results). Both goals can be accomplished. In fact, this text has both theoretical presentations and exercises for the students to explore theoretical concepts, but they are chosen carefully so as not to overwhelm the student or expand the size of a manageable text. Most theoretical exercises investigate a straightforward property or practice using simple theoretical topics (such as a definition) in a proof setting. A corresponding learning objective is for undergraduate readers (even at an introductory level) to gain confidence and practice with manageable and straightforward proofs.

This book uses the Daniell-Riesz approach throughout, but presents it in a way that makes the material understandable to almost a beginning undergraduate mathematician. The joys of this study are numerous. Here are just three:

The mathematics is fun. Presented in as simple a format as possible, it is often surprising. For example, there are two infinities (as sizes of sets of real numbers); in fact, there are *at least* two. There are functions with only two output values that cannot be integrated—at least using Riemann sums. There are spaces of *functions* with a geometry that matches the Euclidean geometry for spaces of *points*.

It is important. Lebesgue's integral opens up modern function theory to the student. His integral is required to understand the current research of function theorists. It provides a gateway into the modern mathematics of functions, both real and complex.

It supports ideas learned in other mathematics courses that describe functions, such as real analysis and complex analysis. Learned before a student takes these courses (or at the same time, or afterwards), Lebesgue's integral becomes another "card in the student's deck" to see what functions are all about.

The last goal for the text is to discuss applications. They are advanced ideas in function theory, but the book does its best to present them in a manageable package—in an introductory fashion that works for undergraduates. Other books (cf. [86]) do a wonderful job of laying out a complete theoretical framework (and giving a full treatment of associated applications), but that task requires a lengthy development impossible here. Such thoroughness is *not* the goal of this text. Instead, *The Lebesgue Integral for Undergraduates* develops the integral's definition, describes its characteristics useful in other contexts within the text, and provides a small taste of many function theoretic topics that evolve out of the integral (though these are mainly limited to real functions). Other mathematicians might choose a different set of topics with persuasive reasons. For example, it is tempting (as an expansion that would serve many students well) to have included more material on complex-valued functions. Or a more general discussion of Banach spaces. Or C^*-algebras. And so on. The topics that made the book's "final cut," such as Hilbert space, are not all encompassing—that was impossible for this project—but they are selectively impressive and enjoyable to teach and learn. Follow-up investigations can always take place. In a forward-looking way, the text concludes by inviting undergraduate students into further study, including the traditional measure-theoretic graduate Lebesgue course. They can see the broad plain of function theory stretching before them, and they can choose to step forth in exploration.

This book explains all the necessary real analysis concepts, such as limits of sequences, continuity, and a set's supremum and infimum values. These prerequisites are grounded in discussions that have clear and interesting payoffs: they *must* be used in order to define the Lebesgue integral and its applications. Finally, the material is modern, which motivates its study as relevant—current function theory research uses it.

After extensive classroom use and student feedback, the text includes many features to enable learning. Each section contains embedded Questions; they help create a "workbook" aspect and make sure ideas are understood. Answers appear at the end of each section. The text also has Reading Questions (a quick sectional review check), about 700 exercises (for practice on both calculational problems and proof writing), solutions to odd-numbered exercises, historical vignettes and Chapter Notes (that include stories of many great mathematicians, celebrating the diversity of mathematical talent), and an application at the end of each chapter (describing an idea with powerful implications—for our physical world, for the way we think about chance, or for an advanced mathematical topic that intrigues us).

Special acknowledgment and thanks go to Stanley Seltzer, Steven Kennedy, Carol Baxter, and many others at the MAA for their supportive professional work to bring this textbook to publication. The (anonymous) reviewers' comments and the copy editor were particularly helpful and dramatically improved the exposition. Many friends, colleagues, and students supported this project. Jay Howard at Butler University provided funding from his office. Others who nourished it at critical stages include Scott Chapman, Carl Cowen, Patricia Johnston, Susan Johnston, John Mugge, James Rovnyak, Steven Seubert, Derek Thompson, John Wilson, Phyllis Cohen, Jeremiah Farrell, and André Wehner. I thank them. Freshmen to seniors at three academic institutions have taken the integration course and have given feedback, improving the explanations considerably. My success at teaching and writing about the Lebesgue integral has occurred because of this support. Any error in this book is a result of my work and is my responsibility, but describing the Lebesgue integral has provided great joy for which I am profoundly grateful.

Introduction

Henri Lebesgue produced one of the greatest advances in function theory when he wrote his dissertation [80] in 1902 at the age of 27. In it, he reworked the framework of the (definite) integral, which is intuitively (and simplistically) thought of as "the area under the functional curve." Prior to his work, mathematicians used a definition of the integral due to Bernhard Riemann, outlined in his paper submitted to join the University of Göttingen faculty in 1854. Riemann defined the integral in terms of a limit of Riemann sums (basically, sums of rectangular areas drawn under the curve).

Lebesgue's formulation is completely different from Riemann's, allowing mathematicians to consider the integral for a more general type of function. Instead of basing his integral on rectangular areas, Lebesgue's approach requires a more abstract foundation—a *Theory of Measure*—creating a new understanding of the sizes of sets. This new "measure theory" then produces the integral for a "simple function." Where Riemann's rectangular areas came from partitioning a function's domain axis, Lebesgue's simple functions effectively partition the range. Lebesgue then uses the integrals of simple functions (in a limiting process) to define the integral for a general class of functions L^1 (pronounced *ell-one*).

In this way, Lebesgue's method describes L^1, the collection of functions that his method can integrate, in terms of measures. Once a full treatment of measure theory is learned, the material is straightforward, at least in its starting framework. Yet undergraduate students everywhere who have tackled Lebesgue's Theory of Measure will agree it is, simply put, tough. The traditional approach is the stuff of graduate school mathematics. What can an undergraduate curriculum do?

Thankfully, a different method that describes the integral, formulated and increasingly taught throughout the 20th century, breaks through the impasse. An outstanding 1973 text by the British mathematician Alan J. Weir [121] explains the technique. Weir uses an approach established by Percy John Daniell [34] in 1917 and refined into a concise description in a 1950s advanced text by Frigyes Riesz and Béla Sz.-Nagy [98]. We call this method the *Daniell-Riesz approach*. It succeeds in

Figure 0.1. Percy John Daniell and Frigyes Riesz.

1

defining Lebesgue's integral differently from Lebesgue. It nearly avoids any dependence on measure theory. Additionally, it develops the integral in terms of step functions (instead of using Lebesgue's more difficult choice of simple functions). Notably, the Daniell-Riesz collection of functions in L^1 turns out to be exactly the same as Lebesgue's class of integrable functions, but the Daniell-Riesz definition is *not* dependent on L^1 membership characterized in terms of measure. More precisely, Lebesgue's notion of L^1 forces a preliminary discussion of the difficult concept of "measurable functions," but the Daniell-Riesz approach does not. In a follow-up volume [120], Weir uses the Daniell-Riesz approach to develop integrals with respect to other measures, besides Lebesgue measure.

This new approach works—and this book now brings it to undergraduates. (Weir wrote at a graduate level.) It presents topics that emphasize the basic definition of the integral along with its many foundational properties and applications. Using the Daniell-Riesz approach, almost any undergraduate interested in functions can access the material. Without the Lebesgue integral, work in functional analysis and such applications as probability theory is restricted. With it, these fields open up to organizational structures such as function spaces and their use. For these reasons, we hope the Daniell-Riesz approach to Lebesgue's integral catches on.

1
Lebesgue Integrable Functions

The integral of a function is as old as the calculus Sir Isaac Newton and Gottfried Leibniz independently formulated. Their description of calculus in the middle to late 1600s came late in the Baroque period, a time when music, art, and literature flourished. Just as advances in the Renaissance led to Baroque cultural and intellectual development, so did advances in the Baroque period (especially its mathematical framework found in calculus) lead to the later Age of Enlightenment.

Development of intellectual endeavors advanced as these broad expressions, and also for specific topics. The integral is a case in point. Newton and Leibniz correctly defined and intuitively understood the (definite) integral as the area under a function's curve. They also understood the relationship between the integral and the derivative: the integral of a (continuous) function f can be found from another function F where the derivative of F equals f. Leibniz put this relationship, which we now call the fundamental theorem of calculus, into the following familiar notation:

$$\int_a^b f(x)\,dx = F(b) - F(a), \text{ where } \frac{dF(x)}{dx} = f(x) \text{ for } x \text{ in } (a, b).$$

This work of Newton and Leibniz was the first grand advancement on the integral (and was perhaps the greatest, as it set the stage for all further development). It was also important, since it led to scientific applications that illuminated the world in which we live. In his seminal treatise *Principia* [88], Newton used the integral[1] to prove Johannes Kepler's first law:

> Planets move in elliptical orbits around the sun.

In fact, in the *Principia* Newton proved all three of Kepler's laws. They followed from his fundamental understanding (in the form of mathematical principles) of the way gravity acts as a force. The work showed the world for the first time that physical nature was not indeterminable or random but was explainable and predictable through mathematics. In rightful praise of the work, Sir Edmund Halley (who used the results in the *Principia* to predict accurately the return

---

[1] Newton has long been properly credited (independently, along with Leibniz) with "inventing" calculus, but he wrote the *Principia* in a format different from the calculus we know today. The Notes section at the end of this chapter discusses this issue.

of the comet now named after him) included a 48-line ode in its publication (which Halley financed at no small expense). The poem included the lines

> *Then ye who now on heavenly nectar fare,*
> *Come celebrate with me in song the name*
> *Of Newton, to the Muses dear; for he*
> *Unlocked the hidden treasuries of the Truth...*

Celebrate, indeed: it was mathematics, including the integral, that helped unlock these Truths on nature's workings. Shift now to modern times: throughout much of contemporary function theory, it is the Lebesgue integral that serves mathematicians well, helping them uncover new discoveries each day.

In the *Principia*, Newton's presentation of the integral is almost unrecognizable in today's mathematical notation. Yet we can understand how it provides a fundamental key to the proof of Kepler's second law, simply from the fact that "area" is the focus of its statement:

The vector that extends from the sun to a moving planet sweeps out area at a constant rate.

At an early date, Leibniz's writings provided notational clarity to the integral in terms of functional expressions. But the precise and rigorous mathematical definition of the integral—the term $\int_a^b f(x)dx$ on the left side of the fundamental theorem's equation—was left unspecified for almost 200 years, until Bernhard Riemann formulated it as limits of finite sums (which we now call Riemann sums). His work was representative of the Age of Enlightenment, using rigor and exactness to explain important ideas so their mathematical underpinnings were clear and unquestioned. We will describe Riemann's integral in Chapter 2. It formulates well Newton's and Leibniz's idea of the integral, as it provides an acceptable definition of the integral for any continuous function, and one for which the above description of the fundamental theorem holds. In fact, it turns out to define very well the area under the curve of any function continuous except at a finite collection of points, or even a "countably infinite" collection of points—a notion we will make precise later in this chapter. Unfortunately, there are many functions we would like to integrate that do not meet this criterion.[2] Their Riemann integrals can be undefined, and so Riemann's definition falls short. By the late 1800s it was clear the Riemann integral had issues. The mathematical concept of the integral needed another step forward in humankind's understanding before it would be fully developed.

That advancement came in 1902, when the French mathematician Henri Lebesgue (pronounced leh-beg, with the same "eh" sound in both syllables) developed a new version of the integral in his doctoral thesis [80]. A well-known type of infinite series, Fourier series, motivated Lebesgue's work. Fourier series were constructed as a sum of a (possibly) infinite number of terms of sine and cosine functions. (We'll explore Fourier series in Chapter 2.) By the late 1800s, mathematicians and physicists realized Fourier series nicely represent bounded functions f. Joseph Fourier worked out the construction of the series in the early 1800s, which came from individually integrating sine and cosine expressions in f. Fourier also showed the usefulness of the series. He used the series representation to solve the famous heat equation, which determines

---

[2] A standard example, integrating over the interval $[0, 1]$, is $f(x) = 1$ if x is rational, and $f(x) = 0$ if x is irrational. Section 2.1 will detail reasons why this function's integral is problematic when using Riemann's definition.

mathematically how, for example, the temperature on a stovetop's metal griddle will change over time and finds that temperature at any spot on the griddle. As he produced his results, Fourier assumed the integrals of the sines and cosines would exist for any bounded function, but that turned out not to be the case—the later-developed Riemann integral failed to be defined in this application for a large class of discontinuous functions. Henri Lebesgue made this discovery as a graduate student and described explicitly which functions Riemann's integral could handle. As a result, it became clear the Riemann integral had limitations when used in Fourier series. Mathematicians needed an integral for a broader class of functions—one that could represent a more complete range of functions as Fourier series, and one that could generally resolve when the Fourier series converged to its represented function.

Lebesgue therefore considered alternative formulations of the definite integral. Realizing the key was the way area was measured, he worked to create an alternative construction of that calculation. In 1901, he began to develop an insightful new "theory of measure" (along with a corresponding way to integrate) in an immediately noteworthy paper *Sur une généralisation de l'intégrale définie* [81]. Lebesgue's eventual formulation (explained in his 1902 dissertation) was impressive. It is extremely difficult even to understand his theory of measure at the level required of the integral. But once the measure theory is understood, the integral follows readily, and Lebesgue used the measure theory to define—in a completely different way—the definite integral. In order for the fundamental theorem to hold, it was a natural requirement that the value obtained from Lebesgue's integral of a function f should agree

Figure 1.1. Henri Lebesgue.

with the value obtained from Riemann's integral of f, and it does (whenever Riemann's integral exists). But Lebesgue's integral turned out to be significantly better. It is well-defined and meaningful for many highly discontinuous functions whose Riemann integrals are not even defined. Many times, when Riemann fails, Lebesgue works.

The mathematical impact of Lebesgue's work is profound. It helps determine when a Fourier series correctly represents its corresponding function (a question that still generates research and that was not resolved satisfactorily in terms of "function spaces" until 1966 by the Swedish mathematician Lennart Carleson [24]). Lebesgue's integral also provides a way to think about "spaces of functions," in much the same way we think about the two-dimensional plane, for example, as a "vector space of points." And in 1933, the Russian mathematician Andrei Kolmogorov showed the Lebesgue integral provides a natural foundation for mathematical constructs in probability theory and statistics. In the end, Lebesgue's broadly formulated measure-theoretic integral (the so-called "Lebesgue-Stieltjes integral") manifests itself as a fully developed version of the integral, in that it turns out to generate an integration tool that works for many applications.

Lebesgue's development of the integral is brilliant. His description of it, however, uses a process that requires a very full understanding of measure theory—a fascinating but abstractly complex set of ideas beyond the scope of a broadly accessible undergraduate course. Fortunately, a different process manages to produce the Lebesgue integral, and we use it here. As it is based primarily on insights from two mathematicians during the 20th century named Percy John

Daniell and Frigyes Riesz, we call this alternative method the *Daniell-Riesz approach*. The technique turns Lebesgue's formulation on its head. Instead of presenting measure theory first, it blazes a straight and direct route to a definition of the Lebesgue integral, nearly sidestepping measure theory altogether. Once the integral is defined, it can then determine the measure of a set. It thereby attains the brilliant heights of a complete and rigorously formal theory of the integral without a single crash against the cliffs of measure theory.

To understand the Daniell-Riesz approach, we'll first need to work through a few advanced but straightforward mathematical concepts from a standard undergraduate curriculum. These include countable infinity, sequences, and limits. Those definitions occur in this chapter—in Sections 1.1 and 1.4. Section 1.3 defines the Lebesgue integral for a particular type of function (a step function), and Section 1.5 expands the definition to a large array of functions. After just these few sections, we'll understand how the Lebesgue integral works, and we'll also have read about and proven a few of the integral's basic properties and its usefulness. (Many more of its powerful properties will come in later chapters.)

Concepts about Lebesgue's general notions of measure theory will come later (and only briefly in this chapter), since they are not required when using the Daniell-Riesz approach in order to understand the integral. That explanation will simply answer a few natural questions we will have (such as "What is the Lebesgue measure of a general set?"). The results will turn out to be natural consequences of this chapter's work. So enjoy—go slowly, realize we are learning a lot of impressive mathematics, and have confidence our understanding of this material will strengthen the more we look at examples, the theory, and applications. This text will present all of its material in a way that should be exciting and fun.

1.1 Two Infinities: Countable and Uncountable

The understanding of infinity took an important leap forward in the late 1860s, when Georg Cantor, who worked in Germany at the University of Halle, made a startling discovery—one unfortunately not well known outside of the mathematics community and one that continues to surprise those who see it for the first time. Asking "How many elements does a set have?" Cantor knew the answer would be either a finite number or an infinite number. But then he discovered there are different sizes of infinity!

Much earlier, in his 1655 paper *De sectionibus conicus*, John Wallis had introduced the symbol "∞" as a representation of infinity. That notation continues to make good sense (and is unaffected by Cantor's later discovery) when discussing, for example, limits. In that vein, it describes when a function or a sequence grows larger without bound. That concept does not refer to an infinite number of elements (an infinite "cardinality") but to an unboundedness of a process. For example, introductory calculus tells us $\lim_{x \to 0} \frac{1}{|x|} = \infty$, which means the function $1/|x|$ grows without bound as values of x get close to 0. It does not say we are somehow "evaluating" the function $f(x) = 1/|x|$ at 0 and obtaining the number ∞. Wallis's use of infinity is clearly important, but Cantor's notion of an "infinite cardinality of sets" is much grander; it describes what is meant by "infinite numbers."

Cantor's insights on the numerical size of a set—its cardinality—came by realizing some infinite sets can be written as infinite sequences, but some cannot. At first, Cantor thought there would be just one numerical value for infinity. But as he tried to prove this, his work led to

1.1 Two Infinities: Countable and Uncountable

surprising conclusions as profound as any mathematics ever produced. Cantor published these results in several papers that together developed a new theory of sets. They explored finite and infinite numbers in terms of representations of sets as sequences (or not), and they developed a new arithmetic for infinite numbers. His results affect almost all areas of mathematics—any that deal with infinite sets. His theory of sets continues to produce new research results, as mathematicians are still discovering properties of set cardinalities.

Cantor used the notation $|S|$ to express the cardinality of a set S, where the term "cardinality" refers to the set's numerical size. The elements in a finite set could be written as a finite sequence (or perhaps more simply put, as a finite list). In that case the cardinality of the set is the length of the sequence. For example, the cardinality of the set $S = \{A, B, C\}$ is 3, since its elements A, B, C can be written as the sequence s_1, s_2, s_3 with $s_1 = A$, $s_2 = B$, and $s_3 = C$. We write $|S| = 3$. Cantor classified finite sets as "countable," a term we continue to use.

Figure 1.2. Georg Cantor.

But in the same way, some infinite sets are countable, since their elements can also be arranged as an (infinite) sequence. Two simple examples are the set of natural numbers $\mathbb{N} = \{1, 2, 3, \ldots\}$ (whose elements are already arranged as a sequence) and the set of whole numbers $\mathbb{Z} = \{\ldots, -2, -1, 0, 1, 2, \ldots\}$. The set $\mathbb{Z}$ is the set of integers (the whole numbers), and it becomes clear $\mathbb{Z}$ is countable when we write its elements as the sequence $0, 1, -1, 2, -2, 3, -3, \ldots$. (Technically, the list forming such a sequence $s_1, s_2, s_3, \ldots$ is allowed to extend infinitely to the right, but not infinitely to both the right and the left.)

Thinking of when infinite sets are countable (when the elements can be written as a sequence $s_1, s_2, s_3, \ldots$) is at the heart of Cantor's insight on the cardinality of infinite sets. We formalize his notion of a countable set in the following definition.

Definition 1.1.1 *A (nonempty) set S is countable when its elements can be arranged in a sequence; that is, when S is of the form*

$$S = \{s_1, s_2, s_3, s_4, \ldots\}.$$

The sequence can be finite or infinite.[3] *A set S is uncountable when it is not countable.*

Note we can show infinite sets are countable by constructing sequences where the elements of the set are allowed to be repeated in the sequence (we can always go back and clean up the sequence by eliminating the repeats). We will often use this approach, not worrying about repeats, to prove that many familiar sets are countable[4].

---

[3] The subscript on each (nonrepeated) element of such a sequence $s_1, s_2, s_3, \ldots$ is the element's *position* in the sequence. That position is an ordinal number, and, for any such sequence element, it must be a natural number in $\mathbb{N}$. When the set S is finite, the last element's position matches the cardinality $|S|$ of the set.

[4] Striking out any repeated element of the set does not change the set—exactly the same elements remain. The only change made from eliminating repeated elements is the description of the set.

We should remind ourselves this definition is for the "cardinality of the set," and so it provides a way to talk about a "countably infinite number" of elements. This concept of infinity therefore progresses beyond that of Wallis's, as it is now in terms of an infinite numerical value—the infinite cardinality (loosely, the size) of a set. We can talk about the number "countable infinity," just as we talk about the number 5 or the number π. As Cantor realized this fact, he saw the need to assign a symbol to the number countable infinity, and he used a variation on the first letter of the Hebrew alphabet: $\aleph_0$ (expressed as "aleph naught"). The notation $\aleph_0$ may seem complicated and peculiar, but it has become a standard label, and we adopt it here.

What are some examples of sets with different sizes? And how hard is it to go about showing an infinite set is countable, according to the definition? The next example illustrates answers to these questions.

Example 1.1.1 Finite sets are obviously countable. The set itself, written in any order, will define an appropriate sequence. For example, the elements of $A = \{w, x, y, z\}$ form a sequence w, x, y, z of length 4, which means $|A| = 4$. We could have written the sequence elements in any order, such as x, w, z, y or z, w, y, x. The point is that such a sequence exists and, no matter what order is used, it always has length 4, making $|A| = 4$.

Working with infinite sets can be more difficult but lots of fun. The set of nonnegative even numbers E^+ is clearly countable, because the set elements are already displayed as a sequence as they are written in the set:

$$E^+ = \{0, 2, 4, 6, 8, 10, \ldots\}.$$

We write $|E^+| = \aleph_0$ and say the set of nonnegative evens is countably infinite. The set of all even numbers $\{\ldots, -6, -4, -2, 0, 2, 4, 6, \ldots\}$ is also countable, since we can write it as a sequence

$$\{0, 2, -2, 4, -4, 6, -6, \ldots\}.$$

In the same way, the set of odd numbers is countably infinite.

A wonderful example comes from the set of rational numbers, which are ratios of whole numbers (integers) with nonzero denominators. For example, $\frac{4}{5}$ is a rational number, since it is the ratio of the integers 4 and 5. The number 7 is rational, since it is the ratio of the integers 7 and 1: $7 = \frac{7}{1}$. The number $\frac{\pi}{3}$ is not rational. The set of rational numbers, denoted $\mathbb{Q}$, is formally defined as $\mathbb{Q} = \{\frac{m}{n} : m, n \in \mathbb{Z} \text{ with } n \neq 0\}$. Numbers not rational are irrational.

The set of rational numbers in the interval $[0, 1)$ can be written as an infinite sequence and therefore has size $\aleph_0$. To see what the sequence is, first form the sequence

$$0, \frac{1}{2}, \frac{1}{3}, \frac{2}{3}, \frac{1}{4}, \frac{2}{4}, \frac{3}{4}, \frac{1}{5}, \frac{2}{5}, \frac{3}{5}, \frac{4}{5}, \ldots.$$

Every rational in $[0, 1)$ is in this sequence, but some of the terms are repeated; for example $\frac{1}{2} = \frac{2}{4}$. But remember—that's OK, since we can go back and clean out the repeats. The result then follows:

$$\text{The set of rationals in } [0, 1) = \left\{0, \frac{1}{2}, \frac{1}{3}, \frac{2}{3}, \frac{1}{4}, \frac{3}{4}, \frac{1}{5}, \frac{2}{5}, \frac{3}{5}, \frac{4}{5}, \frac{1}{6}, \frac{5}{6}, \ldots\right\}.$$ ∎

Question 1.1.1 Show the following sets are countable by writing them as either a finite or infinite sequence. Then use your resulting sequences to determine each set's cardinality.

1. The set of integer odd numbers between -6 and 14.
2. The set of odd numbers.
3. The set of rational numbers in $[1, 2)$. ∎

We mention two important facts. First, the key to proving a set is countable is to write its elements as a sequence. This often requires arranging the set elements in a clever manner—and it must be clear what the pattern of the arrangement is, so it is well-defined. Second, note if we take A as the set of evens and B as the set of odds, then their union (the union of two countably infinite sets) is $\mathbb{Z}$, which is countably infinite ($\mathbb{Z} = \{0, 1, -1, 2, -2, ...\}$). This fact is true in general: the union $A \cup B$ of any two countably infinite sets is always countably infinite. The next theorem formally states this result and presents its straightforward proof (where, again, the key to the proof is to arrange the elements of $A \cup B$ as a sequence).

Theorem 1.1.1 *If A and B are (infinitely) countable sets, then so is their union $A \cup B$.*

Proof. Because A and B are countable, we know we can arrange their elements in sequences:

$$A = \{s_1, s_2, s_3, ...\} \text{ and } B = \{t_1, t_2, t_3, ...\}.$$

Then the elements of $A \cup B$ may also be written as a sequence. (Do you see how?) Doing so completes the proof, and here it is:

$$A \cup B = \{s_1, t_1, s_2, t_2, s_3, t_3, ...\}. \qquad ∎$$

Theorem 1.1.1 and its proof are not surprising or difficult, and the generalization beyond two sets follows naturally. We now present that generalization, whose proof is similar to the case of two sets.

Theorem 1.1.2 The Countable Union Theorem: *The union of a countable number of countable sets is countable.*

Proof. We are looking at a countable number of sets, and so we can write them as a sequence $S_1, S_2, S_3, \ldots$. Each set S_j in this sequence of sets is countable, and so we can describe each one notationally as $S_j = \{s_{j1}, s_{j2}, s_{j3}, ...\}$. Taking their union, $\bigcup_{j=1}^{\infty} S_j$ is a countable set, as shown by our ability to write its elements in the sequence

$$s_{11}, s_{12}, s_{21}, s_{31}, s_{22}, s_{13}, s_{14}, s_{23}, s_{32}, s_{41}, s_{51}, s_{42}, \ldots .$$

The pattern of this sequence may not be obvious, but it does exist, and we present below (in Figure 1.3) a display of its systematic design. Some of the elements might be repeats of each other, but we can always go back and clean out the repeats to get an exact sequence representation of the set elements. ∎

Question 1.1.2 Show the following sets, formed as unions of sets, are countable with size $\aleph_0$. Use either Theorem 1.1.1 or 1.1.2. Also, write the elements of any set you claim is countable as a sequence.

1. $\mathbb{Z} \cup A$, where $A = \{1.5, 2.5, 3.5, \ldots\}$ is the set of numbers of the form $(m + .5)$ with m a positive integer.
2. $A_1 \cup A_2 \cup A_3 \cup A_4 \cup \ldots$, where $A_1 = \{1.1, 2.1, 3.1, \ldots\}$, $A_2 = \{1.11, 2.11, 3.11, \ldots\}$, $A_3 = \{1.111, 2.111, 3.111, \ldots\}$, etc. ∎

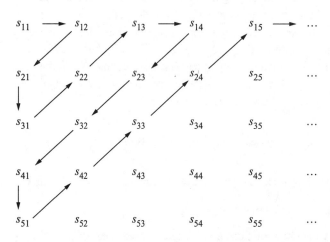

Figure 1.3. Cantor's first diagonalization method.

So how did we arrive at the sequence in the proof of the countable union theorem? We followed Georg Cantor—he came up with the sequence's pattern, using a clever scheme that mathematicians now generally call *Cantor's first diagonalization method*. It works according to the display in Figure 1.3, where the sequences are one below the other. Then the elements in the new sequence are arranged by traveling up and down the diagonals: first s_{11}, then (starting on the next diagonal) s_{12} followed by s_{21}, then back up the next diagonal, and so on.

The countable union theorem is tremendously useful to show many infinite sets are countable. Its most famous use is to show the set of rational numbers $\mathbb{Q}$ is countably infinite. The next question leads you through that proof, whose key construction is to write $\mathbb{Q}$ as a countable union (i.e., a union of a countable number of sets) of countable sets.

Question 1.1.3 This question shows why the set of rational numbers $\mathbb{Q}$ is countably infinite by writing it as a countable union of countable sets.

a. We have seen (in Example 1.1.1) the set of rational numbers contained in the interval $[0, 1)$ is countable. Label this set I_0.
b. Similarly, label I_1 the set of rational numbers contained in the interval $[1, 2)$. Construct a proof that I_1 is countable, based on the proof (in Example 1.1.1) that I_0 is countable.
c. In a similar way, label I_n the set of rational numbers contained in the interval $[n, n+1)$, where n is any integer (note n can be negative). Construct a proof that I_n is countable.
d. Noting the set of integers is countable, recognize the union $\bigcup_{n=-\infty}^{\infty} I_n$ is a countable union. Use the countable union theorem to prove this set is countable.
e. What important set is $\bigcup_{n=-\infty}^{\infty} I_n$? Use your answer with the insights gained from the steps above to construct a formal proof that $\mathbb{Q}$ is countable. ∎

Now we come to the big question: are there sets that are not countable? So far we haven't found one; commonly examined infinite sets such as the rationals and the integers have all turned out to be countably infinite. Maybe that's the case for any infinite set; such a conclusion matches our intuition, doesn't it? Isn't there just one size of infinity? After all, this notion seems to have a lot of supporting evidence. For instance, if we could prove the entire set of real numbers $\mathbb{R}$ is countably infinite, then any infinite subset of it must be, and that would verify infinity as just a single size. Moreover, the rational numbers are dense in $\mathbb{R}$, in the sense that we can find a rational between any two reals. Since the set of rationals is countably infinite, isn't $\mathbb{R}$?

This line of reasoning encouraged Cantor for many years to try to prove $\mathbb{R}$ is countable, but he wasn't successful. Frustrated by this failure, Cantor finally figured out what was going wrong—it wasn't his lack of ability; it was the mathematics. In 1873, he astoundingly proved the set of real numbers is uncountable. Even after he published his work the following year, many members of the mathematical community remained skeptical. The result is so counterintuitive that they suspected there must be problems with his construction or his holistic approach to the subject. To this day, it's sometimes hard to wrap our minds around related results. For example, it turns out there are "more" numbers between 0 and 1 (including both rational and irrational values) than there are positive integers $\{1, 2, 3, \ldots\}$, even though the interval $(0, 1)$ is bounded while $\{1, 2, 3, \ldots\}$ is not. But the proof of the uncountable cardinality of $\mathbb{R}$ is indisputable; the following version of it was Cantor's own simplification of his original proof, and he published it in 1891. It uses a technique "proof by contradiction." That technique first makes an assumption—that the theorem is not true—and then finds a resulting contradiction to some other known fact. Since the contradiction cannot be possible, it must be the case that the assumption made was false, which is equivalent to the theorem being true (and which finishes the proof). One additional technical note: the proof uses decimal expansion of numbers (explained below). That expansion needs to be well-defined, and so if an expansion might end in a series of 9s (such as $1/2 = 0.4999\ldots$), we always choose the expansion that ends in 0s (so we use $1/2 = 0.5000\ldots$). This convention also helps make the reasoning sound (due to a technical reason that does not contribute to the proof's holistic development). Let's see how the proof works.

Theorem 1.1.3 *The set $\mathbb{R}$ of real numbers is uncountable.*

Proof. The result would follow if we could prove the interval $[0, 1)$ is uncountable, since this set is contained in $\mathbb{R}$. That proof's main idea is this: no matter how we might try to list all the numbers in $[0, 1)$ as a sequence, we'll always leave out at least one number (we'll call this number d). So let's assume we have been handed a sequence $\{s_1, s_2, s_3, s_4, \ldots\}$ that (presumably) represents all the numbers in the set $[0, 1)$. How would we construct such a number d in $[0, 1)$ that has been left out of the sequence? Its value will have to be derived from the numbers in our sequence list, as we'll want d to be different from all of them. We'll construct d from the decimal expansions of the numbers. Those numbers are all between 0 and 1; here are three examples of decimal expansions:

$$\frac{1}{2} = 0.50000000\ldots; \quad \frac{1}{3} = 0.33333333\ldots; \quad \frac{\sqrt{2}}{2} = 0.70710678\ldots.$$

We call the values that appear in the decimal expansions the *digits* of the expansion. For example, the first digit in the expansion of $1/2$ is 5. The second digit in the expansion of $1/3$ is 3 (in fact, all the digits are 3). The third digit in the expansion of $\sqrt{2}/2$ is 7.

The number d is going to be a number between 0 and 1, and so it will also have such a decimal expansion. We want d to be different from every number in the sequence, and here's the way we guarantee it is: the first digit in the expansion of d will be different from the first digit in the expansion of the first number in the sequence. The second digit in the expansion of d will be different from the second digit in the expansion of the second number in the sequence. The third digit in the expansion of d will be different from the third digit in the expansion of the third number in the sequence. And so on, all the way along in the expansion of d and all the way along the sequence list. Describing the construction in "one fell swoop," we make d so the nth digit in the expansion of d will be different from the nth digit in the expansion of the nth number in the sequence.

How do we construct d to make that happen? Let d's first digit be one larger than the first digit of the first sequence element; let d's second digit be one larger than the second digit of the second sequence element; let d's third digit be one larger than the third digit of the third sequence element; etc. This works for any digit—except for one technical exception (which we'll handle in a minute) and for 9 (and then adding 1 produces 10, which is not a digit). In this case, we change the digit 9 to the digit 0. Finally, to make sure we don't get a number d that ends in a series of 9's (remember—we never allow that convention), we'll also change the digit 8 to the digit 0.

An example shows how it works. Imagine with me—suppose the sequence had started this way:

$$s_1 = 0.\underline{5}0000000, s_2 = 0.3\underline{3}333333\ldots, s_3 = 0.70\underline{7}106781\ldots, \ldots$$

Then d's expansion would have started $0.648\ldots$. Can you see why? If you can, then you see how d is constructed—the digits marked in bold and underlined, above, were the ones changed (by adding one) to produce the value for d. It should be clear why d is different from each of these three numbers, and (due to its difference in the nth digit) different from every number in the sequence. Finally, this construction works for not only this sequence example, but every time. The inescapable conclusion: no matter how one might try to write the numbers between 0 and 1 in a sequence, there are always values left out. In short, $[0, 1)$ is uncountable. The infinite size of $[0, 1)$ is larger than the infinite size of any countable set, such as $\mathbb{Q}$. That fact, as mentioned earlier, also implies $\mathbb{R}$ is uncountable. ∎

Mathematicians now label $\aleph_1$ (read as "aleph-one") as the cardinality of $\mathbb{R}$, the larger of the two infinities presented here.[5] But Theorem 1.1.3 raises additional questions: we now know there are two sizes of infinity; are there more? Is there an infinite number between $\aleph_0 = |\mathbb{Q}|$ and $\aleph_1 = |\mathbb{R}|$? Cantor was the first to ask this question, in 1877, but he was not able to answer it. Mathematicians expected the answer to be "no," that there is no infinite number between these two values, and that conjecture became known as the continuum hypothesis. Cantor never managed to prove the result, and published his last paper in set theory (on ordinal numbers) in 1897 (see, for example, [21]). The continuum hypothesis remained unsolved for many years, and was identified by David Hilbert in his famous 1900 speech in Paris to the International Congress of Mathematicians (an assembly that meets periodically) as one of the 23 most significant unsolved problems of that time. (Many of the problems remain unsolved.)

---

[5] Actually, $\aleph_1$ is traditionally used for $|\mathbb{R}|$ when assuming the continuum hypothesis. Our use of the symbol $\aleph_1$ is consistent with this assumption.

1.1 Two Infinities: Countable and Uncountable

Two later results finally settled the matter, but in a startling way (and one that illustrated the complexity inherent in Cantor's struggle to prove it). Both results are based on a rigorous list of axioms called the *Zermelo-Fraenkel set axioms*, which were formally proposed in 1922 (after Cantor's death) as standard assumptions that form an "underlying foundation" for mathematical logic. First, in 1940, the famous Austrian mathematician Kurt Gödel proved the continuum hypothesis could not be proven false (in the sense it is inconsistent with the Zermelo-Fraenkel set axioms). Then in 1963, an American, Paul Cohen at Stanford University, proved it could not be proven true (in the sense it is consistent with the Zermelo-Fraenkel set axioms). So the answer to the continuum hypothesis is there is no definitive answer—it cannot be proven true or false. In this way, it is a powerful display of the limitations of mathematics—there exist undecidable theorems that cannot be proved nor disproved. Talk about mind-boggling!

This section will ask you to determine cardinalities of sets, including sets of objects other than real numbers. In many cases, the answer will depend on the continuum hypothesis. One productive way forward is to assume the validity of the continuum hypothesis one way or another—that action will allow us to state a conclusion about cardinalities. So from here on out, we'll take the following stand: the reals will have cardinality $\aleph_1$; infinite sets with cardinality less than $\aleph_1$ will have cardinality $\aleph_0$; and real sets with cardinality larger than $\aleph_0$ will have cardinality $\aleph_1$. Without assuming the continuum hypothesis, our conclusion in many such cases would have to be "we just can't know."

Theorem 1.1.3 gives a characterization of often quite fascinating sets and can provide many curiosities as its implications are considered. Just facts about the rationals and the irrationals are interesting; for example, the rationals are dense in the set of real numbers, but between any two rationals, no matter how close together, there is not only one additional real number, but an uncountable number of reals, whose size $\aleph_1$ is the same as the full set of reals. It's hard to believe there is enough space in the real line to squeeze everything in, and it shows how difficult it is to characterize infinite sets in terms of their uncountability as a sign of how measurably large they are. In 1926 (eight years after Cantor's death), David Hilbert expressed the mathematical community's eventual acceptance and celebration of Cantor's theory and the breadth of complexities it illuminates by famously writing, "No one will drive us from the paradise Cantor created."

We finish this section with examples of infinite sets and their cardinalities, saying why each set has cardinality $\aleph_0$ or $\aleph_1$. In this way, you will get a sense of how to approach various types of sets to determine their cardinalities.

Example 1.1.2 We determine whether each of the following commonly encountered infinite sets are countable or uncountable.

a. We have already shown the set of integers $\mathbb{Z}$ is countably infinite; $|\mathbb{Z}| = \aleph_0$.
b. Question 1.1.3 proved $|\mathbb{Q}| = \aleph_0$.
c. The set of irrationals (those numbers not rational and which we label $\mathbb{I}$) must be of size $\aleph_1$; for if, instead, it was countable, then $\mathbb{R}$ could be (incorrectly) expressed as the union of two countable sets: $\mathbb{R} = \mathbb{Q} \cup \mathbb{I}$.

   How many irrational numbers can you name? Some famous ones are $\sqrt{2}$, π, e, and $\ln(2)$ (though the irrationality of many numbers, such as $\pi + e$, are unknown). Are you surprised to discover there are many more irrationals than there are rationals?

d. The set of linear polynomials over $\mathbb{Q}$ is defined as $S = \{qx + r : q, r \in \mathbb{Q}\}$; i.e., those degree-one polynomials whose coefficients q and r are rationals. $|S| = \aleph_0$, since S can be expressed as the countable union

$$\bigcup_{n=1}^{\infty} \{s_n x + r : r \in \mathbb{Q}\},$$

where the rational numbers q have been written as sequence elements s_n (which we are able to do since $\mathbb{Q}$ is countable). Each of the sets in the union is easily seen to be countable, since they are enumerated in terms of the rational numbers r. The countable union theorem applies.

e. Similarly, the set of polynomials over $\mathbb{Q}$ is countable, where this set is defined formally as $P = \{a_n x^n + a_{n-1} x^{n-1} + \cdots + a_0 : a_n, a_{n-1}, \ldots, a_0 \in \mathbb{Q}\}$; i.e., those nth-degree polynomials with rational coefficients. The proof is more complicated but just the same as the last example's proof, as it follows from writing the set of such polynomials as a countable union of countable sets. Here is a quick description of one way to do that: $P = \bigcup_{n=1}^{\infty} P_n$, where P_n is the set of such nth-degree polynomials. Since $P_n = \cup \cdots \cup \{a_n x^n + a_{n-1} x^{n-1} + \cdots + a_0\}$, where each union is taken over the enumerated set of rationals (one union for the range over $\mathbb{Q}$ for each rational coefficient a_j), $|P| = \aleph_0$ follows from the countable union theorem. ∎

Question 1.1.4 Determine if each set is countable or uncountable. Say why.

1. Any real interval $[a, b]$ (with $a < b$).
2. The set of linear polynomials over $\mathbb{Z}$, defined as $T = \{mx + n : m, n \in \mathbb{Z}\}$.
3. The set of linear polynomials over the irrationals, defined as $T = \{ax + b : a, b$ are irrational$\}$. ∎

Though the ideas in this section are profound, the notion of infinity turns out to be much more involved than what we have presented here. For example, there is a way to add different sizes of infinities together to build up new ones, and such arithmetic operations turn out to be consistent with existing set theory (in the sense of Kurt Gödel's and Paul Cohen's work). Similarly (given the Zermelo-Fraenkel axioms), it can be shown that $\aleph_0$ is the smallest infinite cardinal number, but there is no largest. (Cantor was also the first to prove this.) In particular, there is a cardinal infinity larger than $\aleph_1$, and one larger than that, and so on—an infinity of infinities. They must be sizes of sets of objects different from real numbers, since $|\mathbb{R}| = \aleph_1$. Though these facts (after what you have learned here about countable and uncountable sets) might be somewhat intuitive, their proofs result from careful considerations and are rigorous, and many further explorations on infinite cardinalities exist. They often rely on more advanced concepts in set theory than we have room for here, and they await reading mathematics beyond this text's discussions. Instead, in the next section, we'll use countable and uncountable set sizes to develop important facts needed to define the Lebesgue integral.

Solutions to Questions

Throughout the text, solutions to questions are not always complete in their presentation, but offer quick insight into each question's details.

1.1 Two Infinities: Countable and Uncountable

1.1.1 (1) $A = \{-5, -3, -1, 1, 3, \ldots, 13\}$ has $|A| = 10$. (2) $B = \{\ldots, -3, -1, 1, 3, 5, \ldots\} = \{1, -1, 3, -3, 5, -5, \ldots\}$ has $|B|$ equal to countable infinity. (3) $C = \{1, 1\frac{1}{2}, 1\frac{1}{3}, 1\frac{2}{3}, 1\frac{1}{4}, \ldots\}$ has $|C|$ equal to countable infinity.

1.1.2 (1) $\mathbb{Z} \cup A$ is already written as the union of two countable sets. By Theorem 1.1.1, $|\mathbb{Z} \cup A| = \aleph_0$. $\mathbb{Z} \cup A = \{0, 1.5, 1, 2.5, -1, 3.5, 2, 4.5, -2, \ldots\}$ (2) $|A_k| = \aleph_0$ for each $k = 1, 2, \ldots$. By the countable union theorem, $|\bigcup_{k=1}^{\infty} A_k| = \aleph_0$. As a sequence, $\bigcup_{k=1}^{\infty} A_k = \{1.1, 2.1, 1.11, 1.111, 2.11, 3.1, 4.1, 3.11, 2.111, \ldots\}$.

1.1.3 $I_0 = \{0, \frac{1}{2}, \frac{1}{3}, \frac{2}{3}, \frac{1}{4}, \ldots\}$. $I_1 = \{1, 1\frac{1}{2}, 1\frac{1}{3}, 1\frac{2}{3}, 1\frac{1}{4}, \ldots\}$. $I_n = \{n, n+\frac{1}{2}, n+\frac{1}{3}, n+\frac{2}{3}, n+\frac{1}{4}, \ldots\}$. Each set in the union is countable. By the countable union theorem, the union is countable. $\mathbb{Q} = \bigcup_{n=-\infty}^{\infty} I_n$. Hence $|\mathbb{Q}| = \aleph_0$.

1.1.4 (1) Uncountable. The proof is of the same nature as for the proof of Theorem 1.1.3 to show $[0, 1)$ is uncountable. (2) Countable. $T = \bigcup_{m \in \mathbb{Z}} (\bigcup_{n \in \mathbb{Z}} \{mx + n\})$. Since $|\mathbb{Z}| = \aleph_0$, this represents T as a countable union of countable sets. By the countable union theorem, $|T| = \aleph_0$. (3) Uncountable, since for any fixed irrational a, $|\{ax + b : b \text{ is irrational}\}| = |\{b : b \text{ is irrational}\}|$, which is uncountable. Since T contains this uncountable set, it is also uncountable.

Reading Questions for Section 1.1

1. When is a set S countable? When is it uncountable?

2. Give an example of distinct finite sets X and Y with $|X| = |Y|$.

3. Give an example of nonempty finite sets X and Y with $|X| < |Y|$.

4. State the countable union theorem and give an example of its use to prove an infinite set A is countable.

5. What does the continuum hypothesis say? Is it true or false?

6. Give examples of three countably infinite sets.

7. Give examples of three uncountable sets.

Exercises for Section 1.1

Determine if each mathematical statement in Exercises 1–6 is true or false. Prove each true statement; give a counterexample for each false statement.

1. Any set S contained in an uncountable set T must be uncountable.

2. Any set S containing an uncountable set T must be uncountable.

3. Any set S contained in a countable set T must be countable.

4. Any set S containing a countable set T must be uncountable.

5. If S is countable and T is uncountable, then $S \cup T$ is countable.

6. If S is countable and T is uncountable, then $S \cup T$ is uncountable.

In Exercises 7–9, find the cardinality of each finite set, writing its elements as a sequence as in Definition 1.1.1.

7. $S = \{$ Bob, Chris, Jamie, Rob $\}$.

8. The set of prime numbers smaller than 30.

9. The *power set* of the set $A = \{a, b, c\}$. (The power set of A is the set of all subsets of A, including the empty set and the set A itself. The power set of A is denoted 2^A. The elements of 2^A are sets, not elements of A.)

In Exercises 10–22, determine if each set is countably infinite or uncountably infinite. Write the elements of each countably infinite set as a sequence $\{s_1, s_2, s_3, \ldots\}$, or use the countable union theorem to prove the result.

10. The set of natural numbers $\mathbb{N} = \{n : n = 1, 2, 3, \ldots\}$.

11. The set of prime numbers $P = \{p : p \in \mathbb{N}, $ where p has no factor other than 1 and $p\}$.

12. The set of odds $O = \{\ldots, -3, -1, 1, 3, \ldots\}$.

13. The set of rational numbers contained in an arbitrary, nontrivial closed bounded interval $[a, b]$. (For a pair of real numbers a and b with $a < b$, this is the set $S = \{q : q \in [a, b]$ and $q \in \mathbb{Q}\}$.)

14. The Cartesian product set $\mathbb{N} \times \mathbb{N} = \{(m, n) : m, n \in \mathbb{N}\}$.
    (*Hint*: write the ordered pair elements as a countable union of countable sets.)

15. The Cartesian product set $\mathbb{Q} \times \mathbb{Q} = \{(q, r) : q, r \in \mathbb{Q}\}$.

16. The Cartesian product set $A \times B = \{(a, b) : a \in A$ and $b \in B\}$, where A and B are countable.

17. The Cartesian 3-product set $\mathbb{Q}^3 = \{(q_1, q_2, q_3) : q_1, q_2, q_3 \in \mathbb{Q}\}$.

18. The Cartesian n-product set $\mathbb{Q}^n = \{(q_1, q_2, \ldots, q_n) : q_1, \ldots, q_n \in \mathbb{Q}\}$.

19. For a given value n, the set of nth-degree polynomials with rational coefficients, which is
$$S = \{q_0 + q_1 x + q_2 x^2 + \cdots + q_n x^n : q_0, q_1, \ldots, q_n \in \mathbb{Q}\}.$$

(*Hint*: see the previous problem and think about S in terms of an equivalently sized Cartesian product set.)

20. The set of real zeros for any finite degree polynomial with rational coefficients, which is
$$A = \{x : q_0 + q_1 x + q_2 x^2 + \cdots + q_n x^n = 0 \text{ for some } n \in \mathbb{N}$$
$$\text{and coefficients } q_0, q_1, \ldots, q_n \in \mathbb{Q}\}.$$

Numbers in this set are algebraic. Cantor was the first to prove the set of algebraic numbers is countable. (*Hint*: see the previous problem and realize any nth-degree polynomial has at most n distinct real zeros.)

21. The set of irrational numbers between 0 and 1.

22. The set of irrational numbers between two real numbers a and b, where $a \neq b$.

In Exercises 23–28, determine the cardinality of each set.

23. $A = \{x : x^2 + 4x + 3 = 0\}$.

24. $B = \{t : t$ solves some linear polynomial $mx + n = 0$, where $m, n \in \mathbb{Z}\}$.

25. $B = \{t : t$ solves some linear polynomial $mx + n = 0$, where $m, n \in \mathbb{Q}\}$.

26. $B = \{t : t$ solves some linear polynomial $mx + n = 0$, where $m, n \in \mathbb{R}\}$.

27. $C = \{t : t \in \mathbb{R}$ solves some quadratic polynomial $kx^2 + mx + n = 0$, where $k, m, n \in \mathbb{Z}\}$.

28. $C = \{t : t \in \mathbb{R}$ solves some quadratic polynomial $kx^2 + mx + n = 0$, where $k, m, n \in \mathbb{R}\}$.

Exercises 29 and 30 explore theoretical issues about uncountable sets.

29. This exercise motivates the phrase "Cantor's second diagonalization method," which mathematicians call the process for the construction of r in the proof of Theorem 1.1.3. Work through the following procedure, assuming (incorrectly) that the real numbers between 0 and 1 are listed as the sequence $s_1, s_2, s_3, \ldots$ as in the proof of Theorem 1.1.3. But this time, you choose the sequence: start by choosing at least seven values $s_1, \ldots, s_7$, all in $[0, 1)$, that begin the sequence. For example, you could pick s_1 to be $\pi/4$.
    (a) First, for at least the first seven terms of the sequence, list the decimal expansion of each s_n one under the other:
    $$s_1 = 0.b_{11}b_{12}b_{13}b_{14}b_{15}b_{16}b_{17}\ldots$$
    $$s_2 = 0.b_{21}b_{22}b_{23}b_{24}b_{25}b_{26}b_{27}\ldots$$
    $$s_3 = 0.b_{31}b_{32}b_{33}b_{34}b_{35}b_{36}b_{37}\ldots, \text{ and so on.}$$
    (b) Now circle the nth decimal b_{nn} in the expansion of s_n. For example, you should circle the first decimal b_{11} in s_1 and the second b_{22} in s_2. Comment on the pattern you obtain; what forms the diagonal?
    (c) Use your display from part (b), and the length of the diagonal you obtained, to explain why, in the proof of Theorem 1.1.3, r does not equal any of the sequence elements s_n.

30. Prove if $S \cup T$ is uncountable, then either S or T is uncountable. Generalize this to three sets. What would be the generalization for any finite number of sets?

Advanced Exercises. Exercises 31–34 explore the cardinality of the power set of the natural numbers $\mathbb{N}$. This power set is denoted $2^\mathbb{N}$ and is the set of all subsets of $\mathbb{N}$, including the empty set and $\mathbb{N}$ itself. Notationally, $2^\mathbb{N} = \{S : S$ is a subset of $\mathbb{N}\}$. Notice the elements of $2^\mathbb{N}$ are sets, not natural numbers.

31. The proof that $2^\mathbb{N}$ is uncountable goes by way of contradiction. To that end, assume $2^\mathbb{N} = \{S_1, S_2, S_3, \ldots\}$, where each S_i is a subset of $\mathbb{N}$. Define a set S consisting of the natural numbers n for which $n \notin S_n$. Say why $S \in 2^\mathbb{N}$.

32. Say why $S \neq S_i$ for any $i = 1, 2, 3, \ldots$.

33. Explain why the conclusions of the last two exercises contradict the assumption made in Exercise 31.

34. Use your results of the last three exercises to write a complete, formal proof that $|2^\mathbb{N}| = \aleph_1$.

Find the cardinality of each set in Exercises 35–37.

35. The set A of functions f mapping $\{1, 2\}$ to itself.

36. The set B of functions f mapping $\mathbb{N}$ to $\{1, 2\}$. (*Hint*: use the fact that the power set of $\mathbb{N}$ has cardinality $\aleph_1$. This fact follows easily using the same type of Cantor diagonalization argument that formed the crux of the proof of Theorem 1.1.3.)

37. The set C of functions f mapping $\mathbb{N}$ to $\mathbb{N}$. (*Hint*: the answer is $\aleph_1$.)

1.2 A Taste of Measure Theory

Before Henri Lebesgue developed his integral, he described a new way to define the size of a set A, which he called the measure of A and denoted $m(A)$. Lebesgue's measure theory is fascinating but difficult. It typically starts with four defining (or at least desirable) properties. For example, we would naturally (i) want the measure of any real set to be properly defined. We would also (ii) want the measure of the union of a countable number of disjoint sets to equal the summation of their measures; symbolically, $m(\cup A_n) = \sum m(A_n)$, where the sets are each disjoint from the others. (Sets are disjoint when they do not have any common elements.) We might insist (iii) the measure of an interval should be its length. And we would want (iv) the measure of a set to remain the same when it is shifted along the real line (either to the left or right) by a translation. Yet (regrettably) it turns out to be impossible to satisfy all four fundamental principles. Sets of real numbers just don't universally behave in simple, manageable ways when it comes to examining their measures. This fact forced Lebesgue to be very careful when working with the measure of particular sets. For example, if not all sets have a well-defined measure, then Lebesgue had to think carefully about which sets do and which ones don't. Furthermore, a more general theory of measure is very complicated. For example, the measure-theoretic structure of many sophisticated sets requires a description of "measure" that involves a more abstract concept commonly termed "outer measure." A full treatment of Lebesgue's approach to the integral involves these types of advanced considerations. We will not take on that challenge here.

Thankfully, the Daniell-Riesz approach to the Lebesgue integral, which we employ, does not demand such a full consideration of measure. This section presents the only two concepts, both simple, that the Daniell-Riesz approach requires. Namely, we first define Lebesgue's measure for any interval, which is easy. Then we take up a slightly more difficult task—how to identify sets that have measure zero. We'll describe what is meant by "measure zero" (it is the smallest possible set measurement under Lebesgue's formulation). We'll also begin to see how measure-zero sets can help us understand the behavior of functions. The section will finish with many examples and applications. It should be clear we are not interested here in developing measure theory in its full generality, but only with a few facts necessary to define the Lebesgue integral. Also, since we will be interested in integrating real-valued functions, we will restrict our attention throughout this discussion to sets of real numbers. (Lebesgue developed measure theory for a large variety of sets such as sets of complex numbers or sets in the two-dimensional Cartesian plane of real ordered pairs.)

We start with intervals of real numbers. Not surprisingly, the Lebesgue measure of any interval is defined to be its length. This definition is not dependent on whether the interval

1.2 A Taste of Measure Theory

is open or closed (whether it includes its endpoints or not). Intuitively, you may think of an endpoint—or any single point—as adding no additional measure to a set. This fact is presented in

Definition 1.2.1 *The Lebesgue measure of an interval is its length. Unless specified, we will use the term measure for Lebesgue measure and the notation $m(S)$ for the Lebesgue measure of a set S. So*

$$m([a, b]) = b - a,$$

and this result also holds for open or half-open intervals.

For example, the definition says the measure of the half-open interval $[0, 1)$ is $m([0, 1)) = 1 - 0 = 1$. For the open interval $(1, 7)$, it is $m((1, 7)) = 7 - 1 = 6$. We can think of a set that consists of a single point x as a closed interval $[x, x]$, and then $m(\{x\}) = m([x, x]) = x - x = 0$. It is natural to think instinctively of a single point, by itself, as adding no measure to a set. You can already see some sets have Lebesgue measure zero and some do not.

Question 1.2.1 Find the measure for each of the following sets.
1. $S = [7, 42]$
2. $A = (-3, 4]$
3. $B = (-12, -3)$
4. $C = [6, 21)$
5. $D = [1, 2] \cup (3, 9)$ *Hint*: use the fact that $m(A \cup B) = m(A) + m(B)$ for sets A and B having no elements in common. (Such sets are disjoint.) ∎

There are many sets besides those of a single point that have Lebesgue measure zero, and we investigate large categories of them here. We begin with an appropriate definition of such sets, which will turn out to be a useful help to identify many of them in a concrete way. The definition puts a set's measure-zero calculation in terms of a covering of open intervals of the set. We say a collection of open intervals covers a set S when S is contained in the union of the intervals. In this case, we often say the intervals (collectively) form an *open cover* of S. As we apply the definition in a practical way to sets S that arise in this textbook, we will always expect such a collection to consist of at most a countable number of these open intervals. (Though, technically, one can consider an uncountable cover; we won't do that here.)

Definition 1.2.2 *A set S has Lebesgue measure zero when it can be covered with a sequence of open intervals $I_1, I_2, I_3, \ldots$ whose bounded total measure, defined as $\sum_{n=1}^{\infty} m(I_n)$, is arbitrarily small. In this case, S is a null set and we write $m(S) = 0$.*

What does this definition mean by the phrase "arbitrarily small"? Intuitively it means just what it says: we can find an open cover for a Lebesgue measure-zero set S that has bounded total measure as small a number as we want. More rigorously, mathematicians use the Greek symbol ε for the desired small size. Understanding the size must be positive (as it provides an upper bound on interval lengths), we say for measure zero sets S:

> Given $\varepsilon > 0$, S can be covered with a sequence of open intervals $I_1, I_2, \ldots$, where $\sum_{n=1}^{\infty} m(I_n) \leq \varepsilon$.

This phrase intuitively describes ε as small. Use of the term ε will be familiar if you have seen ε-δ definitions of limits in calculus. In this text, we will use ε many times, including the definition of limits of sequences and functions in Section 1.4. The next example illustrates the definition's straightforward application, as we prove a finite set has measure zero.

Example 1.2.1 If S has a finite number of elements, then it has Lebesgue measure zero. For example, a set $S = \{s_1, s_2, s_3\}$ with three elements has $m(S) = 0$, since, given $\varepsilon > 0$, we may cover S with the open intervals $I_1 = (s_1 - \varepsilon/6, s_1 + \varepsilon/6)$, $I_2 = (s_2 - \varepsilon/6, s_2 + \varepsilon/6)$, and $I_3 = (s_3 - \varepsilon/6, s_3 + \varepsilon/6)$. The intervals are centered at s_k of S (hence S is contained in their union), and each has length $(s_k + \varepsilon/6) - (s_k - \varepsilon/6) = \varepsilon/3$, where $k = 1, 2, 3$. This open cover satisfies Definition 1.2.2, since the union of the intervals has bounded total measure $\varepsilon/3 + \varepsilon/3 + \varepsilon/3 = \varepsilon$. By Definition 1.2.2, $m(S) = 0$.

The same reasoning shows $m(S) = 0$ for any finite set S. For if $S = \{s_1, s_2, \ldots, s_n\}$, then, given $\varepsilon > 0$, we can cover S with the open intervals $I_1 = (s_1 - \varepsilon/2n, s_1 + \varepsilon/2n)$, $I_2 = (s_2 - \varepsilon/2n, s_2 + \varepsilon/2n), \ldots, I_n = (s_n - \varepsilon/2n, s_n + \varepsilon/2n)$. The set S is contained in the union of these intervals, since each element of S is the center of one of them. Each interval has length ε/n, and so the bounded total measure of the n intervals is ε (the sum of the n lengths). ∎

Question 1.2.2 Use ideas presented in Example 1.2.1 to find two open covers for the following sets: one of bounded total measure equal to 1, and one of arbitrarily small bounded total measure ε.

1. The set of positive odd integers between 2 and 12.
2. The set of the first five multiples of π: $S = \{n \cdot \pi \mid n = 1, 2, 3, 4, \text{ or } 5 \}$. ∎

In a manner similar to the technique used in Example 1.2.1, Definition 1.2.2 may be used to show $m(S) = 0$ for any countably infinite set S.

Theorem 1.2.1 *Any countably infinite set of reals S has Lebesgue measure zero.*

Proof. Arrange the elements of the set S as a sequence $a_1, a_2, a_3, a_4, \ldots$, and define, for $\varepsilon > 0$,

$$I_n = \left(a_n - \frac{\varepsilon}{2^{n+1}}, a_n + \frac{\varepsilon}{2^{n+1}}\right).$$

Since each element of S is a center of one of the intervals, S is contained in $\bigcup_{n=1}^{\infty} I_n$, and we calculate the bounded total measure (applying the closed form for a geometric series at the penultimate step):

$$\sum_{n=1}^{\infty} m(I_n) = \sum_{n=1}^{\infty} (a_n + \frac{\varepsilon}{2^{n+1}}) - (a_n - \frac{\varepsilon}{2^{n+1}}) = \varepsilon \cdot \sum_{n=1}^{\infty} (1/2)^n = \varepsilon \cdot \frac{(1/2)}{1 - (1/2)} = \varepsilon.$$

By Definition 1.2.2, $m(S) = 0$. ∎

Theorem 1.2.1 points toward important measure-zero sets—ones we have already seen are countably infinite. The natural numbers $\mathbb{N} = \{1, 2, 3, \ldots\}$, the set of prime numbers $\{2, 3, 5, 7, 11, 13, \ldots\}$, the integers $\mathbb{Z} = \{\ldots, -2, -1, 0, 1, 2, \ldots\}$, any set of rationals contained in an interval $[a, b]$, and the full component of rationals $\mathbb{Q}$ all have Lebesgue measure zero. And so the question immediately arises: are there uncountable sets that have Lebesgue

1.2 A Taste of Measure Theory

measure zero? The answer is emphatically (while perhaps surprisingly) yes! The next question describes a famous set, which originated with Cantor's studies, that is both uncountable and has zero Lebesgue measure. It has lots of details and steps, so this example is more involved than others we have provided to this point. But it is important, as it shows not all Lebesgue measure-zero sets are countable.

Question 1.2.3 This question examines the famous "Cantor set," whose elements are contained in the interval $[0, 1]$. The Cantor set, which we will label C, is constructed iteratively, as it is formed by repeated applications of a procedure. To determine C, first remove the middle third $(\frac{1}{3}, \frac{2}{3})$ from the interval $[0, 1]$, so two intervals remain: $[0, \frac{1}{3}]$ and $[\frac{2}{3}, 1]$. Now remove the middle third from each of these two intervals, so $(\frac{1}{9}, \frac{2}{9})$ and $(\frac{7}{9}, \frac{8}{9})$ are removed. Now four intervals remain: $[0, \frac{1}{9}]$, $[\frac{2}{9}, \frac{1}{3}]$, $[\frac{2}{3}, \frac{7}{9}]$, and $[\frac{8}{9}, 1]$. The Cantor set C is the set of elements remaining after an infinite repetition of this process, removing (successively smaller) middle thirds at each step.

Figure 1.4. The formation of the Cantor set C after four applications of removing middle thirds of remaining intervals. The figure's black line segments are the sections of $[0, 1]$ that are removed. The $2^4 = 16$ grey sections are what remain after four applications, and in which C is contained.

1. The Cantor set C clearly contains the element 0, since it is never removed as a middle third of any subinterval. The element $1/3$ is another element of C, for the same reason. Use the interval endpoints described in the above paragraph to name at least six other elements of C.
2. Because each step removes a middle third of the remaining intervals, it turns out the elements in C are precisely represented as those with a ternary expansion $c = \frac{c_1}{3} + \frac{c_2}{3^2} + \frac{c_3}{3^3} + \cdots$, where each of the terms c_n is either a 0 or a 2. For an element c of the Cantor set, no term has a numerator in the expansion that equals 1, since those terms would imply the element is in a corresponding middle third. A standard format for writing this ternary expansion in decimal form is $c = 0.c_1 c_2 c_3 \ldots [3]$. The bracketed 3 at the end of the expansion indicates it is ternary. When a Cantor set number's ternary expansion does not end in all 0s or all 2s, then it is not the endpoint of some removed middle-third interval. For example,

$$0.02020202\ldots [3] = \frac{0}{3} + \frac{2}{3^2} + \frac{0}{3^3} + \frac{2}{3^4} + \cdots = 2\sum_{k=1}^{\infty}(1/9)^k = 2 \cdot (1/9)/[1 - 1/9] = 1/4$$

is an element of the Cantor set. Use this fact to find at least one other element of C that is not the endpoint of a removed middle-third interval.
3. Use the ternary decimal expansion $0.c_1 c_2 c_3 \ldots [3]$ for values in C to prove C is uncountable, mimicking the proof by way of contradiction of Theorem 1.1.3. First, assume C is countable, so the elements in C can be written as a sequence $a_1, a_2, \ldots$. Then write the elements in their ternary expansion in a unique manner, so that, for example, $a_1 = 0.c_{11} c_{12} c_{13} \ldots [3]$. Now apply Cantor's second diagonalization argument to obtain a value c different from any of the sequence elements. This value c, different from each a_i in the list, will have decimal terms that come from a targeted decimal in each of the ternary expansions and where the targeted decimal is changed from 0 to 2 or from 2 to 0. Argue why this c cannot equal any

of the sequence elements, and state why that fact produces a desired contradiction (which finishes the proof).

4. It turns out (from a property that the Lebesgue measure of a countable union of disjoint sets is the summation of the sets' individual measures) that the Lebesgue measure of C may be calculated as the measure of the interval $[0, 1]$ minus the total measure of the intervals that were removed. At the first step, we removed a single interval of length $1/3$, at the second step, two intervals of total length $2/9$, at the third step, four intervals of total length $4/27$, and (as we now see from the pattern), at the nth step n intervals of total length $2^{n-1}/3^n$. Use this observation to show the bounded total measure of the intervals removed is $\frac{1}{3}\sum_{n=1}^{\infty}(\frac{2}{3})^{n-1}$.

5. A geometric series $\sum_{n=1}^{\infty} r^{n-1}$, with $|r| < 1$ is evaluated in closed form as $1/(1 - r)$. Use this fact and the last result to determine the bounded total measure of the union of the intervals removed in the formation of C.

6. Use your last result to show $m(C) = 0$. ∎

The final topic of this section describes an important term often characterizing functions when thinking of their Lebesgue integral. It turns out sets of measure zero will add nothing to the integral: we can change a function's values on any set of measure zero, and it will not affect the Lebesgue integral of the function. And so any two functions whose range output values agree *everywhere except on a set of measure zero* will have the same Lebesgue integral. We'll state and prove this important fact in the next sections. The following definition formalizes this special relationship between functions.

Definition 1.2.3 *Two functions f and g are said to be equal almost everywhere when $f(x) = g(x)$ for all x in the functions' domains, except possibly for a set of x values that has measure zero. In this case, we write $f = g$ a.e., which is read "f equals g almost everywhere." (The notation a.e. stands for "almost everywhere.")*

One important function to study as we compare Lebesgue's integral to Riemann's is the *Dirichlet function*, which is defined on the interval $[0, 1]$ by

$$f(x) = \begin{cases} 1 & \text{if } x \text{ is rational} \\ 0 & \text{if } x \text{ is irrational.} \end{cases}$$

Can you see why $f(x)$ equals the function $g(x) = 0$ almost everywhere? (The reason is that f differs from $g = 0$ on only the rationals in $[0, 1]$, which form a countable set of measure zero.) So we are able to write $f(x) = g(x) = 0$ a.e., and we realize, from the above discussion, that the Lebesgue integral of f should be the same as the Lebesgue integral of 0. Indeed, we will soon see the Lebesgue integral of f is nicely defined to be 0. (Most mathematicians would intuitively say its integral ought to be 0, since $f = 0$ a.e.) What a contrast to the Riemann integral of f, which will turn out to be undefined! The Riemann integral is not powerful enough to integrate f. The Lebesgue integral is.

Question 1.2.4 For the following pairs of functions f and g, determine the set of points S where $f \neq g$. Then determine $m(S)$ and use your result to decide if f equals g almost everywhere.

1. $f(x) = 1$ and $g(x) = \begin{cases} 1 & \text{if } x \text{ is rational} \\ 0 & \text{if } x \text{ is irrational.} \end{cases}$

1.2 A Taste of Measure Theory

2. $f(x) = 0$ and $g(x) = \begin{cases} 1/n & \text{if } x = m/n \text{ is a rational in lowest terms} \\ 0 & \text{if } x \text{ is irrational.} \end{cases}$
(The function g is sometimes called the "modified Dirichlet function.")
3. $f(x) = x$ for all $x \in [0, 1]$ and $g(x) = 0$.
4. $f(x) = x$ and $g(x) = \begin{cases} 1 & \text{if } x \text{ is rational} \\ x & \text{if } x \text{ is irrational.} \end{cases}$
5. $f(x) = x + 1$ and $g(x) = (x^2 - 1)/(x - 1)$.
6. $f(x) = 0$ and $g(x) = \begin{cases} 1 & \text{if } x \in C \\ 0 & \text{if } x \notin C, \end{cases}$ where C is the Cantor set. ∎

We can make many other observations about function properties that hold almost everywhere. Several of them will be important when considering the integral. For example, we might wonder where a function is continuous. Our notion of a property holding almost everywhere (except on a set of measure zero) immediately leads us to consider if the function is continuous almost everywhere (a.e.). Or if a function is differentiable a.e. Or if a sequence of functions converges to a function limit a.e. Or if a function is bounded a.e. We could go on and on, and such considerations will lead to be useful concepts in the upcoming theory. Acknowledging the importance of laying a sure foundation, each item will be taken up in detail when needed.

With measure theory tasted, we're ready to begin integrating. The next section starts the process, showing how to integrate step functions. They are easy to handle, and they will form a foundation on which the Lebesgue integral will be fully defined.

Solutions to Questions

1.2.1 (1) $m(S) = 42 - 7 = 35$. (2) $m(A) = 7$. (3) $m(B) = 9$. (4) $m(C) = 15$.
(5) $m(D) = 1 + 6 = 7$.

1.2.2 Let $I_k = (k - 1/22, k + 1/22)$ for $k = 2, 3, \ldots, 12$, so $\sum_{k=2}^{12} m(I_k) = 1$. Then, for $\varepsilon > 0$, let $I_k = (k - \varepsilon/22, k + \varepsilon/22)$ for $k = 2, 3, \ldots, 12$. These eleven sets have bounded total measure equal to ε. (2) Let $I_n = (n \cdot \pi - 1/10, n \cdot \pi + 1/10)$ for $n = 1, 2, 3, 4, 5$, so $\sum_{n=1}^{5} m(I_n) = 1$. Then, for any $\varepsilon > 0$, let $I_n = (n \cdot \pi - \varepsilon/10, n \cdot \pi + \varepsilon/10)$ for $n = 1, 2, 3, 4, 5$, so they have bounded total measure ε.

1.2.3 (1) For example, $2/3$, $1/9$, $2/9$, $7/9$, $8/9$, $1/27$. (2) For example, $.20202\ldots[3] = (2/3) \cdot \sum_{k=0}^{\infty}(1/9)^k = (2/3)/(1 - 1/9) = 3/4$. (3) The proof follows the detailed outline in the problem statement. (4) Since all of the intervals removed are discrete (they have no points in common), their total measure equals the stated sum. (5) From the formula in part (4), the bounded total measure is $(1/3)/(1 - 2/3) = 1$. (6) Labeling $B = [0, 1]\backslash C$, since B and C are disjoint and from the result from part (5), $1 + m(C) = m(B) + m(C) = m(B \cup C) = m([0, 1)) = 1$. Hence $m(C) = 0$, even though C is an uncountable set.

1.2.4 (1) S is the set of irrationals. No, f does not equal g a.e. (2) $S = \mathbb{Q}$, so $f = g$ a.e.
(3) $S = (0, 1]$, so $f \neq g$ on a set of measure 1. (4) $S = \mathbb{Q}\{1\}$, so $f = g$ a.e. (5) $S = \{1\}$, so $f = g$ a.e. (6) $S = C$, which has $m(C) = 0$, and so $f = g$ a.e.

Reading Questions for Section 1.2

1. What is the definition of Lebesgue measure for an interval of real numbers with endpoints a and b?

2. When does a set S have Lebesgue measure zero?

3. Give an example of two Lebesgue measure-zero sets, at least one having an infinite number of elements.

4. Give an example of a Lebesgue measure-zero uncountable set.

5. When are two functions f and g equal almost everywhere?

6. What is an example of two functions that are not equal, but are equal almost everywhere?

7. How are the Lebesgue integrals for two functions equal almost everywhere related?

Exercises for Section 1.2

Find the measure of each set in Exercises 1–8. Your solution should indicate reasons for your answers.

1. $[0, 2]$
2. $(1, 3)$
3. $(-3, 21.4]$
4. $[-2\pi, \pi + 6)$
5. $(-\sqrt{5}, \sqrt{3}]$

6. $(-3, 2] \cup [6, 9]$ (*Hint*: the measure of two disjoint sets turns out to be the sum of their measures.)

7. $\{-6, -\sqrt{2}, 3, 7, 31, e, \pi\}$
8. $\{2, 4, 6, 8, 10, \ldots\}$

Use Definition 1.2.2 to show each of the sets in Exercises 9–15 have measure zero.

9. $\{-3, 7, 31\}$
10. $\{-6, -\sqrt{2}, 3, 7, 31, e, \pi\}$
11. $\{-20, -19, -18, \ldots, 0, 1, 2, 3, 4, 5\}$
12. $\mathbb{Z}$
13. $\{2, 4, 6, 8, 10, \ldots\}$
14. $\{\ldots, -4, -2, 0, 2, 4, \ldots\}$
15. The rational numbers contained in $(0, 2]$

In Exercises 16–20, determine the set of real values where the given function f does not equal the function g. Then use Definition 1.2.3 to prove $f = g$ almost everywhere.

16. $f(x) = \begin{cases} 1 & \text{if } x \in (0, 1) \\ 0 & \text{if } x \notin (0, 1), \end{cases}$ and $g(x) = \begin{cases} 1 & \text{if } x \in [0, 1] \\ 0 & \text{if } x \neq [0, 1] \end{cases}$

17. $f(x) = \sqrt{x}, x \in (0, 16]$, and $g(x) = \sqrt{x}, x \in [0, 16]$

18. $f(x) = (x^2 - x)/x$ and $g(x) = x - 1$

19. $f(x) = (x^2 - 4)/(x + 2)$ and $g(x) = x - 2$

20. $f(x) = 0$ for $x \in [0, 1]$ and $g(x) = \begin{cases} 1 & \text{if } x \text{ is rational in } (0, 1) \\ 0 & \text{if } x \text{ is irrational in } (0, 1). \end{cases}$

1.2 A Taste of Measure Theory

Exercises 21–26 deal with the Cantor set C.

21. How can we know that the number $0.200200200\ldots [3] = \frac{2}{3} + \frac{0}{3^2} + \frac{0}{3^3} + \frac{2}{3^4} + \cdots$ is in C?

22. Expressing the infinite series expansion of the previous exercise's number as a geometric series, determine its closed-form value.

23. List four other elements of C in terms of their ternary expansion in decimal form (at least two having a nonterminating expansion), and find each number's closed-form value.

24. Is $0.1202020\ldots [3] = \frac{1}{3} + \frac{2}{3^2} + \frac{0}{3^3} + \frac{2}{3^4} + \cdots$ in C? Why not? What is this number's closed-form value?

25. In the formation of C, what is the sum of the Lebesgue measures $m(I)$, where I is any of the intervals removed in the first four applications of removing middle thirds of what remains of $[0, 1]$ at each step?

26. What is the Lebesgue measure $m([0, 1]\backslash C)$, which is the measure of the set of elements in $[0, 1]$ not in C?

Advanced Exercises. Exercises 27–32 explore modified Cantor sets, which are constructed from a process similar to the one used to form C, but modified so the Lebesgue measure of the resulting set might not be zero. As for C, these sets are uncountable. (Their cardinality is $\aleph_1$, but they can be constructed with non-zero Lebesgue measure.)

27. The Smith-Volterra-Cantor set, which we call SVC (named with Cantor after the Irish mathematician Henry Smith and the Italian Vito Volterra), is constructed in exactly the same way as C, but by removing, at the nth step, middle intervals of width $1/4^n$ instead of middle thirds. Therefore, after one step, the measure of the amount removed is $1/4$, and the intervals $[0, 3/8] \cup [5/8, 1]$ remain. After two steps, the total removed is $1/4 + 2 \cdot 1/16 = 1/4 + 1/8 = 3/8$. Determine an infinite sum that represents the total measure of the intervals removed from $[0, 1]$ in the formation of SVC. Writing it as a geometric series, find its numerical value.

28. Using your result from the last exercise, what is the Lebesgue measure $m(SVC)$?

29. Referring to the definition in Exercise 27, does SVC contain any intervals? Why not? (*Hint*: assume it does. What would happen to that interval in a step of the construction of SVC?)

30. If you removed middle fourths instead of middle thirds to form a generalized Cantor set G, then what would be the Lebesgue measure $m(G)$? (*Hint*: note $G \neq SVC$. After two steps of removing middle fourths, the total measure of the set removed is $1/4 + 2(1/4)(3/8) = 1/4 + 3/4^2$. What is it after three steps? After n steps?)

31. What is the Lebesgue measure of a generalized Cantor set formed by removing, at the nth step, middle intervals of width r^n, where $r < 1/3$?

32. A Cantor set D can be formed on any interval $[a, b]$ (not just on $[0, 1]$) by removing, at the nth step, middle thirds of whatever interval remains.
    (a) Starting with an interval $[a, b]$, what is the first interval removed?
    (b) What is the Lebesgue measure of the resulting set?

(c) Cantor understood that two sets C and D both have cardinality $\aleph_1$ exactly when there is a one-to-one, onto function $f : C \to D$ (so $f(c_1) = f(c_2)$ implies $c_1 = c_2$, and for any $d \in D$, there exists an element $c \in C$ such that $f(c) = d$). If C is the Cantor set (formed on $[0, 1]$) and D the Cantor set formed on an interval $[a, b]$, find such a one-to-one onto function f. Make sure you show why your function has the desired properties.

(d) What is the cardinality $|D|$?

A Lebesgue measure can be defined for higher-dimensional sets. For example, we might look at sets in the two-dimensional plane (whose elements may be viewed either as ordered pairs $(x, y) \in \mathbb{R}^2$ or as complex numbers $z = x + iy \in \mathbb{C}$). Exercises 33–35 deal with this measure.

33. An interval I in $\mathbb{R}^2$ is a rectangle with horizontal and vertical sides, and we define the Lebesgue measure of I as its area. Find the measure of the following intervals.
    (a) $I = \{(x, y) : 1 \leq x \leq 5, 2 < y \leq 10\}$
    (b) $I = \{z = x + iy : 1 \leq x \leq 5, 2 < y \leq 10\}$
    (c) $I = \{(x, y) : |x| < 2, |y| < 4\}$

34. Draw each of the intervals in the last problem on a two-dimensional plane.

35. A set S in $\mathbb{R}^2$ has measure zero according to Definition 1.2.2 (but understood in terms of intervals of $\mathbb{R}^2$ instead of $\mathbb{R}$). In other words, $m(S) = 0$ when it can be covered with a sequence $I_1, I_2, I_3, \ldots$ of open intervals (each of the form $\{(x, y) : a < x < b, c < y < d\}$) whose bounded total measure $\sum_{n=1}^{\infty} m(I_n)$ is arbitrarily small. Show any horizontal line $S = \{(x, y) : y = y_0\}$, where $y_0 \in \mathbb{R}$ is a constant, has $m(S) = 0$.

1.3 Lebesgue's Integral for Step Functions

The Lebesgue integral of a function f is designed to be interpreted as its "area under the curve" (between f and the domain axis). Of course, a function can be so discontinuous or oscillating through so many points that this idea must be stretched quite far to make sense. In addition, there exist generalizations (which we will present in Section 4.3) of the integral for which it is intentionally designed not to be interpreted as area under the curve. Those integrals will be defined with respect to other measures we will construct, not the Lebesgue measure briefly explored in the last section. Those measures can produce different results and will be worth studying for different reasons. In this section's beginning investigations of the integral, though, the functions examined will have integrals that can always be thought of as areas. The values the Lebesgue integral produces will have that familiar understanding, matching what you learned about the definite integral from calculus.

A function whose curve forms simple rectangular areas certainly ought to have its integral defined in terms of them. That's the whole point of this section. Functions whose curves have a finite number of rectangular areas underneath them are step functions. They are constant over intervals. The heights of the rectangles come from the constant function values, and the lengths of the corresponding intervals (over which the function is constant) form the widths. The Lebesgue integral of a step function is defined as the sum of the rectangular areas. It's really simple.

1.3 Lebesgue's Integral for Step Functions

We describe step functions notationally in terms of characteristic functions. A characteristic function on a set S uses the Greek letter chi as its functional symbol $\mathcal{X}_S(x)$, which equals 1 if x is in S and 0 if x is not in S. For example, $\mathcal{X}_{\{3,4,7\}}(x)$ is 1 when x is 3, 4, or 7, and it is 0 otherwise. Similarly, $\mathcal{X}_{[1,2]\cup[4,6]}(x)$ is 1 for x between 1 and 2 or between 4 and 6 (inclusive) and 0 otherwise. You get the idea. When we multiply a characteristic function by a constant, it changes the output value from 1 to that constant value. For example, $5 \cdot \mathcal{X}_{\{3,4,7\}}(x)$ is 5 exactly when x is 3, 4, or 7, and 0 otherwise. The function $\mathcal{X}_{[-3,2]}(x)$ is a characteristic function on the interval $[-3, 2]$, and $\mathcal{X}_{(a,b)}(x)$ is a general characteristic function on an (open) interval (a, b). As we talk about step functions, we will always assume the intervals are bounded. (Hence, for step functions, a and b will always be finite.) Every characteristic function on an interval is of the form $\mathcal{X}_{(a,b)}(x)$, except the interval may contain one or both of its endpoints. We formalize these essential facts about a characteristic function in the following definition.

Definition 1.3.1 *A characteristic function on a set S is* $\mathcal{X}_S(x) = \begin{cases} 1 & \text{if } x \in S \\ 0 & \text{if } x \notin S. \end{cases}$

A characteristic function on a bounded interval I is $\mathcal{X}_I(x) = \begin{cases} 1 & \text{if } x \in I \\ 0 & \text{if } x \notin I, \end{cases}$ *where I takes on one of the forms $I = (a, b)$, $[a, b)$, $(a, b]$, or $[a, b]$ for finite endpoints a and b.*

Characteristic functions on bounded intervals form the essential components of step functions. The simplest step function is a characteristic function on an interval multiplied by a constant. It produces a single rectangular area below its curve, and so its Lebesgue integral needs to be defined as the area of that rectangle. In short, if I is an interval and c is a real constant, we should define the Lebesgue integral of $c \cdot \mathcal{X}_I(x)$ to be $c \cdot$ (length of I). Now remember: the length of the interval is the way we defined its Lebesgue measure, $m(I)$. And so—no big surprise here—we need to define this function's Lebesgue integral as

$$\int c \cdot \mathcal{X}_I(x)\, dx = c \cdot m(I).$$

Any step function f is a finite sum of such terms $c \cdot \mathcal{X}_I(x)$, as the next definition makes formal.

Definition 1.3.2 *A step function f has form* $f(x) = \sum_{j=1}^{n} c_j \cdot \mathcal{X}_{I_j}(x)$, *where each c_j is a real constant and each I_j is a bounded interval.*

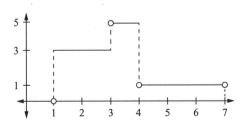

Figure 1.5. The step function $f(x) = 3 \cdot \mathcal{X}_{[1,3]}(x) + 5 \cdot \mathcal{X}_{(3,4]}(x) + 1 \cdot \mathcal{X}_{(4,7)}(x)$.

You can see where the name "step function" comes from when you graph one of them. The graph looks like a series of steps. Sometimes the steps alternate between high and low risers, or wide versus narrow treads, and sometimes they step alternatively up and down (a directional pattern is not necessary), but each such graph visually calls forth a sense of steps. For example, Figure 1.5 displays the graph of $f(x) = 3 \cdot \mathcal{X}_{[1,3]}(x) + 5 \cdot \mathcal{X}_{(3,4]}(x) + 1 \cdot \mathcal{X}_{(4,7)}$. Can you spot the three rectangular regions under its steps?

A general step function's Lebesgue integral is the natural extension of the simpler case (of the constant times a characteristic function on an interval). To integrate

$$f(x) = c_1 \cdot \mathcal{X}_{I_1}(x) + c_2 \cdot \mathcal{X}_{I_2}(x) + \cdots + c_n \cdot \mathcal{X}_{I_n}(x),$$

we sum the n rectangular areas formed, obtaining:

$$\int f(x)\,dx = \int c_1 \cdot \mathcal{X}_{I_1}(x)\,dx + \int c_2 \cdot \mathcal{X}_{I_2}(x)\,dx + \cdots + \int c_n \cdot \mathcal{X}_{I_n}(x)\,dx$$
$$= \sum_{j=1}^{n} \int c_j \cdot \mathcal{X}_{I_j}(x)\,dx.$$

And we know how to integrate each term, since each is a constant times a characteristic function on an interval. Doing so produces the following definition of the Lebesgue integral of any step function.

Definition 1.3.3 *The Lebesgue integral of the step function* $f(x) = \sum\limits_{j=1}^{n} c_j \cdot \mathcal{X}_{I_j}(x)$ *is*

$$\int_{-\infty}^{\infty} f(x)\,dx = \sum_{j=1}^{n} c_j \cdot m(I_j).$$

A comment on notation: we often suppress the dependence of the function on x and write $\int_{-\infty}^{\infty} f(x)\,dx$ as $\int f$. You might notice that this definite integral (as the area under the curve) is taken over all the reals $\mathbb{R}$. We are allowed (but not required) to write it over its integrand values as $\int_{\mathbb{R}} f$. Though the integral is over the whole real line, the step function always equals zero (it vanishes) outside of a finite interval, since all the intervals I_j are bounded. Therefore the integral is always nicely defined as a finite value.

It is not hard to see (and we will prove it before this section is over) that when you add two step functions, you get another step function. That notion is the foundation of an important fact: the collection of step functions forms a linear space. When f and g are step functions, then so is $\alpha f + \beta g$, where α and β are any real constants. The idea of a linear space of functions will be important as we move forward—we will develop the Lebesgue integral on a large space of functions that forms a linear space. (Surely we will want, for example, the sum of two Lebesgue integrable functions to be Lebesgue integrable.) Indeed, we will want this linear space to have other powerful properties associated with it. For example, the property of completeness will guarantee that any Cauchy sequence of functions that are Lebesgue integrable converges to a Lebesgue integrable function. (We will make all these terms precise later, at the time when we explain each property.) For now, the structure of linearity will form the first step toward an ultimate goal: to describe many wonderful relationships between functions that are Lebesgue integrable.

Before more fully discussing some of the important properties of the Lebesgue integral of step functions, we start with an example.

1.3 Lebesgue's Integral for Step Functions

Example 1.3.1 It is easy to calculate the Lebesgue integral of the step function $f(x)$ depicted in Figure 1.5. By Definition 1.3.3,

$$\int_{-\infty}^{\infty} (3 \cdot \mathcal{X}_{[1,3]}(x) + 5 \cdot \mathcal{X}_{(3,4]}(x) + 1 \cdot \mathcal{X}_{(4,7)}(x))\, dx = 3m([1,3]) + 5m((3,4]) + 1m((4,7))$$
$$= 3(3-1) + 5(4-3) + 1(7-4) = 14.$$

We could work in a similar way with $g(x) = 3 \cdot \mathcal{X}_{[1,2]}(x) + \mathcal{X}_{(2,3]}(x) + 2\mathcal{X}_{(2,4]}(x) + 3 \cdot \mathcal{X}_{(3,4]}(x) + 1 \cdot \mathcal{X}_{(4,7)}(x)$. Its Lebesgue integral is

$$\int g = \int (3 \cdot \mathcal{X}_{[1,2]} + \mathcal{X}_{(2,3]} + 2\mathcal{X}_{(2,4]} + 3 \cdot \mathcal{X}_{(3,4]} + 1 \cdot \mathcal{X}_{(4,7)})$$
$$= 3m([1,2]) + m((2,3]) + 2m((2,4]) + 3m((3,4]) + 1m((4,7))$$
$$= 3(2-1) + (3-2) + 2(4-2) + 3(4-3) + (7-4) = 14.$$

The functions f and g are the same (that's why we got the same answer), but represented in two ways. The first writes the function in terms of characteristic functions on intervals that are disjoint; they do not intersect. The second has intervals that intersect, but the constant coefficients at each point sum to the right value. While Definition 1.3.3 requires the integrals to be calculated differently, the answers thankfully always turn out to be the same. ∎

The happy circumstance in the last example is an illustration of a general fact: no matter how you represent a step function, its Lebesgue integral, defined in terms of the representation, turns out to be the same. We prove that result because it is essential to confirm the Lebesgue integral as being "well-defined." You might want to skip ahead and accept it as true without reviewing its proof (which is technical), but you'd miss out—the proof works with step functions in an instructive and interesting way. It starts with the statement and proof of a lemma. (The term "lemma" is used for a straightforward fact, often resulting from detailed mathematical calculations, that helps prove a more powerful result.) We will then use the lemma to prove the main theorem (showing the consistency of the integral across different step-function representations).

Lemma 1.3.1 *A step function $f = \sum_{i=1}^{n} k_i \cdot \mathcal{X}_{J_i}$ can be written as $f = \sum_{j=1}^{m} c_j \cdot \mathcal{X}_{I_j}$, where $I_s \cap I_t = \emptyset$ ($s \neq t$) for any two of the intervals.*[6]

Proof. The proof explicitly describes the construction of the intervals in the second form of f. To that end, examine the n intervals J_i and assume they have r distinct endpoints $p_i, i = 1, 2, \ldots, r$, listed in order. Place the points on the real line and construct the following $m = 2r - 1$ intervals (which form the disjoint I_j's in the second form): r of them "trivial" intervals of the form $[p_j, p_j]$ (which consist of a single point p_j) and $r - 1$ of them the intervals (p_j, p_{j+1}) formed between each pair of adjacent endpoints.

Now for any interval I_j formed in this way, and for any of the original intervals J_i, either $I_j \subseteq J_i$ or $I_j \cap J_i = \emptyset$. Define c_j as the sum of the constants k_i in the original expression that

---

[6] The point of this lemma is to write a step function in a specific way—as a finite sum of constants times characteristic functions of disjoint intervals. This format will be useful; it will allow us to show that the Lebesgue integral of any step function is well-defined.

correspond with those intervals J_i that contain I_j; in other words,

$$c_j = \sum k_i, \text{ where } J_i \cap I_j \neq \emptyset.$$

If there are no such intervals J_i, then $c_j = 0$.

We have now defined disjoint intervals I_j, $j = 1, 2, \ldots, m$ and corresponding constants c_j; we use them to write the function f as $\sum_{j=1}^{m} c_j \cdot \mathcal{X}_{I_j}$. The final step in the proof is to show this has the same value for $f(x)$ as the original representation. But any real value x (between the two extreme points p_1 and p_r) lies in one of the intervals I_s, and then $\sum_{j=1}^{m} c_j \cdot \mathcal{X}_{I_j}(x) = c_s$. From the construction of the constants, this value is $c_s = \sum k_i$, where $I_s \subseteq J_i$. From the construction, the intervals J_i correspond precisely to the characteristic functions having $\mathcal{X}_{J_i}(x) = 1$. Hence the representation $f = \sum_{i=1}^{n} c_i \cdot \mathcal{X}_{I_i}$ produces the same value at x: $f(x) = \sum k_i$, where $I_s \subseteq J_i$. The proof is complete. ∎

Example 1.3.2 We illustrate the construction used in the proof of Lemma 1.3.1 on the step function

$$f = 2 \cdot \mathcal{X}_{[1,3]} + 4 \cdot \mathcal{X}_{[0,2]} + 1 \cdot \mathcal{X}_{(1,4)} + 3 \cdot \mathcal{X}_{(2,6)}.$$

The four intervals used in f's representation overlap. Matching notation in Lemma 1.3.1, the intervals are $J_1 = [1, 3]$, $J_2 = [0, 2]$, $J_3 = (1, 4)$, and $J_4 = (2, 6)$. We now rewrite f so it is constructed from disjoint intervals. From smallest to largest, we list the six endpoints of the four intervals: 0, 1, 2, 3, 4, and 6. We now define six trivial intervals that contain only an endpoint: $I_1 = [0, 0]$, $I_2 = [1, 1]$, $I_3 = [2, 2]$, $I_4 = [3, 3]$, $I_5 = [4, 4]$, and $I_6 = [6, 6]$. We define five open intervals of values between the adjacent endpoints; $I_7 = (0, 1)$, $I_8 = (1, 2)$, $I_9 = (2, 3)$, $I_{10} = (3, 4)$, and $I_{11} = (4, 6)$. Then f is $\sum_{j=1}^{11} c_j \cdot \mathcal{X}_{I_j}$, where c_j is found by adding coefficients from the original representation. For example, $c_1 = 4$, since $x = 0$ (the only value in I_1) is in only one interval $J_2 = [0, 2]$, and that interval has coefficient 4. $c_2 = 2 + 4 = 6$, since $x = 1$ (the only value in I_2) is in J_1 and J_2, and they have coefficients 2 and 4. Similarly, $c_3 = 2 + 4 + 1 = 7$, $c_4 = 2 + 1 + 3 = 6$, $c_5 = 3$, and $c_6 = 0$ (no interval J_i contains $x = 6$).

The other coefficients are defined similarly. The interval $I_7 = (0, 1)$ is contained in J_2 only, and so $c_7 = 4$. Likewise, I_8 is contained in J_1, J_2, and J_3, so $c_8 = 2 + 4 + 1 = 7$. In the same way, $c_9 = 2 + 1 + 3 = 6$, $c_{10} = 1 + 3 = 4$, and $c_{11} = 3$. The function f, in terms of disjoint intervals, is

$$f = 4 \cdot \mathcal{X}_{[0,0]} + 6 \cdot \mathcal{X}_{[1,1]} + 7 \cdot \mathcal{X}_{[2,2]} + 6 \cdot \mathcal{X}_{[3,3]} + 3 \cdot \mathcal{X}_{[4,4]}$$
$$+ 0 \cdot \mathcal{X}_{[6,6]} + 4 \cdot \mathcal{X}_{(0,1)} + 7 \cdot \mathcal{X}_{(1,2)} + 6 \cdot \mathcal{X}_{(2,3)} + 4 \cdot \mathcal{X}_{(3,4)} + 3 \cdot \mathcal{X}_{(4,6)}.$$

That's the expression that the construction in Lemma 1.3.1 produces. It is not the only expression of f in terms of disjoint intervals. In fact, in this case the expression may be simplified: we could combine terms with immediately adjacent intervals that have the same constant coefficient. The disjointness of the intervals is maintained, but in the fewest possible number of terms, and this completes the construction:

$$f = 6 \cdot \mathcal{X}_{[1,1]} + 4 \cdot \mathcal{X}_{[0,1)} + 7 \cdot \mathcal{X}_{(1,2]} + 6 \cdot \mathcal{X}_{(2,3]} + 4 \cdot \mathcal{X}_{(3,4)} + 3 \cdot \mathcal{X}_{[4,6)}.$$

1.3 Lebesgue's Integral for Step Functions

We note, finally, this last representation is unique, in the sense it is the only representation in terms of discrete intervals that cannot be simplified further. In fact, that statement holds for any of the constructions in Lemma 1.3.1—it automatically simplifies to a uniquely written representation in terms of discrete intervals that do not further simplify. The proof of this fact is immediately realized, as this simplification is carried out to obtain the fewest terms. ∎

Whatever representation we use, we want it to produce the same value for $\int f$ as any other representation when we apply Definition 1.3.3. The next theorem proves that happens by equating the integral produced by any representation with the integral produced by the disjoint interval construction in Lemma 1.3.1. But before we present and prove that important result, we illustrate it in detail for the step function we worked with in the last example. The illustration will help us with the general notation and strategy of proof in the theorem.

Example 1.3.3 We show Definition 1.3.3's calculation of the integral of

$$f = 2 \cdot \mathcal{X}_{[1,3]} + 4 \cdot \mathcal{X}_{[0,2]} + 1 \cdot \mathcal{X}_{(1,4)} + 3 \cdot \mathcal{X}_{(2,6)}$$

is identical to Definition 1.3.3's calculation of the integral of

$$f = 4 \cdot \mathcal{X}_{[0,0]} + 6 \cdot \mathcal{X}_{[1,1]} + 7 \cdot \mathcal{X}_{[2,2]} + 6 \cdot \mathcal{X}_{[3,3]} + 4 \cdot \mathcal{X}_{[4,4]}$$
$$+ 0 \cdot \mathcal{X}_{[6,6]} + 4 \cdot \mathcal{X}_{(0,1)} + 7 \cdot \mathcal{X}_{(1,2)} + 6 \cdot \mathcal{X}_{(2,3)} + 4 \cdot \mathcal{X}_{(3,4)} + 3 \cdot \mathcal{X}_{(4,6)}.$$

We first note the integral is based on the measure of the characteristic functions' intervals, and intervals that consist of a single point have measure zero. Therefore, Definition 1.3.3's integral of f as represented immediately above will be the same as when we use the representation

$$f = 4 \cdot \mathcal{X}_{(0,1)} + 7 \cdot \mathcal{X}_{(1,2)} + 6 \cdot \mathcal{X}_{(2,3)} + 4 \cdot \mathcal{X}_{(3,4)} + 3 \cdot \mathcal{X}_{(4,6)}.$$

We need to examine only the $r-1$ components of the function that come from intervals with more than a single point. These are the only ones that contribute to the value of the integral.

We recognize the measure (or length) of each interval J_i in the first representation of f is equal to the sum of the lengths of corresponding intervals I_j in the second. For example, $m[1, 3] = m(1, 2) + m(2, 3)$. Multiplying by the interval's corresponding coefficient produces $2m[1, 3] = 2m(1, 2) + 2m(2, 3)$. The general formula for this relationship is $k_i \cdot m(J_i) = \sum k_i \cdot m(I_j)$, where the sum is taken over all intervals I_j where $I_j \subseteq J_i$. The first representation has four intervals; the four corresponding equations produced in this way are

$$2m[1, 3] = 2m(1, 2) + 2m(2, 3)$$
$$4m[0, 2] = 4m(1, 2) \qquad\qquad + 4m(0, 1)$$
$$1m(1, 4) = 1m(1, 2) + 1m(2, 3) \qquad\qquad + 1m(3, 4)$$
$$3m(2, 6) = \qquad\qquad 3m(2, 3) \qquad\qquad + 3m(3, 4) + 3m(4, 6).$$

Now add down the columns. The sum of the left-hand column gives $\int f$ from Definition 1.3.3 applied to the first representation. We get

$$\int f = 7m(1, 2) + 6m(2, 3) + 4m(0, 1) + 4m(3, 4) + 3m(4, 6).$$

The right-hand side is $\int f$ from Definition 1.3.3 applied to the second representation. ∎

Theorem 1.3.1 *If a step function f has two different representations*

$$f = \sum_{j=1}^{m} c_j \cdot \mathcal{X}_{I_j} = \sum_{i=1}^{n} k_i \cdot \mathcal{X}_{J_i},$$

then the integral is well-defined (in a consistent way) according to Definition 1.3.3:

$$\int f = \sum_{j=1}^{m} c_j \cdot m(I_j) = \sum_{i=1}^{n} k_i \cdot m(J_i).$$

Proof. Lemma 1.3.1 links any representation of f to the one that it constructs in terms of characteristic functions on disjoint intervals. We first show the two representations of f produce an identical calculation of $\int f$. To that end, we assume the first representation $f = \sum_{i=1}^{m} c_j \cdot \mathcal{X}_{I_j}$ is constructed as in Lemma 1.3.1 out of the second representation. Applying Definition 1.3.3 to the second representation, $\int f = \sum_{i=1}^{n} k_i \cdot m(J_i)$. Lemma 1.3.1's construction of the first representation implies $c_s = \sum k_i$, where $I_s \subseteq J_i$.

Following the ideas presented in Example 1.3.3, we get $k_i \cdot m(J_i) = \sum k_i \cdot m(I_j)$, where the sum is taken over the disjoint intervals I_j where $I_j \subseteq J_i$. Summing over the J_i intervals (which corresponds to summing down the $r - 1$ nontrivial interval columns as we did in Example 1.3.3), we get

$$\sum_{i=1}^{n} k_i \cdot m(J_i) = \sum_{i=1}^{n} \sum k_i \cdot m(I_j) = \sum_{j=1}^{r-1} c_j \cdot m(I_j).$$

But that sum of the $r - 1$ terms is the same as the sum over all m terms, since all the other terms in the disjoint representation of f involve intervals that consist of a single point (and therefore have zero measure). We conclude that

$$\sum_{i=1}^{n} k_i \cdot m(J_i) = \sum_{j=1}^{m} c_j \cdot m(I_j).$$

The proof is complete when we realize the term on the right is identical to the integral (from Definition 1.3.3) obtained by simplifying the Lemma 1.3.1 construction to the unique disjoint interval representation—the representation of f that has the fewest terms and intervals that are all disjoint. That simplification (illustrated and discussed in Example 1.3.2) combines terms with immediately adjacent intervals that have the same coefficient. The equivalence to the integral of the simplified representation then follows by observing that, for any coefficient c and immediately adjacent disjoint intervals I_s and I_t, $c \cdot m(I_s) + c \cdot m(I_t) = c \cdot m(I_s \cup I_t)$. The integral based on any representation of f therefore equals the integral calculation for the unique representation of f in terms of disjoint intervals with fewest terms. The proof is complete. ∎

Question 1.3.1 Illustrate the last theorem—prove Definition 1.3.3's calculation of the integral of $f = 3 \cdot \mathcal{X}_{[0,3)} + 2 \cdot \mathcal{X}_{[1,2]} + 1 \cdot \mathcal{X}_{(2,3)}$ is identical to Definition 1.3.3's calculation of the integral of f when written as $f = 3 \cdot \mathcal{X}_{[0,1)} + 5 \cdot \mathcal{X}_{[1,2]} + 4 \cdot \mathcal{X}_{(2,3)}$. Also show the second form

1.3 Lebesgue's Integral for Step Functions

is the unique representation of f from Lemma 1.3.1 that has disjoint intervals with the fewest terms. ∎

The rest of this section discusses some properties of the integral as it is applied to step functions. The complete collection of step functions turns out to form a linear space. That fact follows immediately from straightforward calculations. For example, if the step functions have representations

$$f = c_1 \cdot \mathcal{X}_{I_1} + \cdots + c_n \cdot \mathcal{X}_{I_n} \text{ and } g = s_1 \cdot \mathcal{X}_{J_1} + \cdots + s_m \cdot \mathcal{X}_{J_m}, \text{ then}$$

$$f + g = c_1 \cdot \mathcal{X}_{I_1} + \cdots + c_n \cdot \mathcal{X}_{I_n} + s_1 \cdot \mathcal{X}_{J_1} + \ldots + s_m \cdot \mathcal{X}_{J_m},$$

which is a step function.

The fact that the set of step functions forms a linear space is important. For example, it allows us to discuss the linearity of the integral, which will turn out to be fundamental to the integral's behavior.

Theorem 1.3.2 *For any step functions f and g and constants a and b, the function $a \cdot f + b \cdot g$ is a step function, and*

$$\int (a \cdot f + b \cdot g) = a \cdot \int f + b \cdot \int g.$$

Proof. If $f = \sum_{j=1}^{m} c_j \cdot \mathcal{X}_{I_j}$ and $g = \sum_{j=m+1}^{n} c_j \cdot \mathcal{X}_{I_j}$, then $a \cdot f + b \cdot g = \sum_{j=1}^{n} k_j \cdot \mathcal{X}_{I_j}$, where $k_j = a \cdot c_j$ when $j = 1, 2, \ldots, m$ and $k_j = b \cdot c_j$ when $j = m+1, m+2, \ldots, n$. In other words, $a \cdot f + b \cdot g$ is a step function. Its integral is calculated according to Definition 1.3.3:

$$\int (a \cdot f + b \cdot g) = \int \sum_{j=1}^{n} k_j \cdot \mathcal{X}_{I_j} = \sum_{j=1}^{n} k_j \cdot m(I_j) = \sum_{j=1}^{m} a \cdot c_j \cdot m(I_j) + \sum_{j=m+1}^{n} b \cdot c_j \cdot m(I_j)$$

$$= a \sum_{j=1}^{m} c_j \cdot m(I_j) + b \sum_{j=m+1}^{n} c_j \cdot m(I_j) = a \cdot \int f + b \cdot \int g. \quad \blacksquare$$

Question 1.3.2 Work through each part of this question to illustrate Theorem 1.3.2.

1. Express the step functions $f = 4 \cdot \mathcal{X}_{[0,3)} + 2 \cdot \mathcal{X}_{[1,2]} - 1 \cdot \mathcal{X}_{(2,3)}$ and $g = 2 \cdot \mathcal{X}_{[0,2)} - 4 \cdot \mathcal{X}_{[1,3]}$ using representations that involve only disjoint intervals.
2. Using your results from the last part, find a representation for $6f + 3g$ as a step function.
3. By direct calculations using Definition 1.3.3, show $\int (6f + 3g) = 6\int f + 3\int g$.
4. In the same manner, show $\int (-2f - 5g) = -2\int f - 5\int g$. ∎

This section finishes with a theorem describing one more important property of integrals of step functions. Both this theorem and a version of Theorem 1.3.2 will hold true for general Lebesgue integrable functions, which we present in Section 1.5. You may thereby think of this theorem as a prelude to a more general result.

Theorem 1.3.3 *If f and g are step functions having $f(x) \geq g(x)$ for all real values x, then $\int f \geq \int g$.*

Proof. By Theorem 1.3.2, $(f - g)(x) = f(x) - g(x)$ is a step function. Represent it as $f - g = \sum_{j=1}^{m} c_j \cdot \mathcal{X}_{I_j}$. Since $f(x) \geq g(x)$ for all x values, the coefficients c_j forming the function values must all be nonnegative. By Definition 1.3.3,

$$\int (f - g) = \sum_{j=1}^{m} c_j \cdot m(I_j),$$

which is nonnegative since every term in the sum is nonnegative. (The measure $m(I_j)$ of I_j is its length, which is nonnegative.) Theorem 1.3.2 then implies

$$\int f - \int g = \int (f - g) \geq 0,$$

which proves the result. ∎

Step functions clearly form a special class of functions, and they may feel small in number when compared against the set of functions we will want to integrate (and we will be able to integrate in Section 1.5). But don't let this small army fool you. The collection of step functions will soon form a powerful foundational basis for the Lebesgue integral. Section 1.5 will define integrals of other functions in terms of limits of step-function integrals. In this way, they provide a richness that serves us well—this section's ideas will soon prove extraordinarily useful to define the class of "Lebesgue integrable functions."

Solutions to Questions

1.3.1 From the first representation, $\int f = 3 \cdot 3 + 2 \cdot 1 + 1 \cdot 1 = 12$. From the second, $\int f = 3 \cdot 1 + 5 \cdot 1 + 4 \cdot 1 = 12$. Since the second representation's intervals in each characteristic function are disjoint, and because any such representation must involve at least three terms, it must be the unique construction guaranteed by Lemma 1.3.1.

1.3.2 (1) $f = 4 \cdot \mathcal{X}_{[0,1)} + 6 \cdot \mathcal{X}_{[1,2]} + 3 \cdot \mathcal{X}_{(2,3)}$ and $g = 2 \cdot \mathcal{X}_{[0,1)} - 2 \cdot \mathcal{X}_{[1,2)} - 4 \cdot \mathcal{X}_{[2,3]}$. (2) $6f + 3g = 30 \cdot \mathcal{X}_{[0,2)} + 24 \cdot \mathcal{X}_{[2,2]} + 6 \cdot \mathcal{X}_{(2,3)} - 12 \cdot \mathcal{X}_{[3,3]}$. (3) $\int (6f + 3g) = 30 \cdot 2 + 24 \cdot 0 + 6 \cdot 1 - 12 \cdot 0 = 66$. Since $\int f = 13$ and $\int g = -4$, we get $\int (6f + 3g) = 66 = 6(13) + 3(-4) = 6 \int f + 3 \int g$. (4) The result follows from $-2f - 5g = -18 \cdot \mathcal{X}_{[0,1)} - 2 \cdot \mathcal{X}_{[1,2)} + 8 \cdot \mathcal{X}_{[2,2]} + 14 \cdot \mathcal{X}_{(2,3)} + 20 \cdot \mathcal{X}_{[3,3]}$.

Reading Questions for Section 1.3

1. What is the characteristic function on a set S?

2. What is the form of a characteristic function of an interval?

3. What is the form of a general step function, and what is its associated Lebesgue integral?

4. Give an explicit example of a step function with distinct coefficients over at least three disjoint intervals.

1.3 Lebesgue's Integral for Step Functions

5. How do you calculate the integral of a step function?

6. When f and g are step functions and $a, b \in \mathbb{R}$, what can you say about the function $af + bg$?

7. What can you say about the Lebesgue integrals of step functions f and g when $f \geq g$?

Exercises for Section 1.3

Graph each step function presented in Exercises 1–5 on a separate xy coordinate plane.

1. $f = 4 \cdot \mathcal{X}_{[-2,0]} - 6 \cdot \mathcal{X}_{(0,1]} + 2 \cdot \mathcal{X}_{(1,2]} - 3 \cdot \mathcal{X}_{(3,4)} + \mathcal{X}_{(4,5]}$
2. $g = -.24 \cdot \mathcal{X}_{(-3,-1]} - .60 \cdot \mathcal{X}_{(0,2]} + .25 \cdot \mathcal{X}_{(3,5)} - .30 \cdot \mathcal{X}_{(5,6)} + \mathcal{X}_{(7,8]}$
3. $h = -\pi \cdot \mathcal{X}_{(-3,-1]} + \sqrt{2} \cdot \mathcal{X}_{(0,2]} - \mathcal{X}_{(3,5)}$
4. $s = -10 \cdot \mathcal{X}_{(\sqrt{2},5]} - 6 \cdot \mathcal{X}_{(5,10]} - 5 \cdot \mathcal{X}_{(10,15)} - 3 \cdot \mathcal{X}_{(15,25)} - \mathcal{X}_{(30,40]}$
5. $t = 10 \cdot \mathcal{X}_{(1,5]} + 6 \cdot \mathcal{X}_{(5,10]} + 5 \cdot \mathcal{X}_{(10,15)} + 3 \cdot \mathcal{X}_{(15,25)} + \mathcal{X}_{(30,40]}$

Use the functions defined in Exercises 1–5 to find the integrals listed in Exercises 6–13. Use Definition 1.3.3 to find each answer.

6. $\int f$ and $\int g$
7. $\int h$ and $\int t$
8. $\int (f + g)$
9. $\int (h + s)$
10. $\int (s - h)$
11. $\int 3f$
12. $\int (5g + 2t)$
13. $\int (f + g + h + t)$

Prove each statement in Exercises 14–19 about step functions and their integrals.

14. If f is a step function, then $\int (-f) = -\int f$.

15. The function $f = 0$ is a step function.

16. $\int 0 = 0$.

17. If $a < b < c$ for real values $a, b,$ and c, then $k_1 \mathcal{X}_{(a,c)} + k_2 \mathcal{X}_{(b,c)} = k_1 \mathcal{X}_{(a,b]} + (k_1 + k_2) \mathcal{X}_{(b,c)}$ for any two constants k_1 and k_2.

18. If I and J are intervals with $I \subseteq J$, then $\int \mathcal{X}_I \leq \int \mathcal{X}_J$.

19. If a and b are real constants with $a \leq b$, then $\int a \cdot \mathcal{X}_I \leq \int b \cdot \mathcal{X}_I$ for an interval I.

In Exercises 20–22, express each step function in a representation that involves only disjoint intervals.

20. $f = 2 \cdot \mathcal{X}_{[0,4)} + 3 \cdot \mathcal{X}_{[1,3)} - 4 \cdot \mathcal{X}_{(2,4)}$
21. $g = -2 \cdot \mathcal{X}_{[-1,2)} + 5 \cdot \mathcal{X}_{[1,3]}$
22. $h = 2\pi \cdot \mathcal{X}_{[-4,4)} - 3\pi \cdot \mathcal{X}_{[-\sqrt{2},\sqrt{2})} + 4\pi \cdot \mathcal{X}_{(-1,1)}$

Find each integral's value in Exercises 23–27.

23. $\int (2 \cdot \mathcal{X}_{[0,4)} + 3 \cdot \mathcal{X}_{[1,3)} - 4 \cdot \mathcal{X}_{(2,4)})$
24. $\int (-2 \cdot \mathcal{X}_{[-1,2)} + 5 \cdot \mathcal{X}_{[1,3]})$
25. $\int (2\pi \cdot \mathcal{X}_{[-4,4)} - 3\pi \cdot \mathcal{X}_{[-\sqrt{2},\sqrt{2})} + 4\pi \cdot \mathcal{X}_{(-1,1)})$

26. $\int (10 \cdot \mathcal{X}_{(-10,8)} + 20 \cdot \mathcal{X}_{[-4,4)} - 30 \cdot \mathcal{X}_{[-2,30)} + 40\sqrt{5} \cdot \mathcal{X}_{(0,100)})$

27. $\int (\pi \cdot \mathcal{X}_{(-10,8)} + e \cdot \mathcal{X}_{[-4,4)} - 4\pi^2 \cdot \mathcal{X}_{[-2,30)} + \sqrt{10} \cdot \mathcal{X}_{(0,100)})$

Prove each statement in Exercises 28–37.

28. Define the function $|f|$ by $|f|(x) = |f(x)|$ for all x. Then $|f|$ is a step function whenever f is a step function.

29. $|\int f| \leq \int |f|$ for any step function f. (See the previous exercise's description of $|f|$.)

30. For $a, b \in \mathbb{R}$ with $a \geq b$, $\max\{a, b\} = a$ and $\min\{a, b\} = b$. A standard practice defines the function $\max\{f, g\}$ by $\max\{f, g\}(x) = \max\{f(x), g(x)\}$ for all x, and the function $\min\{f, g\}$ by $\min\{f, g\}(x) = \min\{f(x), g(x)\}$ for all x. Then $\max\{f, g\}$ and $\min\{f, g\}$ are step functions whenever f and g are step functions.

31. Define the function f^+ by $f^+(x) = \begin{cases} f(x) & \text{if } f(x) \geq 0 \\ 0 & \text{if } f(x) < 0 \end{cases}$ for all x. Then f^+ is a step function whenever f is a step function.

32. Define the function f^- by $f^-(x) = \begin{cases} -f(x) & \text{if } f(x) \leq 0 \\ 0 & \text{if } f(x) > 0 \end{cases}$ for all x. Then f^- is a step function whenever f is a step function.

33. $f = f^+ - f^-$.

34. $|f| = f^+ + f^-$.

35. $|f| + |g|$ and $|f + g|$ are step functions whenever f and g are step functions.

36. Define the function fg by $fg(x) = f(x)g(x)$ for all x. Then fg is a step function whenever f and g are step functions.

37. *Schwarz's inequality for step functions*: if f and g are step functions, then $\left(\int fg\right)^2 \leq \int f^2 \cdot \int g^2$. (*Hint*: expand the integral of the step function $(af + bg)^2$, which is nonnegative for all $a, b \in \mathbb{R}$.)

1.4 Limits

To define Lebesgue's integral, we will have to understand limits of terms that occur in various settings: of sequences, of functions, and of sequences of functions. This section defines and describes each type, providing explanations that empower you to follow examples and solve exercises about limits. The material is important mathematically. Its immediate application will be to Lebesgue's integral in the next section. It also comes in handy in other ways—for example, many important characteristics of the real number system are expressed in terms of limits. We'll briefly study one of these important ideas—the axiom of completeness—after looking at limits of sequences.

The concept of limits was applied powerfully (though not always directly or using modern descriptions) in the 17th century and was in place by the time Newton and Leibniz used it as a cornerstone to calculus. They essentially defined the derivative as a limit of the difference

1.4 Limits

quotient as the denominator approaches zero. In 1689, Jacob Bernoulli described the exponential number e as a limit that we now often use as its definition: $e = \lim_{n \to \infty} (1 + 1/n)^n$. But a precise concept of limits was not made rigorous until the early 19th century (around 1820 to be more exact), when European mathematicians began to understand the need for mathematical strictness in the definition of a limit. In France, Augustin Cauchy led this movement. For example, he described (in his famous treatise of 1821) a continuous (increasing) function as one where "an infinitely small increase of the variable produces an infinitely small increase of the function itself" [27, pp. 31–32]. His understanding and rigorous description of many types of limits have become standard mathematical explanations.[7] They are typically defined in terms of arbitrarily small values commonly labeled ε (epsilon) and δ (delta), a notation that Karl Weierstrass first adopted during the 1840s after studying at Germany's University of Münster. The definitions are presented in this section's commentary. We'll start with limits of sequences after first reviewing some notation.

1.4.1 Limits of Sequences

We have already defined a sequence (of real values) as a list of terms in the form $a_1, a_2, a_3, \ldots$, and we can write out sequences longhand in this fashion. A simple example is $S = 1/2, 1/3, 1/4, 1/5, \ldots$. Sequences are often written in terms of their pattern, based on an "index variable" that describes the subscript, say n. You can see that the values for n will be understood as integers (the indexed subscripts in our sequence notation $a_1, a_2, a_3, \ldots$ start at the whole number 1 and increase by one for each term). We use set notation when describing sequences in this way, emphasizing the pattern each term takes in terms of the index n: we write $\{a_n\}_{n=1}^{\infty}$ for $a_1, a_2, a_3, \ldots$. In this way, the description of where the index begins (here at $n = 1$) and ends (here at $n = \infty$) is added to the right set bracket. For example, the pattern associated with $S = 1/2, 1/3, 1/4, 1/5, \ldots$ is in terms of the denominator, which is expressible (in terms of the index) as $n + 1$. It increases by one, just as the index does, for each successive term. Therefore this sequence may be written $S = \{1/(n+1)\}_{n=1}^{\infty}$.

The index does not need to begin at $n = 1$. In fact, we will soon define the limit L of a sequence by how the sequence proceeds, and not on where it starts. We will essentially use only the tail end of the sequence and consider the value L to which those sequence terms eventually approach. In this way, we will often suppress notation that describes where the sequence starts. In fact, we will sometimes suppress the index altogether and just write the sequence as $\{a_n\}$. We can also rewrite the sequence's pattern by reindexing it. Reindexing does not change the sequence (or the pattern on which it is based), as it describes the sequence only with a different beginning index value. A reindexing will require an adjustment in the notation for the pattern, so that the terms of the longhand sequence will still match the terms that the shorthand expression produces. For example, the sequence $S = 1/2, 1/3, 1/4, 1/5, \ldots$ may be thought of as starting at the index $n = 2$ (which most people think is actually a bit more natural). Then the denominators exactly match the index n. This reindexing (starting at $n = 2$) produces $S = \{1/n\}_{n=2}^{\infty}$. It's still the exact same sequence as we discussed before (where we wrote S as $S = \{1/(n+1)\}_{n=1}^{\infty}$). The only change is in a reindexing.

---

[7] Bernhard Bolzano independently developed a modern definition of limit in his work with continuous functions in Prague in 1817, but his ideas went unnoticed until after his death [40].

We are now ready to define rigorously the limit of a sequence.

Definition 1.4.1 *The expression $\lim_{n\to\infty} s_n = L$ means the sequence $\{s_n\}$ approaches L in the following sense: Given $\varepsilon > 0$, there exists a positive real N such that $|s_n - L| < \varepsilon$ whenever $n > N$. In this case, we say L is the limit of the sequence $\{s_n\}$.*

We mention three important insights. First, this definition matches our intuitive idea of a limit. It says that the terms in the sequence get close to L (within any arbitrarily small distance ε) if the term s_n is sufficiently far along in the sequence (where the term's index n is past the value N). Second, the index values for n are integers (typically, positive integers). But in this definition's consideration, the value for N, as the lower bound on the values for n being considered, does not have to be an integer. (Some texts insist $N \in \mathbb{Z}$, but that restriction needs not apply nor do we require it here.) When using the definition to prove a sequence limit is L, it is often best to focus on the term $|s_n - L|$. Start simply: substitute the values for s_n and L. Then simplify the resulting expression as far as possible—enough to bound the simplified result by an expression in N (which is yet to be determined) as you assume $n > N$. Then set this expression in N equal to ε to accomplish two things: it will show $|s_n - L| < \varepsilon$ and you can solve the equation to find an appropriate choice for N (and finish the proof!). The next example shows how it works.

Example 1.4.1 We prove $\lim_{n\to\infty} \frac{1}{n} = 0$.

Proof. Given $\varepsilon > 0$, we focus on the term $|s_n - L|$, substituting values and simplifying as much as possible:

$$|s_n - L| = \left|\frac{1}{n} - 0\right| = \frac{1}{n} < \frac{1}{N}, \text{ whenever } n > N.$$

We were able to remove the absolute value symbols since the index values n may be assumed positive. Now the strategy is to set this expression equal to ε. We get $1/N = \varepsilon$, which determines an appropriate choice of N as $N = 1/\varepsilon$.

Though finding an N that satisfies the definition is the key step in the proof, the last paragraph's logical train of thought does not follow Definition 1.4.1's. To construct a rigorous proof, write more formally. Outline the logic from the calculations. The proof then appears in a beautiful, succinct structure, powerfully using the choice of N. Here's the result for this example:

Given $\varepsilon > 0$, we choose $N = 1/\varepsilon$. Then whenever $n > N$, we have

$$|s_n - L| = \left|\frac{1}{n} - 0\right| = \frac{1}{n} < \frac{1}{N} = \frac{1}{1/\varepsilon} = \varepsilon,$$

and so the limit satisfies Definition 1.4.1. ∎

Example 1.4.2 We use Definition 1.4.1 to show the following limits hold.

1. $\lim_{n\to\infty} \frac{n}{n+1} = 1$.

   Proof. Given $\varepsilon > 0$, choose $N = 1/\varepsilon - 1$ (this choice of N is not obvious but easily becomes clear when simplifying $|s_n - L|$). Finding a common denominator gives

   $$|s_n - L| = \left|\frac{n}{n+1} - 1\right| = \left|\frac{n}{n+1} - \frac{n+1}{n+1}\right| = \left|\frac{n-(n+1)}{n+1}\right| = \left|\frac{-1}{n+1}\right| = \frac{1}{n+1}.$$

1.4 Limits

Whenever $n > N$,

$$|s_n - L| = \frac{1}{n+1} < \frac{1}{N+1} = \frac{1}{(1/\varepsilon - 1) + 1} = \varepsilon,$$

and so the limit satisfies Definition 1.4.1.

2. $\lim\limits_{n\to\infty} (1/9)^n = 0$.

Proof. Given $\varepsilon > 0$, we focus on $|s_n - L|$ to determine an appropriate choice for N:

$$|s_n - L| = \left|\left(\frac{1}{9}\right)^n - 0\right| = \frac{1}{9^n} < \frac{1}{9^N}$$

whenever $n > N$. Setting this simplified expression equal to ε, taking reciprocals and then the logarithm base 9 of both sides gives $\log_9(9^N) = \log_9(1/\varepsilon)$. Then $N\log_9(9) = \log_9(1/\varepsilon)$, or $N = \log_9(1/\varepsilon)$. With an appropriate choice for N determined, we can write the proof into the same logical flow of Definition 1.4.1:

Given $\varepsilon > 0$, we choose $N = \log_9(1/\varepsilon)$. Then whenever $n > N$,

$$|s_n - L| = \left|\left(\frac{1}{9}\right)^n - 0\right| = \left|\left(\frac{1}{9}\right)^n\right| = \frac{1}{9^n} < \frac{1}{9^N} = \frac{1}{9^{\log_9(1/\varepsilon)}} = \frac{1}{1/\varepsilon} = \varepsilon. \quad \blacksquare$$

Question 1.4.1 Provide a proof, following the logical flow of Definition 1.4.1, of the following limits.

1. $\lim\limits_{n\to\infty} \dfrac{3}{2n+1} = 0$
2. $\lim\limits_{n\to\infty} \dfrac{2n}{5n+1} = \dfrac{2}{5}$ ∎

One more important comment describes the completeness of sets. It is easy to see (as the terms could be defined as the expansion of π out to the nth digit) that the following sequence converges[8] to π :

$$3, \ 3.1, \ 3.14, \ 3.141, \ 3.1415, \ 3.14159, \ldots.$$

Each element in the sequence is a rational number (each has a finite decimal expansion), but the limit is not: π is irrational. This says the set of rationals $\mathbb{Q}$ is not complete; there exist convergent sequences of elements of $\mathbb{Q}$ whose limit is outside of the set.

Completeness turns out to be an important concept, and it is one that is difficult to study for many sets, including the set of real numbers $\mathbb{R}$. In the end, we will have to treat the completeness of $\mathbb{R}$ as a fundamental property expressed axiomatically. It might make you uncomfortable to hear that such an important property relies on an axiom—completeness is going to be *assumed* rather than *proven*. Let such discomfort turn into a feeling of reassurance. Mathematicians now know $\mathbb{R}$ is completely characterized by this property along with two others. (Without a thorough description, those are that ">" provides an ordering for real numbers, and an algebraic structure, using multiplication and addition, makes $\mathbb{R}$ what mathematicians call a "field"). Any other set with these three axioms will always act mathematically like the set of real numbers. So the axioms behave as they should—they characterize $\mathbb{R}$ in a fundamental way. We describe this axiomatic

---

[8] A sequence $\{s_n\}$ converges when $\lim_{n\to\infty} s_n$ exists.

approach to completeness by stating it formally, and we do so using a notion of a sequence being bounded above and nondecreasing (a standard approach used in many textbooks).

Axiom of Completeness *A sequence of real numbers that is nondecreasing and bounded above converges to a real number.*

The next question briefly considers how the axiom of completeness plays out for a simple example.

Question 1.4.2 Let $\{s_n\} = \{1, 3/2, 5/3, 7/4, \dots\}$.

1. State why the nondecreasing sequence $\{s_n\}$ is bounded above: provide an upper bound.
2. What does the axiom of completeness then assure us about $\{s_n\}$?
3. What real value is the limit of the sequence $\{s_n\}$? ∎

The axiom of completeness can be phrased in many different ways: the terms of the bounded nondecreasing sequence always have a least upper bound; or any bounded sequence always has a convergent subsequence; or any sequence whose terms get arbitrarily ε-close (where $|a_n - a_m| < \varepsilon$ when n and m are larger than a corresponding N) always converges in limit to a real value. It turns out any one of these descriptions of the axiom of completeness is logically equivalent to any other (cf. [1]). They are essentially all different ways to say intuitively that the set of real numbers has no gaps in it. It should feel comforting to know that fact, as it should feel comforting to know that this property is one of the very defining axioms of the reals. We will discuss some of these issues in more detail in Chapter 2, when we discuss least upper bounds (also called *suprema*) and use them to think about the way the integral is defined.

Question 1.4.3 This question describes $\mathbb{Q}$ as an algebraic field in which addition and multiplication of two rationals result in a rational number. But it also shows $\mathbb{Q}$ is not complete, since the limit of a sequence of rationals is not always a rational number. In this way, $\mathbb{Q}$ has a type of totality (closure) for basic algebraic processes, but lacks another totality (completeness) for the process of taking limits.

1. For rational numbers a/b and c/d (where a, b, c, and d are integers and $b, d \neq 0$), show both the sum $(a/b) + (c/d)$ and the product $(a/b) \cdot (c/d)$ are of the form m/n, where $m, n \in \mathbb{Z}$ and $n \neq 0$. (*Hint*: use the fact that the sum and product of integers is an integer, and the product of two nonzero numbers is nonzero.)
2. Let $a_1 = .1$, $a_2 = .12$, $a_3 = .123$, and so on (here, $a_{11} = .1234567891011$). Show that $\{a_n\}$ is nondecreasing, and that a_n is in $\mathbb{Q}$ for $n = 1, 2, 3, \dots$ but $\lim_{n \to \infty} a_n$ is not. ∎

1.4.2 Functional Limits

We now turn our attention to limits of functions, a topic you have likely already studied in an introductory calculus course. Intuitively, we say a function f approaches a limit L as its domain variable x approaches a value c when f gets close to L as x gets close to c. The corresponding rigorous mathematical definition is essentially due to Augustin Cauchy and Karl Weierstrass, as it uses small positive values δ (delta) and ε (epsilon) to describe the small distances between x and c, and f and L, respectively.

There are important technicalities when using this definition, but we state them here in advance as generally well understood when we formulate both theorems and exercises. We want

1.4 Limits

to avoid getting bogged down into what a function is or becoming overly shackled by these technicalities—they are theoretically crucial but do not need to be the focus of our work. This text will thereby generally assume we are working with functions $f : D \to Y$ defined on a real variable, with domain set D and range Y well understood (either from the context of the problem or from an explicit definition). Any deviation from this structure will be unambiguously stated. As we take functional limits, the variable x will be assumed to approach a value $c \in \mathbb{R}$ so that an open interval containing c exists whose elements, except possibly c itself, are all in D. Precise clarification may be found in a calculus book, or you may wish to refer to more advanced texts such as [1]. We're now ready to state the definition.

Definition 1.4.2 *When L is the limit of f as x approaches c, then we write $\lim_{x \to c} f(x) = L$, which means*

For every $\varepsilon > 0$, there exists $\delta > 0$ such that $0 < |x - c| < \delta$ implies $|f(x) - L| < \varepsilon$.

Just as for limits of sequences, this definition matches our intuitive idea of a limit. It says the function values $f(x)$ get close to L (within any arbitrarily small distance ε) as x is sufficiently close to c (within some small distance δ). Generally speaking, the value of δ may depend on ε, but it cannot depend on the variable x. Analogous to our work with sequence limits, as we prove a limit satisfies this definition, we will focus on the term $|f(x) - L|$. Start simply, substituting values for $f(x)$ and L and then simplifying as far as possible. Look for the term $|x - c|$ as part of the simplified expression, and then apply the fact that $|x - c| < \delta$ (where δ is yet to be determined). Then eliminate the values of x from the simplification by bounding them (from above) by a term involving only δ. Finally, set the resulting expression for δ equal to ε to accomplish two things: it will show $|s_n - L| < \varepsilon$ and you can solve for an appropriate δ in terms of ε, which will finish the proof. The following example shows how it works.

Example 1.4.3 Use the definition to prove $\lim_{x \to 3}(4x + 2) = 14$.

Proof. Given $\varepsilon > 0$, we find a choice for δ by focusing on $|f(x) - L|$:

$$|f(x) - L| = |(4x + 2) - 14| = |4x - 12| = 4 \cdot |x - 3| < 4 \cdot \delta, \text{ whenever } 0 < |x - 3| < \delta.$$

Now we can set this last term equal to ε, which does two things: it automatically results, from the string of equalities and inequalities above, in $|f(x) - L| < \varepsilon$; and it lets us solve for an appropriate choice of δ. We therefore set $4 \cdot \delta = \varepsilon$ and obtain $\delta = \varepsilon/4$. With this choice of δ, we can now write the proof in the same logical flow of Definition 1.4.2.

Given $\varepsilon > 0$, choose $\delta = \varepsilon/4$. Then, whenever $0 < |x - 3| < \delta$,

$$|f(x) - L| = |(4x + 2) - 14| = |4x - 12| = 4 \cdot |x - 3| < 4 \cdot \delta = \varepsilon.$$

The limit therefore satisfies the definition. ∎

You can mimic the last example to find an appropriate choice for δ, whenever proving a correctly evaluated limit of a linear function satisfies the definition. In fact, that value for δ always turns out to be $\varepsilon/|m|$, where m is the slope of the linear function. These facts become apparent in the next question.

As you work each problem involving limits and the definition, make sure your work is neat, organized, and accurate, reflecting properly the roles each item, such as ε, plays in the definition.

The logical procession involved in the definition is precise, and it is easy for beginners to get off track. You may always choose to start your proof with "Given $\varepsilon > 0 \ldots$." That written construction is powerful, since it refers properly to the fact that ε will represent an arbitrarily small distance. It also keeps the procession on track, since the choice of δ and the inequalities to follow will almost always have to depend on the value of ε.

Question 1.4.4 Show each limit satisfies Definition 1.4.2.

1. $\lim_{x \to 1}(5x - 8) = -3$ 2. $\lim_{x \to 3}(-4x + 2) = -10$ 3. $\lim_{x \to 2}(x^2 - 4)/(x - 2) = 4$ ∎

Each limit in the last question involved linear terms (the last one simplified to a linear expression when $x \neq 2$). Limits involving nonlinear expressions can be more complicated when using the definition. Finding a bound (involving only δ or constants) may get out of hand if the only strategy is to apply directly the assumption $|x - c| < \delta$. Thankfully, straightforward methods exist to make proofs of such limits manageable. Some authors call a very useful approach the "restriction technique"; you might wish to read about that process in detail in [47, Section 4.1] or [66, Section 4.3]. We present the restriction technique here in a quick example. Our hope is that it illustrates the process well.

Example 1.4.4 Show $\lim_{x \to 2}(x^3 - 8) = 0$.

Solution. Given $\varepsilon > 0$, we examine $|f(x) - L| = |(x^3 - 8) - 0| = |x^3 - 8| = |x - 2| \cdot |x^2 + 2x + 4|$ (from the algebraic formula to factor a difference in cubes). The definition assumes the first term in the product is less than δ, and so that term is easily handled. But the second term is a mess. To obtain an upper bound for it, use the "restriction technique," which proceeds as follows.

The Restriction Technique: Use the interpretation of δ as the distance between x and $c = 2$, along with the fact that we have control over δ (we are, after all, going to choose an appropriate value for δ). Assume x is within, say, 1 unit of $c = 2$, so $1 \leq x \leq 3$. This restriction on x is equivalent to stating $\delta \leq 1$. For the restricted x values, because the second term $|x^2 + 2x + 4|$ increases, it is between $1^2 + 2 \cdot 1 + 4$ and $3^2 + 2 \cdot 3 + 4$, or between 7 and 19. In any event, the term is always less than or equal to 19 when $\delta \leq 1$, which provides the bound we need to handle the problem easily.

It means, whenever $\delta \leq 1$ and $|x - 2| < \delta$, that $|f(x) - L| = |x - 2| \cdot |x^2 + 2x + 4| \leq |x - 2| \cdot 19 < \delta \cdot 19$. We now set $\delta \cdot 19 \leq \varepsilon$ to obtain the desired inequality $|f(x) - L| < \varepsilon$.

The only task that remains is to determine precisely an appropriate choice of δ. The above analysis requires two things: $\delta \leq 1$ and $\delta \cdot 19 \leq \varepsilon$ (which is equivalent to $\delta \leq \varepsilon/19$). We get both requirements by choosing δ to equal the minimum of the two values: $\delta = \min\{1, \varepsilon/19\}$. The proof is complete. ∎

Question 1.4.5 Use the restriction technique to show each limit satisfies Definition 1.4.2.

1. $\lim_{x \to 3} x^2 = 9$ 2. $\lim_{x \to 2}(x^2 + x) = 6$ 3. $\lim_{x \to 3}(x^3 - 7) = 20$ ∎

We now give a short description of continuity, an extraordinarily important and distinctive property that some functions have. Continuity will turn out to ensure various nice properties of integrals. For example, every continuous, bounded function will have a well-defined Lebesgue

1.4 Limits

integral over a finite range of integration. Intuitively, a function is continuous over a domain set when it is unbroken—you can graph the function without ever "letting your pencil leave the graph paper." We also define continuity at a single value c: a function is continuous at c when its limit there equals $f(c)$. (This statement implicitly states that both $f(c)$ and the limit exist.) Therefore, continuity at c means these values exist and $\lim_{x \to c} f(x) = f(c)$. The next definition formalizes this concept. As in Definition 1.4.2, it assumes f is a function with domain D and there is an open interval containing c that is in D.

Definition 1.4.3 *A function f is continuous at $x = c$ when, given $\varepsilon > 0$, there exists $\delta > 0$ such that $|f(x) - f(c)| < \varepsilon$ whenever $|x - c| < \delta$ (and x is in the domain D). When f is continuous at every value c in a set S, then we say f is continuous on S. When f is continuous on its entire domain, then we say f is a continuous function.*

You can see the definition for continuity at c generally matches the definition of limit as x approaches c, where $f(c)$ is substituted for L. One subtle but important difference (due to the fact that $f(c)$ exists) is that continuity's definition examines x such that $|x - c| < \delta$, not $0 < |x - c| < \delta$. In practice, proving continuity at c is similar to proving a limit, as the next example makes plain.

Example 1.4.5 We examine the function $f(x) = 5x + 2$ in terms of continuity properties. First, we can show f is continuous at a particular point such as $c = 1$. Here, $f(1)$ exists and equals $5 \cdot 1 + 2 = 7$. To show the corresponding limit exists and equals $\lim_{x \to 1}(5x + 2) = 7 = f(1)$, assume $\varepsilon > 0$, and choose $\delta = \varepsilon/5$. Then, whenever $|x - 1| < \delta$, we have

$$|f(x) - f(1)| = |(5x + 2) - 7| = |5x - 5| = 5 \cdot |x - 1| < 5 \cdot \delta = 5 \cdot (\varepsilon/5) = \varepsilon,$$

and so Definition 1.4.2 is satisfied.

We can also show f is continuous over a given set, such as over the interval $[1, 2]$. For if we are given any value c in $[1, 2]$, then f satisfies the definition of continuity at c: given $\varepsilon > 0$, we choose $\delta = \varepsilon/5$. Then, whenever $|x - c| < \delta$, we have

$$|f(x) - f(c)| = |(5x + 2) - (5c + 2)| = |5x - 5c| = 5 \cdot |x - c| < 5 \cdot \delta = 5 \cdot (\varepsilon/5) = \varepsilon.$$

The function f is therefore continuous on $[1, 2]$.

A similar proof would show f is continuous at any value c in $\mathbb{R}$ (indeed, the last proof never used the fact that c was in $[1, 2]$). We conclude f is a continuous function: it is continuous at every point in its domain $D = \mathbb{R}$. ∎

Question 1.4.6 A. Show each function below is continuous at $x = 1$.

1. $f(x) = 7x - 4$
2. $g(x) = 1/x$ (*Hint*: show the limit of $g(x)$ as x approaches 1 equals $g(1)$, using the restriction technique.)

B. Now show each function below is continuous.

1. $h(x) = 3x + 7$
2. $s(x) = x^2$ (*Hint*: use the restriction technique to show the limit of $s(x)$ as x approaches any real value c equals $s(c)$.) ∎

Functions may be continuous at some points but not at others. For example, the piecewise defined function $g(x) = \begin{cases} 1 & \text{if } x \geq 3 \\ 0 & \text{if } x < 3 \end{cases}$ is not continuous at $c = 3$. Its limit does not exist there, since its left- and right-hand limits do not agree. But g is continuous at every other real value. The Dirichlet function[9]

$$d(x) = \begin{cases} 1 & \text{if } x \text{ is rational} \\ 0 & \text{if } x \text{ is irrational} \end{cases}$$

is not continuous at any value, since any group of numbers close to any value c would have some numbers that $d(x)$ maps to 1 and some that $d(x)$ maps to 0. So no limit for d exists at any c.

The examples are relevant to integration theory. The function g turns out to have well-defined Riemann and Lebesgue integrals over any finite interval of integration. But the Dirichlet function $d(x)$ never has a well-defined Riemann integral, even though we can see $d(x)$ differs from 0 on only a set of measure zero. Lebesgue's integral comes to the rescue, and the Lebesgue integral of $d(x)$ is well-defined and equals 0. Chapter 2 will rigorously show the number of places where f is discontinuous can play havoc with the Riemann integral. In fact, a description of a function's points of discontinuity will turn out to characterize precisely when it is Riemann integrable. But the Lebesgue integral is robust in terms of the function's continuity—it will often nicely exist, such as for the Dirichlet function, even when the function is nowhere continuous!

1.4.3 Limits of Function Sequences

We finish this section with a description of convergent sequences of functions. The situation arises as a special case of a convergent sequence, but where the elements of the sequence are functions evaluated at a specific point x. Such sequences are of the form

$$f_1(x), f_2(x), f_3(x), f_4(x), \ldots .$$

The definition of this sequence's convergence to a limit, say $f(x)$, will mimic the definition of sequence convergence from Section 1.4.1, and we present it here.

Definition 1.4.4 *For a specific x, the expression $\lim\limits_{n \to \infty} f_n(x) = f(x)$ means the sequence $\{f_n(x)\}$ approaches the value $f(x)$ in the following sense: Given $\varepsilon > 0$, there exists $N > 0$ such that $|f_n(x) - f(x)| < \varepsilon$ whenever $n > N$. Then we say $f(x)$ is the limit of the function sequence $\{f_n(x)\}$. When convergence happens over a domain set of x-values D, then the corresponding set of limit values $f(x)$ defines a function f on D. We then write $\lim\limits_{n \to \infty} f_n = f$ and say f is the pointwise limit of f_n.*

Definition 1.4.4 thinks of the elements of the sequence as the numbers obtained when evaluating the functions f_n at a specific x. At least that's the most fundamental way to understand the definition—as a sequence of numbers. But we can then consider the set of sequences we get when looking at a collection of x values throughout some well-defined set D (perhaps a common domain set for the f_n functions). Then the sequence $\{f_n(x)\}$ is a sequence of *functions*,

---

[9] This function is named after the German mathematician Peter Gustav Lejeune Dirichlet, who, among other great accomplishments, proved in 1825 when 20 years old the case $n = 5$ of Fermat's last theorem.

1.4 Limits

defined in terms of range values $f(x)$ by the limits at each x. In this way, the sequence of functions f_n can be thought of as approaching some pointwise limit function f. Plus, it's useful. For example, Section 1.5 will use the next example to evaluate the Lebesgue integral $\int_0^1 x \, dx$.

Example 1.4.6 We define a sequence of functions on the interval $[0, 1]$.

$$\text{Let } f_1(x) = \begin{cases} 0 & \text{if } x \in [0, \tfrac{1}{2}) \\ 1/2 & \text{if } x \in [\tfrac{1}{2}, 1], \end{cases} \qquad f_2(x) = \begin{cases} 0 & \text{if } x \in [0, \tfrac{1}{4}) \\ 1/4 & \text{if } x \in [\tfrac{1}{4}, \tfrac{1}{2}) \\ 1/2 & \text{if } x \in [\tfrac{1}{2}, \tfrac{3}{4}) \\ 3/4 & \text{if } x \in [\tfrac{3}{4}, 1], \end{cases} \quad \text{etc.}$$

The nth function in the sequence is piecewise defined on 2^n intervals of equal width $1/2^n$, and it outputs the range value $k/2^n$ on the kth interval, where k runs from 0 to $2^n - 1$. Parts 1–3 below examine the sequence.

1. Examine $x = 1/3$. Find $\lim_{n \to \infty} f_n(\tfrac{1}{3})$.

Solution. Evaluating f_n at $x = 1/3$ produces the sequence

$$\{f_n(1/3)\}_{n=1}^\infty = \left\{0, \frac{1}{4}, \frac{2}{8}, \frac{5}{16}, \frac{10}{32}, \ldots\right\},$$

which approaches $1/3$ in limit. In fact, we can show $\lim_{n \to \infty} f_n(1/3) = 1/3$ satisfies the definition. We will use the following key observation: since the sequence function's output follows from which interval $1/3$ falls into, the nth term in the sequence is $f_n(1/3) = k/2^n$, where $k \in \{0, 1, \ldots, 2^n - 1\}$ and k is determined by $k/2^n \leq 1/3 < (k+1)/2^n$. Proceeding now with the definition, given $\varepsilon > 0$, we choose $N = \log_2(1/\varepsilon)$. Since $k/2^n \leq 1/3 < (k+1)/2^n$,

$$|f_n(1/3) - 1/3| = \left|\frac{k}{2^n} - \frac{1}{3}\right| \leq \left|\frac{k}{2^n} - \frac{k+1}{2^n}\right| = \frac{1}{2^n} < \frac{1}{2^N} = \varepsilon, \text{ whenever } n > N.$$

The proof that the limit satisfies Definition 1.4.4 is complete.

2. For any $x \in [0, 1]$, $f_n(x) = k/n$ when x is in the interval $[\tfrac{k}{n}, \tfrac{k+1}{n})$ for some $k = 0, 1, 2, \ldots, 2^n - 1$. Also, $f_n(1) = (2^n - 1)/2^n$. Use these facts to show $|f_n(x) - x| < 1/2^n$.

Solution. The proof is similar to the case in part 1, when $x = 1/3$. When $x \in [\tfrac{k}{2^n}, \tfrac{k+1}{2^n})$, then $|f_n(x) - x| = |k/2^n - x|$. But because $x \in [\tfrac{k}{2^n}, \tfrac{k+1}{2^n})$, it can be no farther away from the term $k/2^n$ than at the right endpoint of this interval. There, the distance between $k/2^n$ and $(k+1)/2^n$ is $1/2^n$, and so $|k/2^n - x| < 1/2^n$.

3. Use the result in part 2 to prove $\lim_{n \to \infty} f_n(x) = x$ for any $x \in [0, 1]$. Thus $f(x) = x$ can be thought of as the pointwise limit of the sequence of functions $\{f_n(x)\}$.

Solution. A given $x \in [0, 1)$ must be in an interval $[\tfrac{k}{2^n}, \tfrac{k+1}{2^n})$ for some $k = 0, 1, 2, \ldots, 2^n - 1$. Given $\varepsilon > 0$, we choose $N = \log_2(1/\varepsilon)$. Then for $n > N$, part 2 shows $|f_n(x) - x| < 1/2^n < 1/2^N = \varepsilon$, and the limit satisfies Definition 1.4.4 for all $x \in [0, 1)$. The special case $x = 1$ is handled similarly. ∎

As mentioned, the next section will use the last example to evaluate the Lebesgue integral $\int_0^1 x\,dx$. The example will also illustrate a general method to find the Lebesgue integral of a broad class of functions, including continuous functions. Why will it work so well? Because the Daniell-Riesz approach to the Lebesgue integral of f will look for a nondecreasing sequence of step functions f_n whose limit is f (at least almost everywhere). The Lebesgue integral of f will then be the limit of the Lebesgue integral of f_n. That will be made precise in the next section. The example just discussed found such a sequence for the function $f(x) = x$ on the interval $[0, 1]$. (Each f_n is a step function because it is constant over each interval of width $1/2^n$.) The Daniell-Riesz approach will use the integrals of the step functions to calculate the Lebesgue integral $\int_0^1 x\,dx$.

The following question gives you some experience to mimic the kinds of calculations in Example 1.4.6. Several exercises at the end of this section give you further practice with similar function sequences.

Question 1.4.7 Examine a variation on Example 1.4.6. On $[0, 2]$, define the sequence of functions

$$g_1(x) = \begin{cases} 0 & \text{if } x \in [0, 1) \\ 1 & \text{if } x \in [1, 2], \end{cases} \qquad g_2(x) = \begin{cases} 0 & \text{if } x \in [0, \tfrac{1}{2}) \\ 1/2 & \text{if } x \in [\tfrac{1}{2}, 1) \\ 1 & \text{if } x \in [1, \tfrac{3}{2}) \\ 3/2 & \text{if } x \in [\tfrac{3}{2}, 2], \end{cases} \qquad \text{etc.}$$

Each function $g_n(x)$ is piecewise defined on the 2^n subintervals of $[0, 2]$ that have length $1/2^{n-1}$. Additionally, the function values $g_n(x)$ equal the minimum x value in the subinterval, so each $g_n(x)$ value is of the form $k/2^{n-1}$ for some $k = 0, 1, 2, \ldots, 2^n - 1$.

1. Overlay the graphs of g_1 and g_2 on the same display. Also plot $g(x) = x$, $x \in [0, 2]$ on it.
2. For $x \in [0, 2]$, why is $\{g_n(x)\}$ a nondecreasing sequence of values?
3. For $x \in [0, 2)$, show $|g_n(x) - x| < 1/2^{n-1}$.
4. Use your result from part 3 to show $\lim_{n \to \infty} g_n(x) = g(x) = x$ for any $x \in [0, 2]$. ∎

Solutions to Questions

1.4.1. (1) Given $\varepsilon > 0$, choose $N = (1/2)(3/\varepsilon - 1)$. Then, whenever $n > N$, $|3/(2n+1) - 0| < 3/(2N+1) = \varepsilon$.   (2) Given $\varepsilon > 0$, choose $N = (1/5)[2/(5\varepsilon) - 1]$. Then, whenever $n > N$, $|2n/(5n+1) - 2/5| = |\tfrac{2}{5(5n+1)}| < \tfrac{2}{5(5N+1)} = \varepsilon$.

1.4.2. (1) Since $(2n-1)/n < 2$, an upper bound is 2.   (2) It converges.   (3) $\lim_{n\to\infty}(2n-1)/n = 2$: Given $\varepsilon > 0$, choose $N = 1/\varepsilon$. Then, whenever $n > N$, $|(2n-1)/n - 2| = 1/n < 1/N = \varepsilon$.

1.4.3. (1) $a/b + c/d = (ad+bc)/(bd)$ and $(a/b)\cdot(c/d) = (ac)/(bd)$. If $b, d \neq 0$, then (by the hint) they are of the form m/n, where $m, n \in \mathbb{Z}$ and $n \neq 0$.   (2) For any term, $a_n = a_{n-1} + n \cdot 10^{-k}$ for some positive integer k. Since $n \cdot 10^{-k} > 0$, we may conclude $a_n > a_{n-1}$; so the sequence is nondecreasing. Since a_n has a terminating decimal expansion, it is in $\mathbb{Q}$. Since $\lim_{n\to\infty} a_n$ has a nonterminating, nonrepeating decimal expansion, it is not.

1.4 Limits

1.4.4. (1) Given $\varepsilon > 0$, choose $\delta = \varepsilon/5$. Then, whenever $0 < |x - 1| < \delta$, we have $|5x - 8 - (-3)| = 5|x - 1| < 5\delta = \varepsilon$. (2) Given $\varepsilon > 0$, choose $\delta = \varepsilon/4$. Then, whenever $0 < |x - 3| < \delta$, we have $|-4x + 2 - (-10)| = 4|x - 3| < 4\delta = \varepsilon$. (3) Given $\varepsilon > 0$, choose $\delta = \varepsilon$. Then, whenever $0 < |x - 2| < \delta$, we have $|(x^2 - 4)/(x - 2) - 4| = |x - 2| < \delta = \varepsilon$.

1.4.5. (1) Given $\varepsilon > 0$, we restrict x so $\delta \leq 1$ and $2 \leq x \leq 4$. Then $x + 3 \leq 7$, and, for these x values and whenever $0 < |x - 3| < \delta$, we have $|x^2 - 9| = |x + 3| \cdot |x - 3| < 7 \cdot \delta$, which is less than or equal to ε when we choose $\delta = \min\{1, \varepsilon/7\}$. (2) Given $\varepsilon > 0$, we restrict x so $\delta \leq 1$ and $1 \leq x \leq 3$. Then $x + 3 \leq 6$. Now choose $\delta = \min\{1, \varepsilon/6\}$. Whenever $0 < |x - 2| < \delta$, we get $|(x^2 + x) - 6| = |x + 3| \cdot |x - 2| < 6 \cdot \delta \leq \varepsilon$. (3) Given $\varepsilon > 0$, choose $\delta = \min\{1, \varepsilon/37\}$. Whenever $0 < |x - 3| < \delta$, we get (from the restriction) $x^2 + 3x + 9 \leq 4^2 + 12 + 9 = 37$, and $|(x^3 - 7) - 20| = |x - 3| \cdot |x^2 + 3x + 9| < 37 \cdot \delta \leq \varepsilon$.

1.4.6. (A1) Clearly $f(1)$ exists and equals 3. Given any $\varepsilon > 0$, choose $\delta = \varepsilon/7$. Whenever $|x - 1| < \delta$, we have $|f(x) - f(1)| = |(7x - 4) - (7 \cdot 1 - 4)| = 7|x - 1| < 7\delta = \varepsilon$. (A2) Clearly $g(1)$ exists and equals 1. Given $\varepsilon > 0$, we restrict x so $\delta \leq 1/2$ and $1/2 \leq x \leq 3/2$. Then $1/x \leq 2$, and, for these x values and whenever $|x - 3| < \delta$, we have $|g(x) - g(1)| = |1/x - 1| = |x - 1| \cdot |1/x| < 2 \cdot \delta$, which is less than or equal to ε when we choose $\delta = \min\{1/2, \varepsilon/2\}$. (B1) Given nonzero $x = c$ and $\varepsilon > 0$, note that $h(c)$ exists and choose $\delta = \varepsilon/3$. Whenever $|x - c| < \delta$, we have $|h(x) - h(c)| = |(3x + 7) - (3c + 7)| = 3|x - c| < 3\delta = \varepsilon$. (B2) Given $x = c$ and $\varepsilon > 0$, note that $s(c)$ exists and choose $\delta = \min\{c/2, 2\varepsilon/5c\}$. Whenever $|x - c| < \delta$, we have $|s(x) - s(c)| = |x^2 - c^2| = |x + c| \cdot |x - c| < (5c/2) \cdot \delta \leq \varepsilon$.

1.4.7. (2) For $x \in [0, 2)$ and $n \in \mathbb{N}$, it is the case that $k/2^{n-1} \leq x < (k+1)/2^{n-1}$ for some $k = 1, 2, \ldots, 2^n$, which would imply $g_n(x) = k/2^{n-1}$. But, now considering $g_{n+1}(x)$, $j/2^n \leq x < (j+1)/2^n$, and then $g_{n+1}(x) = j/2^n$. Additionally, either $j = 2k$ or $j + 1 = 2(k + 1)$. (Either the left endpoint of this interval equals the left endpoint of the interval just described for g_n, or the corresponding right endpoints are equal.) In the first case, $g_{n+1}(x) = j/2^n = (2k)/2^n = k/2^{n-1} = g_n(x)$. In the second case, $g_{n+1}(x) = j/2^n = [2(k+1) - 1]/2^n < [2k]/2^n = k/2^{n-1} = g_n(x)$. Hence the sequence is nondecreasing. (3) The key fact to use is $|g_n(x) - x| = |\frac{k}{2^{n-1}} - x| < 1/2^{n-1}$. (4) Assume $x \in [0, 2)$. Given $\varepsilon > 0$, we choose $N = \log_2(1/\varepsilon) + 1$. Then for any $n > N$, part (1) shows $|g_n(x) - x| < 1/2^{n-1} < 1/2^{N-1} = \varepsilon$, and the limit satisfies Definition 1.4.4 for all $x \in [0, 2)$. The special case $x = 2$ works similarly.

Reading Questions for Section 1.4

1. Give an example of a sequence displayed in longhand notation.

2. Give an example of a sequence displayed in set notation, emphasizing the pattern.

3. Give an example of a sequence displayed in set notation, and then employ a change in its index, showing how the notation of the sequence pattern changes, as well as the indexed first term of the sequence.

4. What do we mean by the expression $\lim_{n \to \infty} s_n = L$?

5. What do we mean by the expression $\lim_{x \to c} f(x) = L$?

6. When is a function f continuous at a domain value $x = c$?

7. When is a function f continuous on a set S?

8. When is a function f continuous?

9. For a sequence $f_n(x)$ of functions evaluated at a point x, what do we mean by $\lim_{n \to \infty} f_n(x) = f(x)$?

Exercises for Section 1.4

In Exercises 1–8, use Definition 1.4.1 to show each of the limits hold.

1. $\lim_{n \to \infty} 11/(3n) = 0$

2. $\lim_{n \to \infty} \dfrac{2}{n+6} = 0$

3. $\lim_{n \to \infty} \dfrac{2n-3}{n+1} = 2$

4. $\lim_{n \to \infty} \dfrac{2n-3}{5n+1} = 2/5$

5. $\lim_{n \to \infty} \dfrac{2n-3}{n^2+3} = 0$ (*Hint*: the numerator is less than $2n$, and the denominator is greater than n^2.)

6. $\lim_{n \to \infty} (1/3)^n = 0$

7. $\lim_{n \to \infty} (5/9)^n = 0$

8. $\lim_{n \to \infty} r^n = 0$, for any real constant r with $|r| < 1$

In Exercises 9–16, use Definition 1.4.2 to show each limit holds. Apply the restriction technique when necessary.

9. $\lim_{x \to 1/2} (2x + 5) = 6$

10. $\lim_{x \to -5} (4x + 3) = -17$

11. $\lim_{x \to 2} (-5x + 12) = 2$

12. $\lim_{x \to 3} (3x^2 + 1) = 28$

13. $\lim_{x \to 2} (x^3 + 7) = 15$

14. $\lim_{x \to 1} (x^2 + x) = 2$

15. $\lim_{x \to 2} 1/x = 1/2$

16. $\lim_{x \to 9} (\sqrt{x} + 9) = 12$

In Exercises 17–25, use Definition 1.4.3 to discuss continuity.

17. Prove $f(x) = 8x + 2$ is continuous at $x = 4$.

18. Prove $f(x) = 8x + 2$ is continuous on the interval $[1, 4]$.

19. Prove $f(x) = 8x + 2$ is a continuous function.

20. Prove any linear function $f(x) = mx + b$, where $m, b \in \mathbb{R}$ is continuous.

21. Prove $f(x) = x^2$ is continuous at $x = 2$.

22. Prove $f(x) = x^2$ is a continuous function.

23. Prove $f(x) = 1/x$ is continuous at $x = 2$.

24. Prove $f(x) = 1/x$ is continuous on the set $S = [2, 5]$.

25. Identify a value x_0 where $f(x) = 1/x$ is not continuous. Say why.

1.5 The Lebesgue Integral and L^1

In Exercises 26–30, use Definition 1.4.4 to show each limit holds.

26. $\lim_{n \to \infty} f_n(1) = 3$, where, on the interval $[0, 3]$, $\{f_n\}$ is the sequence of functions

$$f_1(x) = \begin{cases} 0 & \text{if } x \in [0, \frac{3}{2}) \\ 9/2 & \text{if } x \in [\frac{3}{2}, 3], \end{cases} \quad f_2(x) = \begin{cases} 0 & \text{if } x \in [0, \frac{3}{4}) \\ 9/4 & \text{if } x \in [\frac{3}{4}, \frac{3}{2}) \\ 9/2 & \text{if } x \in [\frac{3}{2}, \frac{9}{4}) \\ 27/4 & \text{if } x \in [\frac{9}{4}, 3], \end{cases} \quad \text{etc.}$$

In general, f_n is defined over 2^n subintervals of $[0, 3]$, where the function value of x in the subinterval $[3(k-1)/2^n, 3k/2^n)$ for $k \in \{1, \ldots, 2^n\}$ is $f_n(x) = 9(k-1)/2^n$.

27. $\lim_{n \to \infty} f_n(2) = 6$, where $\{f_n\}$ is as in Exercise 26.

28. $\lim_{n \to \infty} f_n(x) = 3x$ for any $x \in [0, 3]$, where $\{f_n\}$ is as in Exercise 26. As you prove this result, feel free to mimic the technique used in Example 1.4.6.

29. $\lim_{n \to \infty} g_n(3/2) = 9/4$, where, on the interval $[0, 2]$, $\{g_n\}$ is the sequence of functions

$$g_0(x) = 0, \, g_1(x) = \begin{cases} 0 & \text{if } x \in [0, 1) \\ 1 & \text{if } x \in [1, 2], \end{cases} \quad g_2(x) = \begin{cases} 0 & \text{if } x \in [0, \frac{1}{2}) \\ 1/4 & \text{if } x \in [\frac{1}{2}, 1) \\ 1 & \text{if } x \in [1, \frac{3}{2}) \\ 9/4 & \text{if } x \in [\frac{3}{2}, 2]. \end{cases} \quad \text{etc.}$$

In general, g_n is defined on 2^n subintervals of $[0, 2]$, where the function value of x in the subinterval $[(k-1)/2^{n-1}, k/2^{n-1})$ is $g_n(x) = ((k-1)/2^{n-1})^2$.

30. $\lim_{n \to \infty} g_n(x) = x^2$ for any $x \in [0, 2)$, where $\{g_n\}$ is as in Exercise 29. As you prove this result, feel free to mimic the technique used in Example 1.4.6.

31. Show the sequences $\{f_n\}$ from Exercise 26 and $\{g_n\}$ from Exercise 29 are each nondecreasing. In other words, show (for any x in the domain of the functions in the sequence) $f_n(x) \le f_{n+1}(x)$, $n = 1, 2, 3, \ldots$, and similarly for g_n. (*Hint*: use the fact that for any x, $f_n(x)$ is the minimum value of the interval I_n that contains x. Argue that $I_{n+1} \subseteq I_n$, and hence $f_{n+1}(x) \le f_n(x)$.)

1.5 The Lebesgue Integral and L^1

Section 1.3 showed how to integrate step functions, and we learned the collection of step functions forms a linear space. This section describes the Daniell-Riesz approach to the Lebesgue integral for a broader category of functions—the set of "Lebesgue integrable functions." They form a linear space of functions L^1 that has many important properties. We'll explore a couple of those properties at the end of this section and look at a full array of them throughout the rest of the book.

The Daniell-Riesz approach uses the integrals of step functions to develop the space L^1 in two steps. First, it describes the Lebesgue integral for more than just step functions—for a space of functions L^0. (There is no standard name for this collection of functions. We hope L^0 will seem appropriate and catch on.) Second, the Daniell-Riesz approach broadens the L^0 category into the fully blossomed space of Lebesgue integrable functions. That expansion into L^1 will turn out to be an easy step after L^0 is developed and discussed. We split this section into two subsections devoted, respectively, to the Daniell-Riesz description of L^0 and of L^1.

1.5.1 The Space L^0

The Daniell-Riesz development of the Lebesgue integral depends on concepts of convergence described in Section 1.4. Here's the way it works: if we want to integrate a function f, then we look for a sequence of step functions $\{\phi_n(x)\}$ that converges pointwise to $f(x)$ and for which the corresponding sequence of Lebesgue integrals $\{\int \phi_n\}$ converges. There are a couple of fine points involved. First, the sequence of step functions doesn't have to converge to $f(x)$ at every value x—it can fail to converge on a set of measure zero. In short, the sequence of step functions $\{\phi_n\}$ converges almost everywhere. Second, the sequence of step functions must be nondecreasing: $\phi_n(x) \leq \phi_{n+1}(x)$ for almost all x. (We'll often suppress the notation for x and write $\phi_n \leq \phi_{n+1}$.)

When we have such a sequence $\{\phi_n\}$, then we define the Lebesgue integral of f to be the limit of the step-function integrals: $\int f = \lim_{n \to \infty} \int \phi_n$. The next definition encapsulates this idea.

Definition 1.5.1 *Suppose $\{\phi_n(x)\}$ is a nondecreasing sequence of step functions that converges pointwise almost everywhere to a function $f(x)$. Then $\int f \equiv \lim_{n \to \infty} \int \phi_n$. For such functions f having a finite integral,[10] we say f belongs to the space L^0 (pronounced "L-naught").*

This definition outlines the definite integral of f as opposed to the general antiderivative (the indefinite integral). We will eventually show that you may interpret this integral value as the area under the curve, just as you interpreted—in your calculus course—the definite integral (the Riemann integral). From calculus, you might be familiar with the notation $\int_a^b f(x)\,dx$, and this notation may be used here. (We will indicate how as we progress through this section.) But the definition works, beautifully, in a general way. For example, it does not require f to equal zero outside of a fixed interval $[a, b]$, nor does it even require f to be bounded above by some upper bound $M < \infty$. (Although we will often integrate, for example, functions that vanish outside a fixed interval, that characterization is not a requirement.)

This collection of functions L^0 is quite large. Trivially, any step function f is the limit of the sequence of step functions $\{f, f, f, \ldots\}$. It thus includes all step functions, but lots more. It will also turn out to include all functions continuous on a finite interval that vanish off the interval. (Their integrals are then guaranteed to be finite.) Unfortunately, it will turn out not to be quite full enough to be a linear space of functions (a fact we will prove later in this section and that will serve as the main reason we will want to expand the space to L^1). But we will be able to show L^0 is almost linear, in the sense that if $\phi_n \to f$ and $\psi_n \to g$, then $(\alpha \cdot \phi_n + \beta \cdot \psi_n) \to (\alpha \cdot f + \beta \cdot g)$ for any nonnegative constants α and β.

We illustrate this important definition with some examples. For simplicity's sake, they will all deal with polynomial functions f. The challenge when using Definition 1.5.1 to calculate integrals of such functions f from scratch is to find a nondecreasing sequence of step functions

---

[10] It's important to note the limit defining the integral always exists, either as a finite value or ∞, because the sequence of terms $\int \phi_n$ is nondecreasing. Therefore the limit is of terms that can only increase, either up to some finite limit, or without bound.

1.5 The Lebesgue Integral and L^1

that converges (at least almost everywhere) to f. Once that's done, the sequence of integrals of the step functions is easy to calculate, and the limit of the sequence can be straightforward. We'll see how it works.

Example 1.5.1 Let $f(x) = x$ if $0 \leq x \leq 1$, and $f(x) = 0$ otherwise. We define an increasing sequence of step functions $\phi_n(x)$ that converges to $f(x)$. Each of them can equal 0 for x not in $[0, 1]$, since $f(x) = 0$ for such x. For values of x between 0 and 1, the functions ϕ_n are defined as

$$\phi_1(x) = \begin{cases} 0 & \text{if } x \in [0, \tfrac{1}{2}) \\ 1/2 & \text{if } x \in [\tfrac{1}{2}, 1], \end{cases} \qquad \phi_2(x) = \begin{cases} 0 & \text{if } x \in [0, \tfrac{1}{4}) \\ 1/4 & \text{if } x \in [\tfrac{1}{4}, \tfrac{1}{2}) \\ 1/2 & \text{if } x \in [\tfrac{1}{2}, \tfrac{3}{4}) \\ 3/4 & \text{if } x \in [\tfrac{3}{4}, 1], \end{cases} \text{ etc.}$$

Just as in Example 1.4.6, each function ϕ_n is piecewise defined on 2^n subintervals having equal length $1/2^n$. The function values $\phi_n(x)$ are equal to the minimum x value in that subinterval, and so each $\phi_n(x)$ value is of the form $(k-1)/2^n$ for some $k = 1, 2, 3, \ldots, 2^n$. For example, the third sequence function (which we won't write out completely, as it needs to be defined on eight subintervals) looks like

$$\phi_3(x) = \begin{cases} 0 & \text{if } x \in [0, \tfrac{1}{8}) \\ 1/8 & \text{if } x \in [\tfrac{1}{8}, \tfrac{1}{4}) \\ \ldots & \ldots \\ 7/8 & \text{if } x \in [\tfrac{7}{8}, 1]. \end{cases}$$

Figure 1.6 displays the graphs of f, ϕ_1, ϕ_2, and ϕ_3. You can see the convergence happening.

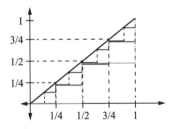

Figure 1.6. $f(x) = x$, $0 \leq x \leq 1$, with $\phi_1(x)$ (in grey), $\phi_2(x)$ (thick grey), and $\phi_3(x)$ (black).

Example 1.4.6 showed why $\{\phi_n\}$ converges to f at every x value. We can also show the sequence of step functions is nondecreasing. For if x is any value in $[0, 1]$, then $\phi_n(x)$ is defined as the minimum value $(k-1)/2^n$ of the interval I_n in which x is an element. This holds true for any f_n, $n = 1, 2, 3, \ldots$. Each I_n is constructed so $I_{n+1} \subseteq I_n$. (In fact, I_{n+1} is exactly half the size of I_n, because the piecewise defined function ϕ_{n+1} is defined according to 2^{n+1} subintervals across $[0, 1]$, double the 2^n number of subintervals that define ϕ_n.) Hence the minimum value of I_{n+1} must be greater than or equal to the minimum value of I_n (by the containment $I_{n+1} \subseteq I_n$), and that means $\phi_{n+1}(x) \geq \phi_n(x)$. In other words, the sequence $\{\phi_n\}$ is nondecreasing.

We therefore have a nondecreasing sequence of step functions $\{\phi_n\}$ that converges pointwise almost everywhere to a function $f(x)$. In fact, the convergence is everywhere. Applying

Definition 1.5.1 (and Definition 1.3.3 on the integral of a step function),

$$\int f = \lim_{n\to\infty} \int \phi_n = \lim_{n\to\infty} \sum_{k=1}^{2^n} \frac{k-1}{2^n} \cdot m(I_n) = \lim_{n\to\infty} \sum_{k=1}^{2^n} \frac{k-1}{2^n} \cdot \frac{1}{2^n} = \lim_{n\to\infty} \frac{1}{2^{2n}} \sum_{k=1}^{2^n} (k-1).$$

A formula for the sum of the first 2^n positive integers evaluates the summation as

$$\sum_{k=1}^{2^n} k - \sum_{k=1}^{2^n} 1 = \frac{2^n(2^n+1)}{2} - 2^n = \frac{2^n(2^n-1)}{2}.$$

Therefore $\int f = \lim_{n\to\infty} \frac{2^n(2^n-1)}{2 \cdot 2^{2n}} = \lim_{n\to\infty} \frac{\mathbf{1} \cdot 2^{2n} - 2^n}{\mathbf{2} \cdot 2^{2n}}$, which is easily evaluated as the ratio of its boldfaced terms: the coefficients on the highest power of 2^n (which is the second power) in the numerator and denominator. Hence the Lebesgue integral is $\int f = 1/2$.

This section later develops notation to write this result as $\int_0^1 x\, dx = 1/2$, which matches the result familiar to all of us from calculus: the area under the curve $f(x) = x$ from $x = 0$ to 1 is $1/2$. ∎

The last example used a technique that will become routine when using Definition 1.5.1. That technique will be a crucial part of what we will call the "standard construction" of the Daniell-Riesz approach. It is used to construct a nondecreasing sequence of step functions converging almost everywhere to a function f in L^0.

> The **Standard Construction** will apply when f equals zero off an interval $[a, b]$. Whenever defining the sequence of step functions $\{\phi_n\}$, divide $[a, b]$ into 2^n subintervals, doubling the number of subintervals for each subsequent function in the sequence. The subintervals then formulate the pieces I_k over which ϕ_n is defined. Each I_k turns out to stretch between endpoints $a + (b-a)(k-1)/2^n$ and $a + (b-a)k/2^n$, for $k = 1, 2, 3, \ldots, 2^n$ (an index over k that includes all the 2^n subintervals). When f is continuous, the construction automatically creates a nondecreasing sequence $\{\phi_n\}$, by defining $\phi_n(x)$ to be the minimum value of f on the corresponding subinterval I_k in which x falls.[11]

You can imagine this standard construction will be particularly useful (for example, when f is continuous on $[a, b]$), as it will always result in a desired nondecreasing step-function sequence. Chapter 2 will fully outline and use the standard construction to prove all continuous functions are in L^0. The next question provides guided practice with it.

Question 1.5.1 Examine the function $f(x) = x^2$ when $0 \leq x \leq 1$ (and $f = 0$ for all other x values). This question finds $\int f$, which will be the area under x^2 from 0 to 1.

1. Construct a nondecreasing sequence of step functions $\{\phi_n\}$. Use the standard construction to form the nth function in the sequence: partition the interval $[0, 1]$ (as in Example 1.5.1) into 2^n subintervals of the form $I_k = [(k-1)/2^n, k/2^n)$, where $k = 1, 2, 3, \ldots, 2^n$. What is the minimum value of f on I_k? Define $\phi_n(x)$ so it has a constant value equal to that minimum.
2. Exhibit explicitly the formula for the functions ϕ_1 and ϕ_2 as you defined them in part 1.

---

[11] I_k is generally not closed, and so the minimum may have to be considered over the interval I_k but also including both endpoints.

3. Superimpose the graphs of the functions $f(x)$, $\phi_1(x)$, $\phi_2(x)$, and $\phi_3(x)$ on the same xy axis.
4. Prove $\lim_{n \to \infty} \phi_n(x) = f(x)$. (*Hint*: for x in I_k,

$$|f(x) - \phi_n(x)| < k^2/2^{2n} - (k-1)^2/2^{2n} < 1/2^{n-1}.)$$

5. Use the formula

$$\sum_{k=1}^{2^n} k^2 = \frac{2^n(2^n + 1)(2 \cdot 2^n + 1)}{6}$$

to show $\int \phi_n(x) = \frac{2 \cdot 2^{2n} - 3 \cdot 2^n + 1}{6 \cdot 2^{2n}}$.
6. Use the result from part 5 along with Definitions 1.5.1 and 1.3.3 to show $\int f = 1/3$. In notation we will soon develop fully, this says that $\int_0^1 x^2 \, dx = 1/3$. ∎

Three important considerations exist. First, the convergence in Definition 1.5.1 of the sequence $\{\phi_n(x)\}$ to $f(x)$ needs happen only almost everywhere. That fact means $\int f$ will be the same as the integral of any function g to which the functions ϕ_n converge almost everywhere. Here's the powerful upshot: if $f(x) = g(x)$ except on a set of measure zero (i.e., for almost every x), then $\int f = \int g$.

Second, we are here interested only in finite integrals. Hence the only sequences of step functions $\{\phi_n\}$ to consider are those for which all of the integrals $\int \phi_n$ are bounded. We may therefore assume, for all $n = 1, 2, \ldots$, that $\int \phi_n < K$ for some real-valued upper bound K. But it turns out any nondecreasing sequence bounded above must converge. For a function f that we want to integrate, we therefore look only for nondecreasing sequences $\{\phi_n\}$ of step functions that converge almost everywhere to f and for which $\int \phi_n$ converges to a finite value.

The third issue is also crucial, but we will have to worry carefully about it before it is resolved well. We need to make sure Definition 1.5.1 is not problematic. At this point, its consistency is not yet clear. Namely, what if we had two different nondecreasing sequences of step functions, say ϕ_n and φ_n, that both converged to the same function f almost everywhere? How do we know the two integral sequences $\int \phi_n$ and $\int \varphi_n$ will converge to the same value? If they don't, then the definition could produce many different values for $\int f$, and then the integral would not be well-defined.

Thankfully, everything works out well. We formulate the result in the following theorem.

Theorem 1.5.1 *Suppose $\{\phi_n\}$ and $\{\varphi_n\}$ are two nondecreasing sequences of step functions, both converging almost everywhere to a function f and whose integrals are bounded above. Then*

$$\lim_{n \to \infty} \int \phi_n = \lim_{n \to \infty} \int \varphi_n.$$

In this way, Definition 1.5.1 provides a consistent formulation of $\int f$.

Proof. The result stems from the following fact, whose proof we will work through in Question 1.5.2: Suppose f and g are in L^0 with $f \geq g$ almost everywhere. Also suppose $\{\phi_n\}$ satisfies Definition 1.5.1's required nondecreasing sequence for f, and $\{\varphi_n\}$ does so for g, so $\phi_n \to f$

a.e., $\varphi_n \to g$ a.e., $\int \phi_n \to \int f$, and $\int \varphi_n \to \int g$. Then

$$\int f = \lim_{n\to\infty} \int \phi_n \geq \lim_{n\to\infty} \int \varphi_n = \int g.$$

Letting $g = f$ throughout the statement of this fact forces the last inequality to be equality, which proves the theorem. ∎

How do we prove the fact used in Theorem 1.5.1's proof? The next question works through the details. The insights of the proof of Theorem 1.5.1 and the structure and details presented throughout Question 1.5.2 follow Weir [121, p. 36]. Working through Question 1.5.2 (it is straightforward and its answers are presented at the end of the section) finishes the proof that the Lebesgue integral is well-defined.

Question 1.5.2 Let f, g, $\{\phi_n\}$, and $\{\varphi_n\}$ be as in the setting in the proof of Theorem 1.5.1: f and g are in L^0 with $f \geq g$ a.e., $\phi_n \to f$ a.e., $\varphi_n \to g$ a.e., $\int \phi_n \to \int f$, and $\int \varphi_n \to \int g$.

1. Fix a sequence function φ_k and show the sequence $\{\phi_n - \varphi_k\}_{n=1}^{\infty}$ is nondecreasing a.e.. Also show its limit is a.e. nonnegative: take n to infinity and apply the facts $\phi_n \to f$, $f \geq g$, and $g \geq \varphi_k$ (each almost everywhere).
2. For a general function h, define the function $h^-(x) = h(x)$ where $h(x) \leq 0$ and $h^-(x) = 0$ where $h(x) > 0$. Use the definition to consider the sequence $\{(\phi_n - \varphi_k)^-\}_{n=1}^{\infty}$ for fixed k. Show it is a sequence of nonpositive step functions converging, as n goes to infinity, almost everywhere to zero. Conclude $\int (\phi_n - \varphi_k)^- \to 0$ as $n \to \infty$. Though the last step is intuitive, it is not trivial to prove (see, for example, [121, Lemma 1, p. 34]).
3. Refer correctly to theorems in Section 1.3 to show $\int (\phi_n - \varphi_k)^- \leq \int (\phi_n - \varphi_k)$ and $\int (\phi_n - \varphi_k) = \int \phi_n - \int \varphi_k$.
4. Use the results of parts 2 and 3 to show $\lim_{n\to\infty} \int \phi_n - \int \varphi_k \geq 0$.
5. Taking the limit as k goes to infinity in part 4, show

$$\lim_{n\to\infty} \int \phi_n - \lim_{k\to\infty} \int \varphi_k \geq 0,$$

and state why the result $\int f = \lim_{n\to\infty} \int \phi_n \geq \lim_{n\to\infty} \int \varphi_n = \int g$ follows. ∎

1.5.2 The Space L^1

Why do we need a space of integrable functions that is even larger than L^0? The answer is in terms of properties we want the space to have, and L^0 comes close but doesn't quite make it. For example, we want any convergent sequence of functions in L^0 to converge to a function in L^0, but that doesn't happen. More fundamentally, it turns out L^0 is not even linear—there are functions f and g and constants α and β for which f and g are in L^0 but $\alpha f + \beta g$ is not. The next example illustrates this with $\alpha = 1$ and $\beta = -1$.

Example 1.5.2 We construct a function $f - g$ not in L^0 even though f and g are. The choice of functions is due to Weir [121, p. 43].

Let f be the characteristic function of the interval $(0, 1)$, so $f = \mathcal{X}_{(0,1)}$. Because f is a step function, we know it is in L^0, and it is easy to see that $\int f = 1$. We now define g. First, list the rationals in $(0, 1)$ as $\{a_1, a_2, a_3, \ldots\}$. (Such a list exists because $\mathbb{Q}$ is countable; see Question 1.1.3.) Then define intervals I_j centered at a_j by $I_j = (a_j - 1/2^{j+2}, a_j + 1/2^{j+2})$.

1.5 The Lebesgue Integral and L^1

Direct calculation, as in the proof of Theorem 1.2.1, shows the union $\bigcup_{j=1}^{\infty} I_j$ has bounded total measure $\sum_{j=1}^{\infty} m(I_j) = 1/2$. Let g be the characteristic function of $\bigcup_{j=1}^{\infty} I_j$, so $g = \mathcal{X}_{(\bigcup_{j=1}^{\infty} I_j)}$. Because g is a limit of a nondecreasing sequence of step functions ($g = \lim_{n\to\infty} \mathcal{X}_{(\bigcup_{j=1}^{n} I_j)}$), because its integral is finite ($\mathcal{X}_{(\bigcup_{j=1}^{n} I_j)} \leq \sum_{j=1}^{n} \mathcal{X}_{I_j}$), and since $\int g = \lim_{n\to\infty} \int \mathcal{X}_{(\bigcup_{j=1}^{n} I_j)} \leq \lim_{n\to\infty} \int \sum_{j=1}^{n} \mathcal{X}_{I_j} = \sum_{j=1}^{\infty} m(I_j) = 1/2$, we see $g \in L^0$.

If L^0 were a linear space, then $\int f - g$ would satisfy $\int f - \int g = \int \mathcal{X}_{(0,1)} - \int g \geq 1 - 1/2 = 1/2$. But then the function $f - g$ can't be in L^0, as Definition 1.5.1 would imply its integral would be less than or equal to 0. For Definition 1.5.1 would require a sequence of step functions $\{\phi_n\}$ nondecreasing almost everywhere up to its limit $f - g$ (and so every one of the ϕ_n would have to be less than or equal to $f - g$ a.e.). But $f - g$ equals zero or one—and so any such step function would have to be a.e. less than or equal to zero, except at those numbers between 0 and 1 not in $\bigcup_{j=1}^{\infty} I_j$ (where it equals 1). Take x to be a value where $f(x) - g(x) = 1$ and where ϕ_n is less than or equal to $f - g$, and look at a nontrivial (positive length) interval I containing x. It contains a rational a_j (as enumerated above in $(0, 1)$, since the rationals are dense in the reals). Hence I contains a nontrivial interval $A = I_j \cap I \cap (0, 1)$, where I_j is as defined in the preceding paragraph, on which $f - g = 1 - 1 = 0$ is identically zero. If $c \cdot \mathcal{X}_I$ is part of ϕ_n's unique step-function representation as described in Lemma 1.3.1, then c had better be less than or equal to zero, or ϕ_n would be larger than $f - g$ at any value in the interval A. Hence $\phi_n \leq 0$ a.e. for all $n = 1, 2, \dots$.

By Theorem 1.3.3, this implies $\int f - g = \lim_{n\to\infty} \int \phi_n \leq \lim_{n\to\infty} \int 0 = 0$. So if there did exist such a sequence $\{\phi_n\}$, it would violate $\int f - g \geq 1/2$. There's a contradiction, leading to the conclusion that no such sequence exists. Thus $f - g$ cannot be in L^0, and therefore L^0 cannot be a linear space. ∎

What can we do now we know L^0 is not linear? A wonderful answer happens to be remarkably simple: we expand L^0 in an extraordinarily basic way to include all possible linear combinations. Fortuitously, that process then creates the linear space we want and need. That space is L^1, and the expansion of L^0 to it turns out to remedy the lack of linearity. The next definition describes the expansion to L^1.

Definition 1.5.2 *The space L^1 of Lebesgue integrable functions consists of any function f of the form $f = g - h$, where g and h are in L^0. Its integral is*

$$\int f = \int g - \int h.$$

The space L^1 contains all of the functions in L^0, since we can take h to be identically zero in the definition. The Lebesgue integral on L^1 has many wonderful properties, which we will study in the upcoming chapters. The one we prove first is in many ways the most important and basic: L^1 is linear.

Theorem 1.5.2 *L^1 is a linear space of functions. Namely, for two functions f and u in L^1 and constants α and β, $\alpha f + \beta u$ is in L^1, and $\int \alpha f + \beta u = \alpha \int f + \beta \int u$.*

Proof. Since f and u are in L^1, by Definition 1.5.2, there are functions g, h, v, and w in L^0 with $f = g - h$ and $u = v - w$.

Now assume a special case: that α and β are both nonnegative (other cases are handled below). Then

$$\alpha f + \beta u = \alpha g - \alpha h + \beta v - \beta w = (\alpha g + \beta v) - (\alpha h + \beta w).$$

It turns out each function in the parentheses is in L^0 (the first of two key facts that we will show in a moment), and so, by Definition 1.5.2, $\alpha f + \beta u$ is in L^1 with its integral defined as

$$\int (\alpha f + \beta u) = \int (\alpha g + \beta v) - \int (\alpha h + \beta w).$$

Using the second key fact that we will show below, $\int (\alpha g + \beta v) = \alpha \int g + \beta \int v$, and, similarly, $\int (\alpha h + \beta w) = \alpha \int h + \beta \int w$. Putting the equalities together,

$$\int (\alpha f + \beta u) = \alpha \int g + \beta \int v - \alpha \int h - \beta \int w$$

$$= \alpha \left(\int g - \int h \right) + \beta \left(\int v - \int w \right) = \alpha \int f + \beta \int u,$$

where the last equality follows from Definition 1.5.2.

So what are the two key facts mentioned, and why are they true? First, for functions s and t in L^0 and nonnegative constants a and b, $as + bt$ is in L^0. The second fact identifies the corresponding integral: $\int (as + bt) = a \int s + b \int t$. The facts follow from an examination of the increasing sequences of step functions: $\{\phi_n\}$ converging almost everywhere to s and $\{\varphi_n\}$ converging almost everywhere to t. They generate an increasing sequence $\{a\phi_n + b\varphi_n\}$ converging almost everywhere to $as + bt$. After all, the limit of the sum is the sum of the limits, and the limit of a constant times a sequence is the constant times the limit of the sequence. The detail that the sequence $\{a\phi_n + b\varphi_n\}$ is increasing results from the fact that a and b are nonnegative. And then $\int (as + bt) = \lim_{n \to \infty} \int (a\phi_n + b\varphi_n) = \lim_{n \to \infty} a \int \phi_n + \lim_{n \to \infty} b \int \varphi_n = a \int s + b \int t$. These results then apply, and the first case, as dealt with in the above paragraph, is proved.

The three other cases follow similarly: (i) When α and β are both nonpositive, write

$$\alpha f + \beta u = \alpha g - \alpha h + \beta v - \beta w = ((-\alpha)h + (-\beta)w) - ((-\alpha)g + (-\beta)v).$$

(ii) When α is nonnegative and β is nonpositive, write

$$\alpha f + \beta u = \alpha g - \alpha h + \beta v - \beta w = (\alpha g + (-\beta)w) - (\alpha h + (-\beta)v).$$

(iii) When α is nonpositive and β is nonnegative, write

$$\alpha f + \beta u = \alpha g - \alpha h + \beta v - \beta w = ((-\alpha)h + \beta v) - ((-\alpha)g + \beta w).$$

In each case, the two facts cited and proved in the preceding paragraph now apply to the structure that was formed, and the result follows. ∎

A corollary (a statement proved almost immediately from a theorem preceding it) that follows Theorem 1.5.2 shows the Lebesgue integral for any L^1 function is well-defined.

1.5 The Lebesgue Integral and L^1

Corollary 1.5.1 *If f in L^1 has two different representations $f = g - h$ and $f = v - w$, where $g, h, v,$ and w are in L^0, then $\int g - \int h = \int v - \int w$. In this way, Definition 1.5.2 provides a consistent formulation of $\int f$.*

Proof. The result follows almost immediately from the two facts given within the proof of Theorem 1.5.2, applying them to both sides of the resulting linear equation of L^1 functions $g + w = v + h$. Then $\int g + \int w = \int v + \int h$, which proves the corollary.

We have now developed a well-defined Lebesgue integral for functions in the linear space L^1. The following question asks about elements of this space. ∎

Question 1.5.3 Find each of the integrals listed, using the functions $f(x) = x$, where $x \in (0, 1)$ (and 0 otherwise); $g(x) = x^2$, where $x \in (0, 1)$ (and 0 otherwise); and $h(x) = 1 + 1/j$, whenever $x \in (\frac{1}{j+1}, \frac{1}{j}]$, $j = 1, 2, 3, \ldots$ (and 0 otherwise).

1. $\int 5f$ 2. $\int -3g$ 3. $\int (4g - 3f)$
4. $\int h$ 5. $\int (h - f)$ 6. $\int (g + 2h)$

The last theorem in this section is now easily proved and will be used in Chapter 2 to prove an important result in a comparison between the Lebesgue and Riemann integrals.

Theorem 1.5.3 *If $f \in L^1$ and $f \geq 0$, then $\int f \geq 0$.*

Proof. $f = g - h$ with $g, h \in L^0$, and so $g \geq h$ almost everywhere. Applying Definition 1.5.2 and the result of Question 1.5.2, $\int f = \int g - \int h \geq 0$. ∎

This chapter finishes with a clarification of an issue mentioned earlier: we can use the notation $\int_a^b f(x)\, dx$ in a manner consistent with how it is used, say, in a calculus course. Suppose we are given a function f and want to know its definite integral, but only if the function is restricted to an interval (a, b). Then we can define a related function g as the function equal to f on (a, b) but vanishing (equaling zero) everywhere else. When g turns out to be in L^1, then we define $\int_a^b f(x)\, dx = \int g$, and we write $f \in L^1(a, b)$.

Example 1.5.3 We find $\int_1^4 x\, dx$ in order to show $f(x) = x$ is in $L^1(1, 4)$.

Solution. The integral's value will be $\int g$, where $g(x) = x$ for $x \in (1, 4)$ and $g(x) = 0$ for all other x. Using the standard construction, much as we did in Example 1.5.1, we form step functions ϕ_n by partitioning the interval $[1, 4)$ into 2^n subintervals I_k. Here $I_k = [1 + (4 - 1)(k - 1)/2^n, 1 + (4 - 1)k/2^n)$, where $k = 1, 2, \ldots, 2^n$. The measure of each subinterval I_k is its length $m(I_k) = 3/2^n$. For $x \in I_k$, we define $\phi_n(x) = 1 + 3(k - 1)/2^n$, which is the minimum value of $g(x) = x$ on the subinterval (of course it is also the left endpoint of the interval, since the function $g(x) = x$ is increasing on $[1, 4]$). We also realize, of course, that $\phi_n(x) = 0$ when $x \notin [1, 4]$.[12]

---

[12] For simplicity's sake, the value $x = 4$ has been left out of the discussion. We can always define $\phi_n(4) = 4$.

The standard construction works much as it did in Example 1.5.1: it is not difficult to prove the sequence $\{\phi_n\}$ is nondecreasing and $\phi_n \to g$. Using Definition 1.5.1,

$$\int \phi_n = \sum_{k=1}^{2^n}(1 + 3(k-1)/2^n) \cdot (3/2^n) = (3/2^n)\left(\sum_{k=1}^{2^n} 1\right) + (9/2^{2n})\left(\sum_{k=1}^{2^n}(k-1)\right).$$

Applying the closed-form expressions for the summations and simplifying,

$$\int \phi_n = \frac{3}{2^n} \cdot 2^n + \frac{9}{2^{2n}} \cdot \frac{2^n(2^n-1)}{2} = 3 + \frac{9 \cdot 2^n - 9}{2 \cdot 2^n}.$$

Evaluating the limit as n goes to infinity, $\int_1^4 x\, dx = \int g = 3 + 9/2 = 15/2$. Since the integral is finite, $f(x) = x$ is in $L^1(1,4)$. ∎

Question 1.5.4 Use the standard construction to find each of the following integrals.

1. $\int_0^1 x^2\, dx$ 2. $\int_1^2 x\, dx$ 3. $\int_{-2}^1 x^2\, dx$

4. $\int_0^b x\, dx$, where b is a positive real value ∎

In the same way, we can use traditional notation for the integral of a function f over the whole real line. There are actually three ways to denote this integral:

$$\int f = \int_{\mathbb{R}} f = \int_{-\infty}^{\infty} f(x)\, dx.$$

It is not appropriate to mix notations. For example, the third type of notation is the only one where the independent variable and the term dx appear; they always must appear together (by convention). Chapter 2 will show how this notation is useful when using techniques of integration, such as integration by substitution. In the same way, if I is an interval of points stretching between a and b, then we may write

$$\int_a^b f(x)\, dx = \int_I f.$$

Notational issues of when the variable x and the symbol dx occur are identical here to those just described.

At the end of this first chapter, we have a strong working definition of Lebesgue's integral applied to a broad class of functions—one that forms a linear function space, and one that will have many beautiful and useful analytic properties. We describe and study those properties in the upcoming chapters.

Solutions to Questions

1.5.1 (1) Let $\phi_n(x) = [(k-1)/2^n]^2$ for $x \in [(k-1)/2^n, k/2^n)$, where $k = 1, 2, \ldots, 2^n$. (And $\phi_n(x) = 0$ otherwise.)

(2) $\phi_1(x) = \begin{cases} 0 & \text{if } x \in [0, \frac{1}{2}) \\ 1/4 & \text{if } x \in [\frac{1}{2}, 1], \end{cases}$ and $\phi_2(x) = \begin{cases} 0 & \text{if } x \in [0, \frac{1}{4}) \\ 1/16 & \text{if } x \in [\frac{1}{4}, \frac{1}{2}) \\ 1/4 & \text{if } x \in [\frac{1}{2}, \frac{3}{4}) \\ 9/16 & \text{if } x \in [\frac{3}{4}, 1]. \end{cases}$ (3) The graphs are step functions. (4) Given $\varepsilon > 0$, choose $N = \log_2(1/\varepsilon) + 1$. Now apply the hint. (5) $\int \phi_n = \sum_{k=1}^{2^n}[(k-1)/2^n]^2 \cdot m([(k-1)/2^n, k/2^n)) = \sum_{k=1}^{2^n}[(k-1)/2^n]^2 \cdot$

1.5 The Lebesgue Integral and L^1

$(1/2^n) = (1/2^{3n}) \sum_{k=1}^{2^n}(k^2 - 2k + 1)$. Now apply $\sum_{k=1}^{2^n} 2k = 2^n(2^n + 1)$ and the given formula, and simplify. (6) $\int f = \lim_{n\to\infty} \int \phi_n = \lim_{n\to\infty} \frac{2 \cdot 2^{2n} - 3 \cdot 2^n + 1}{6 \cdot 2^{2n}} = 2/6 = 1/3$.

1.5.2 (1) For any x and any φ_k, since $\{\phi_n\}$ is a nondecreasing sequence a.e., we must have $\phi_n(x) - \varphi_k(x) \le \phi_{n+1}(x) - \varphi_k(x)$. Hence the sequence $\{\phi_n - \varphi_k\}$ is also nondecreasing a.e. Furthermore, $\lim_{n\to\infty}(\phi_n(x) - \varphi_k(x)) = f(x) - \varphi_k(x) \ge g(x) - \varphi_k(x) \ge 0$ a.e. (2) By the definition of h^-, the functions $(\phi_n(x) - \varphi_k(x))^-$ are nonpositive and are step functions because $\phi_n(x)$ and $\varphi_k(x)$ are. But by part (1), the set of x values where $\phi_n(x) - \varphi_k(x) < 0$ reduces to a set of measure zero in the limit. Therefore $\lim_{n\to\infty}(\phi_n(x) - \varphi_k(x))^- = 0$ a.e.
(3) For any n and k, $(\phi_n(x) - \varphi_k(x))^- \le \phi_n(x) - \varphi_k(x)$. By Theorem 1.3.3, $\int(\phi_n(x) - \varphi_k(x))^- \le \int \phi_n(x) - \varphi_k(x)$. By Theorem 1.3.2, $\int(\phi_n(x) - \varphi_k(x)) = \int \phi_n(x) - \int \varphi_k(x)$.
(4) $0 = \lim_{n\to\infty} \int (\phi_n(x) - \varphi_k(x))^- \le \lim_{n\to\infty} \int (\phi_n(x) - \varphi_k(x)) = \lim_{n\to\infty} \int \phi_n(x) - \int \varphi_k(x)$. (5) $0 \le \lim_{k\to\infty}[\lim_{n\to\infty} \int \phi_n(x) - \int \varphi_k(x)] = \lim_{n\to\infty} \int \phi_n(x) - \lim_{k\to\infty} \int \varphi_k(x)$. Adding $\lim_{k\to\infty} \int \varphi_k(x)$ to both sides and applying Definition 1.5.1, the result follows.

1.5.3 (1) $\int 5f = 5 \int f = 5 \cdot (1/2) = 5/2$. (2) $-3 \int g = -3 \cdot (1/3) = -1$.
(3) $4 \int g - 3 \int f = 4/3 - 3/2 = -1/6$. (4) $\int h = \sum_{j=1}^{\infty}(1 + 1/j)[1/j - 1/(j+1)] = \sum_{j=1}^{\infty} \frac{j+1}{j} \cdot \frac{1}{j(j+1)} = \sum_{j=1}^{\infty} 1/j^2$. The sum converges (it is a p-series with $p = 2$). Actually, Chapter 3 will show its exact value is $\pi^2/6$. (5) $\int h - \int f = \pi^2/6 - 1/2$.
(6) $\int g + 2 \int h = 1/3 + \pi^2/3$.

1.5.4 In each solution, n is assumed to be a positive integer. (1) Define ϕ_n on 2^n equal-sized subintervals of $[0, 1)$ as $\phi_n(x) = ((k-1)/2^n)^2$ for x in the kth subinterval, where $k = 1, 2, \ldots, 2^n$. Then $\int_0^1 x^2 = \lim_{n\to\infty} \int \phi_n = \lim_{n\to\infty} \sum_{k=1}^{2^n}((k-1)/2^n)^2 \cdot (2/2^n) = \lim_{n\to\infty} \frac{(2^n-1)2^n(2 \cdot 2^n-1)}{6 \cdot 2^{3n}} = 1/3$. (2) Define ϕ_n on 2^n equal-sized subintervals of $[1, 2)$ as $\phi_n(x) = (k-1)/2^n + 1$ for x in the kth subinterval, where $k = 1, 2, \ldots, 2^n$. Then $\int_1^2 x\, dx = \lim_{n\to\infty} \int \phi_n = \lim_{n\to\infty} \sum_{k=1}^{2^n}[\frac{k-1}{2^n} + 1] \cdot (1/2^n) = \lim_{n\to\infty} \frac{(2^n-1)2^n}{2 \cdot 2^{2n}} + 1 = 3/2$. (3) Define ϕ_n on 2^n equal-sized subintervals of $[-2, 1)$ as $\phi_n(x) = [3(k-1)/2^n - 2]^2$ for x in the kth subinterval, where $k = 1, 2, \ldots, 2^n$. Then $\int_{-2}^1 x^2\, dx = \lim_{n\to\infty} \int \phi_n = \lim_{n\to\infty} \sum_{k=1}^{2^n}(\frac{3(k-1)}{2^n} - 2)^2 \cdot (3/2^n) = 3$.
(4) Define ϕ_n on 2^n equal-sized subintervals of $[0, b)$ as $\phi_n(x) = b(k-1)/2^n$ for x in the kth subinterval, where $k = 1, 2, \ldots, 2^n$. Then $\int_0^b x\, dx = \lim_{n\to\infty} \int \phi_n = \lim_{n\to\infty} \sum_{k=1}^{2^n}[\frac{b(k-1)}{2^n}] \cdot (b/2^n) = \lim_{n\to\infty} \frac{b^2(2^n-1)2^n}{2 \cdot 2^{2n}} = b^2/2$.

Reading Questions for Section 1.5

1. When is a sequence of step functions $\{\phi_n\}$ nondecreasing?

2. When is a function f in the space L^0?

3. For a function $f \in L^0$, does the nondecreasing sequence of step functions $\{\phi_n\}$, described in Definition 1.5.1, have to converge to f throughout the real line $\mathbb{R}$? Where might the convergence fail?

4. Is L^0 a linear space?

5. When is a function f in L^1?

6. What does the notation $\int_a^b f(x)\,dx$ mean?

Exercises for Section 1.5

Use the standard construction with Definition 1.5.1 to evaluate the given Lebesgue integral in Exercises 1–15.

1. $\int f$, where $f = x$ when $x \in [1, 2]$ and $f = 0$ otherwise

2. $\int g$, where $g = x^2$ when $x \in [0, 2]$ and $g = 0$ otherwise

3. $\int h$, where $h = x - 1$ when $x \in [1, 2]$ and $h = 0$ otherwise

4. $\int f$, where $f = 3$ when $x \in (-3, 2]$ and 0 otherwise (*Hint*: is f a step function?)

5. $\int_0^1 (x+2)\,dx$
6. $\int_3^5 x\,dx$
7. $\int_{-1}^5 3x\,dx$
8. $\int_0^1 2\,dx$
9. $\int_0^2 x^2\,dx$
10. $\int_0^2 7x^2\,dx$
11. $\int_1^4 (x^2 + 1)\,dx$
12. $\int_0^2 (2t + 1)\,dt$

13. $\int_0^x (2t + 1)\,dt$, for a real x

14. $\int_I f$, where $f = 2x^2 + x - 1$ and $I = [1, 2]$

15. $\int_0^1 2x^3\,dx$ $\left(\text{Hint}: \sum_{i=1}^{2^n} i^3 = \left(\frac{2^n(2^n + 1)}{2}\right)^2.\right)$

Use the standard construction and Definition 1.5.1 to evaluate the Lebesgue integral in Exercises 16–20. Since each function f to be integrated is decreasing from left to right over the interval $[a, b]$ of integration, you will want to divide $[a, b]$ into subintervals I_k of the form $I_k = (a + (b - a)(k - 1)/2^n, a + (b - a)k/2^n]$. Then the minimum of f on each I_k will exist—it is f's value at the interval's right endpoint and determines ϕ_n.

16. $\int_0^1 (1 - x)\,dx$
17. $\int_1^5 -3x\,dx$
18. $\int_{-1}^0 x^2\,dx$
19. $\int_{-3}^{-1} x^2\,dx$
20. $\int_0^2 (4 - x^2)\,dx$

Use the standard construction and Definition 1.5.1 to evaluate the Lebesgue integral in Exercises 21–25. Since each function f to be integrated is in some places decreasing and in others increasing over the interval $[a, b]$ of integration, you will want to think carefully about the form of the subintervals I_k that divide $[a, b]$. In each case, make sure the minimum of f on each I_k will exist to help determine ϕ_n.

21. $\int_{-1}^1 |x|\,dx$
22. $\int_0^2 (x - 1)^2\,dx$
23. $\int_{-1}^1 x^2\,dx$
24. $\int_0^2 (x^2 - 2x)\,dx$
25. $\int_{-1}^2 (4 - x^2)\,dx$

Use Definition 1.5.1 and/or Definition 1.5.2 to evaluate the Lebesgue integral in Exercises 26–30. The solutions do not necessarily follow from an application of the standard construction.

26. $\int f$, where $f(x) = j/(j+1)$ whenever $x \in (\frac{1}{j+1}, \frac{1}{j}]$, $j = 1, 2, 3, \ldots$ (and 0 otherwise)

27. $\int -f$, where f is as in the preceding exercise

28. $\int f$, where $f(x) = j$ whenever $x \in (\frac{1}{(j+1)!}, \frac{1}{j!}]$, $j = 1, 2, 3, \ldots$ (and 0 otherwise)[13]

29. $\int \mathcal{X}_\mathbb{I}$, where $\mathbb{I}$ is the set of irrationals in $(0, 1)$

30. $\int f$, where $f = \sum_{n=0}^{\infty} (1/2)^n \mathcal{X}_{[n,n+1)}$

In Exercises 31–35, prove each fact about the Lebesgue integral.

31. If f is in L^1 and a function s equals f almost everywhere, then s is in L^1 and $\int f = \int s$. (*Hint*: write f as $f = g - h$ for g and h in L^0, and then write s as $s = g - (h + f - s)$. Then argue that $h + f - s \in L^0$, since it equals h almost everywhere.)

32. If $\phi(x)$ is a step function, then (i) $\phi(x + k)$ is a step function for any real constant k, and (ii) $\int_{-\infty}^{\infty} \phi(x) \, dx = \int_{-\infty}^{\infty} \phi(x + k) \, dx$.

33. If $f(x) \in L^1$, then $\int_{-\infty}^{\infty} f(x) \, dx = \int_{-\infty}^{\infty} f(x + k) \, dx$ for any real constant k. (*Hint*: use the result of the preceding exercise.)

34. If $\phi(x)$ is a step function, then (i) $\phi(k \cdot x)$ is a step function for any nonzero real constant k, and (ii) $\int_{-\infty}^{\infty} \phi(x) \, dx = |k| \cdot \int_{-\infty}^{\infty} \phi(k \cdot x) \, dx$.

35. If $f(x) \in L^1$ and k is any nonzero real constant, then $\int_{-\infty}^{\infty} f(x) \, dx = |k| \cdot \int_{-\infty}^{\infty} f(k \cdot x) \, dx$. (*Hint*: use the result of the preceding exercise.)

36. If $\phi(x)$ is a step function in L^0, then so is $|\phi(x)|$, and $|\int_\mathbb{R} \phi| \leq \int_\mathbb{R} |\phi|$. (*Hint*: use the triangle inequality for n values, which states $|a_1 + a_2 + \ldots + a_n| \leq |a_1| + |a_2| + \ldots + |a_n|$ for any $a_1, a_2, \ldots, a_n \in \mathbb{R}$.)

Advanced Exercises. Prove each statement in Exercises 37 and 38.

37. The function f defined as $f(x) = j$ whenever $x \in (\frac{1}{j+1}, \frac{1}{j}]$, $j = 1, 2, 3, \ldots$, and $f(x) = 0$ otherwise, is not in L^0 or L^1.

38. For real values $a \leq b \leq c$ and a function f, if f is in $L^1(a, b)$ and $L^1(b, c)$, then $f \in L^1(a, c)$, and $\int_a^b f(x) \, dx + \int_b^c f(x) \, dx = \int_a^c f(x) \, dx$.

Notes for Chapter 1

Remarks are in order to support this chapter's claim that Isaac Newton used the integral to prove Kepler's laws. His development of the integral began at Trinity College, Cambridge, where he matriculated as a student in 1661 and started attending the lectures of Isaac Barrow in 1663. Barrow held the original appointment to the now-famous (as Newton later graced it) Lucasian chair at age 33, was a member of the Royal Society, and successfully worked to find equations of tangents to certain curves and areas under them. To do so he used analytic methods that were precursors to both differential and integral calculus. Newton served as Barrow's

---

[13] This exercise follows [93, p. 95].

student assistant, helping him prepare two important works for publication. Barrow's *Lectiones Opticae* and *Lectiones Geometricae* appeared in 1669 and 1670, respectively (see [20, p. 146]). By 1665, Newton had mastered René Descartes' analytic geometry (representing points as ordered pairs and geometric curves as resulting equations), developed an ability to differentiate and integrate certain non-polynomial curves such as $\sqrt{1-x^2}$ (using infinite binomial series representations)[14], and recognized that finding tangents and areas under curves are inverse problems. This last accomplishment arguably served as the first systematic formulation of the fundamental theorem of calculus, though Barrow himself had understood the inverse relationship in a geometric, nonanalytic format [118]. In 1669, Newton took his observations to Barrow in the form of a tract, *De Analysi per Aequationes Numeri Terminorum Infinitas* ("On Analysis by Infinite Series"), which Barrow helped distribute to others. The *De Analysi* was never published but can be regarded as the first methodical description of calculus (Newton called it the "method of fluxions") issued to the mathematical public.

Figure 1.7. Isaac Newton and Isaac Barrow.

In a peculiar and unfortunate detour away from the mathematical development of the time, in 1670 Newton rejected Descartes' approach to geometry and his analytic notation. Instead (perhaps to continue to keep his method of fluxions more or less secretive [116, p. 588]) he began using a wholly Euclidean description. For example, in the *Principia* Newton gave pictorial descriptions of circles as opposed to ones in terms of their equations $(x-a)^2 + (y-b)^2 = r^2$. As a result, Newton's use of calculus in the *Principia* is practically unrecognizable and was at times unclear in its explanations. Ironically, Newton praised the ancient Greek geometers' methods as "more elegant by far than the Cartesian one." But we modern readers know how much more difficult Newton's task in the *Principia* turned out to be from this choice, as he had to explain the geometry of planetary motion without the mathematical tool provided by Cartesian equations and its analytic algebra. Imagine working with calculus without ever using analytic formulas or functional notation! It is therefore difficult to uncover in a practical way Newton's use of calculus in the *Principia*, and many historians approach the work as a purely Euclidean geometric one. Still, its elements of calculus often lie exposed. Many of its proofs involve limits, though not using terminology we would today call apt, and of course not rigorous—Augustine Cauchy's treatment came 150 years later. Furthermore, the *Principia* gives a Euclidean description of definite integrals in terms of inscribed and circumscribed rectangles. It analyzes ratios of numerators and denominators that (in modern language) both tend toward 0; Newton called such limits "ultimate ratios." Using modern notation, the

---

[14] Papers exist "in Newton's handwriting bearing dates 1665 and 1666 in which the method [of fluxions] is described" [116, p. 583].

limits of ratios can often be expressed as $\lim_{t \to 0}(f(x + h) - f(x))/h$, where h is some function of time t that approaches 0 as t does.

Niccolò Guicciardini's text [49] performs many of the translations necessary to go from Newton's Euclidean geometry description to today's analytic geometry (calculus) description. For example, Newton's proof of Lemma 2 in Book 1 of the *Principia* calculates the area under a curve using limits of sums of areas of inscribed or circumscribed parallelograms. His technique is fundamentally the same notion as Riemann's sums (for continuous functions) that we teach in a modern calculus course. Describing the limiting process, Newton states, "... if the breadth of those parallelograms is diminished, and their number is augmented continually, [then] the ultimate ratios, which the inscribed figure..., the circumscribed figure..., and the curvilinear figure..., have to each other, are ratios of equality." This passage of the *Principia* shows unequivocally that Newton obtains a definite integral through a limit of sums of parallelogram areas.

Newton uses this approach in his analysis of Kepler's laws, which this chapter has put forth as the first great physical application of the definite integral. Similar to his treatment of Lemma 2, Newton's proofs of Kepler's laws are without question purely geometric in the sense of his rejection of Cartesian analysis. His corresponding explanations are necessarily geometric and without a rigorous analytic base. But the *Principia*'s famous proofs of Kepler's laws, such as (in the *Principia*'s language),

Proposition 1: The areas which bodies made to move in orbits describe by radii drawn to an unmoving centre of forces lie in unmoving planes and are proportional to the times,

are proved when "Newton takes a limit" [49, p. 48] of polygonal areas determining curvilinear ones. Newton goes beyond, say, an ancient Greek (e.g., Archimedean) type of calculation in such limits, since he applies it systematically to general curves and "was aware that a translation into calculus (i.e., into the analytical method of fluxions) [for many of his proofs] was feasible" [49, p. 59]. Finally, in many fundamental cases, such as his proof of Proposition 6, his limits are examples of ultimate ratios. In such ways, both differential and integral calculus are involved in this greatest work of Newton's. As the American mathematician Frederick Rickey put it in an award-winning article [97], the *Principia* "is packed with the informal geometrical ideas of the new analysis, the calculus.... The geometric ideas of the calculus are used constantly in the *Principia*, but the algebraic notations are not."

The ideas set forth in the *Principia* showed mathematicians in the late 1600s and early 1700s the power of Newton's methods and indicated to them mathematics' ability to explain fully the physical properties of nature. Around 1710, Jacob Hermann, who promoted Newton's methods in Italy, Germany, and Russia, formulated the methods in terms of analytic geometry and put them into a calculus format [56]. The reader may wish to read others of Newton's works on the calculus, including [87], into which the *De Analysi* evolved, which likewise demonstrated the fundamental theorem of calculus and which was published posthumously. Newton also delivered this manuscript to Barrow by 1671, but could not, as David Burton [20, p. 376] puts it, obtain publication due to a lack of market for mathematical texts.

We also mention expository work on Lebesgue's 1902 advancement and his measure-theoretic approach, appearing in many admired books. Widely read authors of graduate texts, in their order of production from the 1950s, include Tom Apostol [3], Robert Bartle [10], Halsey Royden [105], Walter Rudin [106], and Frank Burk [18]. In the 1990s and beyond, a score of authors approached the Lebesgue integral by linking Lebesgue's measure theory to the study of probability theory. Patrick Billingsley [12], Malcolm Adams with Victor Guillemin [2], and Marek Capinski with Peter Kopp [22] are examples. All the texts require a course in real analysis, are used in many colleges and universities at the advanced undergraduate or graduate level, and come highly recommended for those students who want to delve into the integral using Lebesgue's original measure-theoretic approach. Many are written at a level of sophistication beyond much of the commentary in this text. Michael Carter and Bruce van Brunt's volume [26] does away with as much abstraction as possible, defining (as they put it) Lebesgue's integral "as an extension of the Riemann integral," with the intention of introducing the integral to undergraduates for applied purposes.

The timeline development of the Daniell-Riesz approach is explained in the Introduction. The 1955 book by Frigyes Riesz and Béla Sz.-Nagy [98] defines the Lebesgue integral as in Section 1.5, and it derives the convergence properties that Chapters 2 and 3 will explain. It was the only book of any note to present the Daniell-Riesz approach until Alan J. Weir wrote his 1973 breakthrough text [121], which expanded the discussion with a collection of detailed examples, theorems, and proofs. When read in conjunction with his follow-up 1974 volume [120], the reader understands the Daniell-Riesz approach in its broadest generality. For example, the second volume includes the descriptions of integrals based on measures other than Lebesgue measure. We will present some of that material in Chapter 4. A broadly inclusive graduate text on applied operator theory first published in 1971 by Arch Naylor and George Sell [86] describes the integral via the Daniell-Riesz approach but only referentially, in an appendix. Hilary Priestley's 1997 text [93] is one other that uses the Daniell-Riesz approach; its intended prerequisite matches that for any of these books—at least a first course in real analysis.

Several mathematicians have recently written marvelous descriptions of the history behind the integral. William Dunham produced a beautifully written historical account of calculus and many of the mathematicians involved in its development in [37]. This 2004 book, *The Calculus Gallery*, includes a final chapter on Henri Lebesgue. Other narratives on the development of analysis and that contain material on Lebesgue include Hans Niels Jahnke [61], John Stillwell [113], and Carl Boyer and Uta Merzbach [14]. David Bressoud [17] explains Lebesgue measure and the integral with historical descriptions throughout. Thomas Hawkins [55] relates Lebesgue's early work on his integral to the 19th-century discoveries that motivated his efforts. The authors give details of how the integral developed and connect it to the broad, fascinating overview of the history of mathematics. Finally, a delightfully detailed 1986 BBC Two video production *The Birth of Calculus* is impressive and fascinating. Available on the web, it displays handwritten manuscripts by Newton from 1665 and 1666, and personal papers of Leibniz from 1675.

2
Lebesgue's Integral Compared to Riemann's

This chapter compares the Lebesgue integral with one developed earlier by Bernhard Riemann. The conceptual development of the two integrals will be different. Chapter 1 introduced the Lebesgue integral of a function f in L^1, which was formed from a limit of integrals of a non-decreasing, almost-everywhere convergent sequence of step functions. Avoiding the Lebesgue integral due to its perceived measure-theoretic complexity, a standard calculus course instead teaches students about the Riemann integral of a continuous function f over an interval $[a, b]$, defining it as a limit of Riemann sums. A calculus class first describes the Riemann process as subdividing (or partitioning) $[a, b]$ into n subintervals. Using them and increasing n, the process then takes a limit of sums of circumscribed or inscribed rectangular areas to get the Riemann integral of f. The Riemann process works well and is straightforward when f is a bounded continuous function, or when f has, for example, a finite number of discontinuities. The goal is, intuitively, to find the area under f between a and b.

The Riemann integration explained in an introductory calculus course is not as general a process as the more complete theory of integration Riemann developed in the 1850s. His integral will handle a slightly broader class of functions than continuous ones, and we will present it in Section 2.1. After Riemann produced his method, mathematicians realized it worked for some functions (including those continuous on $[a, b]$) but not for others. (We will give examples in Section 2.1.) But until Henri Lebesgue found the answer in 1902, the mathematics community was unable to determine generally which functions the Riemann integral could handle and which ones it could not. His characterization of Riemann integrable functions is surprisingly simple—it involves only the function's continuity. Section 2.1 will explain that famous result.

Your calculus course showed you the Riemann integral is useful, and we want to reassure you this textbook does not suggest you forget any of that wonderful mathematics. You've built up a lot of intuition from calculus, working with the Riemann integral definition and the many exercises you've solved with calculations using the fundamental theorem of calculus. In fact, you can use your intuition and understanding of the Riemann integral as you explore properties of the Lebesgue integral. The great news is that whenever the Riemann integral of a function f exists, then so will the Lebesgue integral of f, and they will have the same value. Section 2.1 will prove that. Of course you'd expect their values to agree when both are defined, since both of them, in an intuitive sense, find the area under the function's curve.

Then why is the Lebesgue integral so important and necessary, and why do mathematicians consider it better, in an overall sense, than the Riemann integral? For both the Lebesgue and the Riemann integration processes, there are examples (in fact large classes) of functions for which the integral is not defined. But the Lebesgue integral is defined for many more functions than is the Riemann integral. So many more that we will be able to use the Lebesgue integral to describe "complete spaces of functions," where the integral can be applied (in a way that is linked to the space's definition) to all the functions in the space. This ability will let us think about limits and other characterizations of functions in a new way. Using the integral in this way will also provide us with a new understanding of functions: functions are members of the function space just as ordered pairs, for example, are members of a space of points. Finally, it will allow us to study important objects associated with functions, such as operators, which map one function to another. An example of an operator is the derivative, which maps one function (such as $f(x) = x^2$) to another (in this case, $f'(x) = 2x$). But we're getting ahead of ourselves; that material will be developed in the last chapter after we develop many important properties of function spaces and the integral.

This chapter will show the equivalence of the Riemann and Lebesgue integrals of a function f whenever they are both properly defined. It will provide many examples of the broader usefulness of the Lebesgue integral. That discussion comes in Section 2.1, using the standard construction from Section 1.5. In Section 2.2 we will begin to exhibit properties of the Lebesgue integral that go beyond what the Riemann integral can do, starting with the monotone convergence theorem. That section will also describe some features common to both integrals, such as the fundamental theorem of calculus. Integration techniques will be familiar to you from your calculus courses. Section 2.3 continues the discussion of important convergence theorems for the Lebesgue integral, properties that go beyond the reach of the Riemann integral. The mathematics from this section lies at the heart of why the Lebesgue integral is so powerful, as it will lead to results on spaces of functions in Chapter 3. Section 2.4 describes an immediate application of the properties, formulating a Fourier series representation for a function f, and then using the Lebesgue integral to determine when the series properly converges to f. We hope this chapter shows the usefulness of the Lebesgue integral and how it builds upon the Riemann integral. By the end of it, you should begin to feel empowered to investigate modern topics in function theory research, a journey that can provide great enjoyment and rich intellectual rewards.

2.1 The Riemann Integral

In 1853, at the age of 27, Bernhard Riemann developed a first complete rigorous definition of the definite integral. A shy but brilliant mathematician and a student of Carl Friedrich Gauss, Riemann had completed his dissertation in 1851 and was ready to join the faculty at the University of Göttingen. By German academic tradition, that appointment required an *Habilitationsshcrift*—a presentation to the university community on significant new mathematical work. Riemann's paper evolved from his interest in the work of Joseph Fourier, who had developed trigonometric (Fourier) series to solve the famous heat equation, a differential equation that mathematically describes the way heat flows through objects. Along with many other mathematical function theorists, Riemann had begun examining Fourier series, but was studying their solution to a different but similar differential equation, the wave equation. This equation

2.1 The Riemann Integral

mathematically describes the way objects vibrate when, for example, they produce sound. By the 1850s, mathematicians knew that not all solutions to the heat or wave equations could be described as Fourier series, but they were unable to determine precisely when such solutions could be represented this way. Fourier had done very little work on this problem and more or less casually (and incorrectly) had stated that any bounded function had a Fourier series representation.

Riemann correctly perceived the mathematical community's difficulty in solving this Fourier representation problem was due to poor foundational understanding of the integral. The Fourier series representation was formulated from an integration process. Though Newton and Liebniz had developed the fundamental theorem of calculus over two centuries before, and Augustin Cauchy had advanced a mathematically scrupulous description of it in the 1820s, by the 1850s mathematicians still had not produced a rigorously complete definition of the integral—no satisfying foundational analysis of it existed. Riemann changed that in his *Habilitationsshcrift*. He started by defining an integration process that was consistent with the fundamental theorem of calculus, and he used it to say when a function

Figure 2.1. Bernhard Riemann.

f had a well-defined integral over an interval $[a, b]$. (The definition therefore also implicitly determined when the Riemann integral of f was not defined.) In the *Habilitationsshcrift* paper, Riemann immediately applied his definition of the integral to the representation problem, correctly producing a necessary condition for when a Riemann integrable function had a Fourier series. A necessary condition is one that guarantees a theorem's conclusion, and so the condition that Riemann found was one that guaranteed when a Riemann integrable function could be written as a trigonometric series. This theorem on Fourier series was a major breakthrough, and it was coupled with a new understanding of the integral—two extraordinary results in one.

In the same way as Lebesgue's, the Riemann integral may be interpreted (whenever the concept makes sense) as finding the area under a functional curve. As is customary, we think of area below the horizontal axis as negative. As taught in a calculus course, it is eventually possible to describe the Riemann integral over infinite domains of integration (through a limiting process that produces an improper Riemann integral) or for many functions unbounded over an interval $[a, b]$. But a rigorous definition of the Riemann integral always starts with the scenario to which we restrict our attention here: where the area is found (when it exists) for a bounded function over a finite closed interval $[a, b]$. To find this definite integral, Riemann looked at what we now commonly call Riemann sums, which are sums of rectangular areas that produce an approximation of the area under the curve. The formation of the sums will be crucial to the formal definition of the Riemann integral. In your introductory calculus course, you found Riemann sums for a few simple cases, along with their resulting Riemann integrals. But a calculus course's definition of the Riemann integral is too simplistic for our general purposes, as it applies only to curves that are continuous (except possibly at discrete points). Riemann's definition—the one we will explore in this section—can handle a slightly larger class of functions than continuous ones because it formulates the Riemann sums in a more general way than was described in calculus. This section develops that full generality. As we proceed, we will be able to think of Riemann sums as elementary integrals of step functions (as we presented in Section 1.3). The formulation will allow us to describe the Riemann integral in language parallel to language we used to describe Lebesgue's. The description will use the standard construction of Section 1.5, but to attain the full generality of Riemann's integration process, we will need to generalize that construction, phrasing it in terms of two important terms: a function's supremum and infimum. We start there.

2.1.1 The Supremum and Infimum

The full definition of the Riemann integral uses the supremum and infimum values of a set. These terms are also known as the set's least upper bound and greatest lower bound, respectively. Suprema and infima values (the plurals of supremum and infimum) are important in many areas of analysis. Our application will be immediate, since the definition of the Riemann integral is one example of where they arise.

Definition 2.1.1 *A real value M is an upper bound for a set S of real numbers when $s \leq M$ for every element s in S. When a set of real numbers S has an upper bound, then the supremum of S, labeled $\sup S$ (or l.u.b. for the least upper bound), satisfies the properties*

(i) $\sup S$ is an upper bound for S;
(ii) if M is any other upper bound for S, then $\sup S < M$.

Similarly, a real value m is a lower bound for a set S when $s \geq m$ for every element s in A. When a set of real numbers S has a lower bound, then the infimum of S, labeled $\inf S$ (or g.l.b. for the greatest lower bound), satisfies the properties

(i) $\inf S$ is a lower bound for S;
(ii) if m is any other lower bound for S, then $\inf S > m$.

Some sets have a supremum and some do not. Whenever a set, such as the interval $[2, \infty)$, does not have any upper bound, then the supremum of the set is not defined. Similarly, the infimum is not defined for a set such as $(-\infty, 2]$ that has no lower bound. As mentioned in Section 1.4.1, the axiom of completeness is assumed to imply that any bounded set has a supremum (and an infimum). When the supremum $\sup S$ is defined and is a member of the set S, then it is the maximum of S, but that membership is not guaranteed. For example, the supremum of $S = (0, 1)$ is 1, which is not in S. Similarly, $\inf S$ does not have to be a member of S, but it equals the minimum of S when it is.

Example 2.1.1 We prove the set $A = [0, \infty)$ does not have an upper bound: given any positive real value M, we see $M + 1$ is in A but $M + 1$ is not less than or equal to M (which implies M cannot be an upper bound). Therefore, $\sup A$ does not exist. On the other hand, the sets $B = [0, 2]$ and $C = (0, 2)$ have as upper bounds any real number M greater than or equal to 2, since every member x of B and C has $x \leq 2 \leq M$. The least of these upper bounds (which is by definition the supremum) is 2, since

(i) 2 is an upper bound; and
(ii) if M is any other upper bound, then $2 < M$. In particular, if some other upper bound M were less than 2, then M would have to be larger than the set element 1, and then $x \equiv M + (2 - M)/2$ would be an element of B (and C) since it is less than 2 but larger than (the supposed upper bound) M.

2.1 The Riemann Integral

The set B contains the supremum (2 is the maximum of B) while C does not (the set C has no maximum). In the same way, A, B, and C each have infimum equal to 0 (the greatest of their lower bounds), since

(i) 0 is a lower bound; and
(ii) if m is any other lower bound, then $0 > m$.

The sets A and B contain their infimum (0 is the minimum of A and of B) while C does not. ∎

It's not difficult to prove general facts about suprema and infima, many of which are extremely useful when proving theorems about functions and sets. Later in this section, we will cite the two facts in the next theorem as we investigate the theory of the Riemann integral.

Theorem 2.1.1 *For any real-valued sets A and B with $A \subseteq B$, $\sup A \leq \sup B$ and $\inf A \geq \inf B$.*

Proof. We prove the first fact; the proof of the second fact is similar. Since $\sup B$ is an upper bound for B and since $A \subseteq B$, $\sup B$ must also be an upper bound for A. As for any upper bound for A, $\sup B$ either equals $\sup A$, or the second part of Definition 2.1.1 implies it is greater than $\sup A$. In any case, $\sup A \leq \sup B$. ∎

The next example and question work with straightforward sets, finding their suprema and infima.

Example 2.1.2 The supremum of a finite set is its maximum, and the infimum is its minimum. Hence the infimum and supremum of $A = \{0, \frac{1}{3}, \frac{2}{3}, 1\}$ are 0 and 1, respectively. Such a rule does not exist for infinitely countable sets; for example, the supremum of $B = \{0, \frac{1}{2}, \frac{2}{3}, \frac{3}{4}, \ldots\}$ is 1, which is not an element of B, while its infimum 0 is in B. Similarly, the infimum of $C = \{1, \frac{1}{2}, \frac{1}{3}, \frac{1}{4}, \ldots\}$ is 0, which is not an element of C, while its supremum 1 is in C. ∎

Question 2.1.1 Find the supremum and infimum of the following sets. Also determine when they are elements of the set.

(a) $A = (-40, 12]$
(b) $D = \{0, \frac{1}{2}, \frac{3}{4}, \frac{5}{6}, \frac{7}{8}, \ldots\}$
(c) $B = (0, 1) \cup [3, 5)$
(d) $E = \{x^2 : x \in [1, 3)\}$
(e) $C = \{0, 1, 3, 5\}$
(f) $F = \{\sin x : x \in [0, \pi/2)\}$ ∎

The section will soon use supremum and infimum to define the Riemann integral. They help us understand the Lebesgue integral, too. In Section 1.5, a standard construction integrated a continuous function f over $[a, b]$. It used the minimum value of f on each of 2^n subintervals to determine the value of a step function $\phi_n(x)$, where x was an element of the subinterval. Using the infimum instead of the minimum, the construction generalizes to situations where f might not be continuous everywhere, as long as we assume, for example, that $|f(x)|$ is bounded above by some constant M. The standard construction then produces a sequence $\{\phi_n\}$ of step functions that is increasing and convergent. And, when f is continuous almost everywhere, the sequence $\{\phi_n\}$ will converge to f almost everywhere. These facts are precisely those needed in order to conclude f is in L^0.

We recall the beginning step in the standard construction, which was the formation of subintervals. When integrating f over $[a, b]$ in examples in Section 1.5, we used step functions whose 2^n subintervals were of the form

$$I_k = [a + (b - a)(k - 1)/2^n, a + (b - a)k/2^n), k = 1, 2, 3, \ldots 2^n.$$

For example, we might try to find the integral $\int_2^5 f(x)\,dx$ by forming an increasing sequence of step functions $\{\phi_n(x)\}$ that converge almost everywhere to f (so f would be in L^0). The examples in Section 1.5 typically constructed the subintervals for ϕ_n from the above-described format. As an example, you can see ϕ_2 would then have the following four (as $4 = 2^2$) subintervals

$$[2, 2\tfrac{3}{4}), [2\tfrac{3}{4}, 3\tfrac{1}{2})[3\tfrac{1}{2}, 4\tfrac{1}{4}), [4\tfrac{1}{4}, 5).$$

The standard construction of the step function follows from the standard format of the 2^n subintervals.[1] Its most general framework is formulated in terms of the infimum, and the next definition enumerates its steps.

Definition 2.1.2 The Standard Construction. *When attempting to form either the Riemann integral or the Lebesgue integral $\int_a^b f$ for a function $f \in L^0$, the following standard construction can be useful. Define a sequence of step functions $\{\phi_n\}$, where*

1. *Each ϕ_n is piecewise defined over $[a, b]$ using 2^n subintervals of the form*

$$I_k = [a + (b - a)(k - 1)/2^n, a + (b - a)k/2^n), k = 1, 2, 3, \ldots, 2^n.$$

2. *For $x \in I_k$, the step function's value is the infimum of f over the subinterval:*

$$\phi_n(x) \equiv \inf\{f(t) : t \in I_k\}.[2]$$

3. *In summary, $\phi_n(x) \equiv \sum_{k=1}^{2^n} m_k \cdot \mathcal{X}_{I_k}(x)$, where $m_k = \inf\{f(t) : t \in I_k\}$ and I_k is as in part 1.*
4. *Finally, we define the standard construction's dual sequence $\{\psi_n\}$, consisting of step functions*

$$\psi_n(x) \equiv \sum_{k=1}^{2^n} M_k \cdot \mathcal{X}_{I_k}(x), \text{ where } M_k = \sup\{f(t) : t \in I_k\} \text{ and } I_k \text{ is as in part 1.}$$

In the next subsection, we will see how the standard construction and the dual sequence create sequences of step functions that have just the right necessary properties when using the definition of the Riemann integral. The next two theorems present several useful facts.

Theorem 2.1.2 *As defined in the standard construction, the sequences $\{\phi_n\}$ and $\{\psi_n\}$ are nondecreasing and nonincreasing, respectively.*

---

[1] The subinterval format in the standard construction always leaves out the right endpoint. When integrating over an interval $[a, b]$, therefore, the value $x = b$ will always be omitted from the set of subintervals. Because it is a single point, its omission will have no impact on the value of the integral, and we can, indeed, always conveniently define $\phi_n(b) = f(b)$ to ensure each step function in the standard construction is well-defined at b.

[2] Remember: as we study Riemann integrals, we assume f is bounded on $[a, b]$, and so the infima will exist.

2.1 The Riemann Integral

Proof. For $x \in [a, b]$ and index value n, suppose $x \in I_k$, where I_k is one of the subintervals used to form ϕ_n. Also assume x is in some subinterval, say I_j, used to form ϕ_{n+1}, the next function in the sequence. In this case, by the standard construction I_j is exactly half the size of I_k and, more importantly, $I_j \subseteq I_k$. Correspondingly, $\{f(t) : t \in I_j\} \subseteq \{f(t) : t \in I_k\}$. Theorem 2.1.1 then implies $\inf\{f(t) : t \in I_j\} \geq \inf\{f(t) : t \in I_k\}$, which means $m_j \geq m_k$, or $\phi_{n+1}(x) \geq \phi_n(x)$. Thus the sequence $\{\phi_n\}$ is nondecreasing. In the same way in a similar proof, the dual sequence $\{\psi_n(x)\}$ is nonincreasing. ∎

Theorem 2.1.3 *For a real number x, let $\{J_n\}$ be a sequence of bounded intervals containing x with $J_1 \supseteq J_2 \supseteq \ldots$, where x is not an endpoint of any of them and where $\lim_{n \to \infty} m(J_n) = 0$. For a bounded function f, define $g(x) = \lim_{n \to \infty} (\inf\{f(t) : t \in J_n\})$ and $h(x) = \lim_{n \to \infty} (\sup\{f(t) : t \in J_n\})$. Then (1) the limits exist and are independent of the choice of sequence $\{J_n\}$ and (2) $g(x) \leq f(x) \leq h(x)$.*

Proof. First, suppose $|f| < M$. By Theorem 2.1.1, the sequence $\{\inf\{f(t) : t \in J_n\}\}_{n=1}^{\infty}$ is nondecreasing and bounded above by M. Hence, by the axiom of completeness, it converges to some value, which we call $g(x)$. If $\{K_n\}$ is any other sequence of bounded intervals with the same properties as $\{J_n\}$, then we can similarly define $g_1(x) = \lim_{n \to \infty} (\inf\{f(t) : t \in K_n\})$. For $n \geq 1$, since x is not an endpoint of J_n, it must have distance $\delta_n > 0$ from the nearest endpoint of J_n. Let m be so large that $m(K_m) < \delta_n$. Then $K_m \subseteq J_n$, and Theorem 2.1.1 implies $\inf\{f(t) : t \in J_n\} \leq \inf\{f(t) : t \in K_m\} \leq g_1(x)$. Therefore $g(x) = \lim_{n \to \infty} (\inf\{f(t) : t \in J_n\}) \leq g_1(x)$. Interchanging the roles of J_n and K_m proves $g_1(x) \leq g(x)$. Hence $g_1(x) = g(x)$, the limit is independent of the choice of sequence, and $g(x)$ is well-defined. A similar proof works the same way for $h(x)$.

Part 2 follows from $g(x) = \lim_{n \to \infty} \{\inf\{f(t) : t \in J_n\}\} \leq \lim_{n \to \infty} f(x) \leq \lim_{n \to \infty} \sup\{f(t) : t \in J_n\} = h(x)$. ∎

As we saw in Section 1.5, the standard construction also forms a sequence useful for determining the Lebesgue integral of many functions in L^0. Furthermore, the standard construction and its dual sequence will allow us to show fairly easily the equivalence between the Riemann integral and the Lebesgue integral of a function f, whenever the Riemann integral is defined. Finally, the sequences in Definition 2.1.2, coupled with the facts in the last theorem, will form the key to determining when the Riemann integral of a function exists. Let's see what the definition of the Riemann integral is all about and how it works.

2.1.2 The Definition of the Riemann Integral

We now formulate the Riemann integral "$R\text{-}\int_a^b f(x)\, dx$." (We add the notation "$R\text{-}$" to distinguish the Riemann integral from the Lebesgue integral.) As we proceed, the functions f we consider will always be bounded. In other words, over f's domain we will assume $|f(x)| \leq M$ for some positive real number M. We will also assume, for the sake of convenience, $f(x) = 0$ when x is not in $[a, b]$. (If this is not the case, then we will redefine the function values so $f(x) = 0$ off $[a, b]$, a modification that will cause no change in any of the results that follow.)

For such a function f, there exist step functions $\phi(x)$ and $\psi(x)$ such that

$$\phi \leq f \leq \psi.$$

For example, we may choose $\phi(x) = -M\mathcal{X}_{[a,b]}(x)$ and $\psi(x) = M\mathcal{X}_{[a,b]}(x)$. Another set of functions, more complicated but that flows naturally out of our work with the standard construction, are "inscribed" and "circumscribed" Riemann sum approximations to the area under the curve. Using the standard construction and the dual sequence from Definition 2.1.2, these are

(i) $\phi(x) = \sum_{k=1}^{2^n} m_k \mathcal{X}_{I_k}(x)$, where $m_k = \inf\{f(t) : t \in I_k\}$ and

(ii) $\psi(x) = \sum_{k=1}^{2^n} M_k \mathcal{X}_{I_k}(x)$, where $M_k = \sup\{f(x) : x \in I_k\}$.

For this choice of step functions, too, $\phi(x) \leq f(x) \leq \psi(x)$ for $x \in [a,b]$, since $m_k \leq f(x) \leq M_k$ for any x in the partition's interval I_k and, by the functions' definitions, $\phi(x) = m_k$ and $\psi(x) = M_k$.

So how is the Riemann integral of f defined? To formulate the definition, we use the Lebesgue integrals of such step functions as those constructed in Definition 2.1.2. (As mentioned in the last paragraph, they essentially form Riemann sum approximations.) The relationship between the sizes of the step-function integrals will be helpful as we proceed, and Theorem 1.3.3 compared the two:

When $\phi(x) \leq f(x) \leq \psi(x)$ for step functions ϕ and ψ, then the corresponding Lebesgue integrals satisfy $\int \phi \leq \int \psi$.

The Riemann integral of such a function f is determined from two values, the "lower Riemann integral of f" and the "upper Riemann integral of f." The lower Riemann integral of f is the supremum of the set of integrals of those step functions $\int \phi$ with $\phi \leq f$. Intuitively, the lower Riemann integral is as close as you can get to f's area under the curve by taking integrals of step functions less than or equal to f. The upper Riemann integral of f is the infimum of the set of all such $\int \psi$; it is intuitively as close as you can get to f's area under the curve by taking integrals of step functions greater than or equal to f. (These are Lebesgue integrals of step functions, which form Riemann sums.) In many situations, the lower Riemann integral will equal the upper Riemann integral. When that happens, we declare f to be Riemann integrable, and we define the integral of f to be the lower and upper integrals' common value. We formally present these ideas in the following definition, which (along the way) also introduces corresponding notation.

Definition 2.1.3 *In this definition, we continue to assume f is bounded and $f = 0$ when $x \notin [a,b]$.*

(A) The lower Riemann integral of f on $[a,b]$, denoted $\underline{\int_a^b} f$, is the supremum of the set of all integrals of step functions bounded above by f:

$$\underline{\int_a^b} f = \sup\left\{\int \phi \ : \ \phi \text{ is a step function, and } \phi \leq f\right\}.$$

(B) Similarly, the upper Riemann integral of f on $[a,b]$, denoted $\overline{\int_a^b} f$, is the infimum of the set of all integrals of step functions bounded below by f:

$$\overline{\int_a^b} f = \inf\left\{\int \psi \ : \ \psi \text{ is a step function, and } f \leq \psi\right\}.$$

2.1 The Riemann Integral

(C) Because $\phi \leq f \leq \psi$ in (A) and (B), it is always the case that $\underline{\int_a^b} f \leq \overline{\int_a^b} f$. When $\underline{\int_a^b} f = \overline{\int_a^b} f$, then f is Riemann integrable on $[a, b]$, and the Riemann integral of f, denoted $R\text{-}\int_a^b f(x)\, dx$, is equal to the common value:

$$R\text{-}\int_a^b f(x)\, dx = \underline{\int_a^b} f = \overline{\int_a^b} f.$$

(D) When the lower Riemann integral for f does not equal its upper Riemann integral, then f is not Riemann integrable.

We mention three facts. First, the standard construction forms a sequence of step functions for the lower Riemann integral's set—the one over which the supremum is taken. But this set contains many more step functions besides those coming from the standard construction. In the same way, the standard construction's dual functions are in the upper Riemann integral's set—over which the infimum is taken—but that set contains many more step functions besides just these.

Second, it is easy to see that the Riemann integral of a step function f exists and matches the Lebesgue integral. Put in step-function notation as in Definition 1.3.2, $f(x) = \sum_{j=1}^n c_j \cdot \mathcal{X}_{I_j}(x)$ (whose nonzero values are assumed to live on the interval $[a, b]$) has Riemann integral $R\text{-}\int_a^b f = \sum_{j=1}^n c_j \cdot m(I_j)$. This calculation follows quickly from two facts. First, f is itself a member of the set $\{\phi : \phi \text{ is a step function and } \phi \leq f\}$, and so its Lebesgue integral $\int f$ is in the set $S = \{\int \phi : \phi \text{ is a step function, and } \phi \leq f\}$ and serves as the set's supremum. Second, and in the same way, f is in the set $\{\psi : \psi \text{ is a step function and } f \leq \psi\}$, and so $\int f = \inf\{\int \psi : \psi \text{ is a step function, and } f \leq \psi\}$. Putting these facts together,

$$R\text{-}\int_a^b f(x)\, dx = \underline{\int_a^b} f = \overline{\int_a^b} f = \sum_{j=1}^n c_j \cdot m(I_j) = \int f.$$

Third, not all functions are Riemann integrable; in fact, many fairly straightforward ones are quickly seen not to satisfy Definition 2.1.2. The most famous of these is the Dirichlet function, which we discussed in Section 1.4 and is defined for $x \in [0, 1]$ by

$$D(x) = \begin{cases} 1 & \text{if } x \text{ is rational} \\ 0 & \text{if } x \text{ is irrational.} \end{cases}$$

The Dirichlet function is not Riemann integrable over, for example, the interval $[0, 1]$. For in this case as we form a step function, any subinterval $[x_i, x_{i+1}]$ of $[0, 1]$ will contain both a rational number (where $D = 1$) and an irrational number (where $D = 0$). But that means $\underline{\int_0^1} D = 0$ (since the largest step function that is bounded above by D must be $\phi = 0$), while $\overline{\int_0^1} D = 1$ (since the smallest step function that is bounded below by D must be $\psi = 1 \cdot \mathcal{X}_{[0,1]}$). Therefore the lower and upper Riemann integrals do not agree. The unfortunate conclusion is that the Riemann integral fails for $D(x)$, since it is undefined.

In sharp contrast, the Lebesgue integral of the Dirichlet function $D(x)$ exists. Why? Because $D(x)$ is equal almost everywhere to the function $f(x) = 0$: you can see the only place where $D(x)$ is not identically zero is on the rationals between 0 and 1, and the set of rationals has measure zero. Since functions equal almost everywhere have the same Lebesgue integral, the Lebesgue integral of $D(x)$ is equal to the Lebesgue integral of $f(x) = 0$, which is 0.

An interesting variation on the Dirichlet function is Thomae's function, explored in this section's Exercise 35. Before that, the next question provides practice with Definition 2.1.2, discussing Riemann integrals of step functions and functions analogous to the Dirichlet function.

Question 2.1.2 Find the Riemann integral $R\text{-}\int_0^5 f$ of the following functions, or use the definition to explain why the Riemann integral does not exist.

1. $f = 3 \cdot \mathcal{X}_{[0,1]} - 2 \cdot \mathcal{X}_{[4,5]}$
2. $f = \sum_{j=1}^{4} [j/(j+1)] \cdot \mathcal{X}_{[j,j+1]}$
3. $f(x) = \begin{cases} 1 & \text{if } x \text{ is rational} \\ 0 & \text{if } x \text{ is irrational} \end{cases}$
4. $f(x) = \begin{cases} 1 & \text{if } x \in [-1, 4] \\ 0 & \text{otherwise} \end{cases}$ ∎

2.1.3 Darboux's Theorem

This subsection works with the definition of the Riemann integral, characterizing it so we will be able to directly calculate as easily as possible the Riemann integral of some standard functions. The main result is Darboux's theorem, named after the famous French mathematician Gaston Darboux, who earned his Ph.D. in 1866 and taught in Paris through the early part of the 20th century. Darboux supervised six Ph.D. students, a fairly small number for an accomplished researcher. But all six went on to what can be considered extraordinary accomplishments in function theory; they are Émile Borel, Elie Cartan, Édouard Goursat, Émile Picard, Thomas Stieltjes, and Stanislaw Zaremba. All but Zaremba have well-known theorems or mathematical objects named after them, and Zaremba was in other ways among the most accomplished of the group, as the author of numerous admired textbooks and the cofounder and first president of the Polish Mathematical Society.[3]

Fundamental to the development of Darboux's theorem and other concerns of the Riemann integral will be the following alternative characterization of Definition 2.1.1. It provides a different but equivalent definition for the supremum and infimum of a real-valued set.

Figure 2.2. Darboux and his Ph.D. students. From left to right: Gaston Darboux, Émile Borel, Elie Cartan, Édouard Goursat, Émile Picard, Thomas Stieltjes, and Stanislaw Zaremba.

---

[3] More information on Darboux's students (and other mathematicians' advising lineage) can be found at the Mathematical Genealogy Project [84], an Internet resource founded by retired mathematician (from the University of Minnesota at Mankato) Harry Coonce.

2.1 The Riemann Integral

> **Alternative Characterization of the Supremum and Infimum**
>
> A number s equals $\sup S$ exactly when
>
> 1. s is an upper bound for S, and
> 2. given $\varepsilon > 0$, there exists a number $x \in S$ such that $x > s - \varepsilon$.
>
> Similarly, a number i equals $\inf S$ exactly when
>
> 1. i is a lower bound for S, and
> 2. given $\varepsilon > 0$, there exists a number $y \in S$ such that $y < i + \varepsilon$.

We can see why this characterization is equivalent to Definition 2.1.1. In the case of the supremum, suppose s is as in the alternate characterization: given $\varepsilon > 0$, there always exists a number $x \in S$ such that $x > s - \varepsilon$. Then any value less than s (which can be written as $s - \varepsilon$ for some ε) must always be smaller than some value x in S. But that means any value less than s cannot serve as an upper bound of S. In conclusion, s must be the least upper bound in the sense of Definition 2.1.1. Conversely, let's suppose s is a supremum for S as in Definition 2.1.1. That means s is an upper bound for S and $s < M$ for any other upper bound M. Then, given $\varepsilon > 0$, $s - \varepsilon$ cannot be an upper bound, and so must be smaller than some element x in S. In conclusion, s satisfies the alternative characterization. The situation for the infimum works similarly; see Exercise 37.

We now begin characterizing the Riemann integral in a way that will allow us to calculate it for several standard functions, starting with a useful theorem.

> **Theorem 2.1.4** *A function f is Riemann integrable if and only if there exist step functions ϕ and ψ with $\phi \leq f \leq \psi$ so $\int (\psi - \phi)$ is arbitrarily small.*

Proof. This theorem is an if and only if statement, and so we must prove both of the statement's directions (sufficiency and necessity). Begin by assuming f is Riemann integrable (the sufficiency direction). Then $R\text{-}\int_a^b f(x)\,dx = \sup \{\int \varphi : \varphi \text{ is a step function and } \varphi \leq f\}$. Applying the alternative characterization of the supremum, this means, given $\varepsilon > 0$, there exists a step function ϕ such that $\int \phi > R\text{-}\int_a^b f(x)\,dx - \varepsilon$. From the upper Riemann integral (and the alternative characterization for the infimum), there similarly exists a step function ψ such that $\int \psi < R\text{-}\int_a^b f(x)\,dx + \varepsilon$. Subtracting one inequality from the other,

$$\int (\psi - \phi) = \int \psi - \int \phi < \left(R\text{-}\int_a^b f(x)\,dx + \varepsilon \right) - \left(R\text{-}\int_a^b f(x)\,dx - \varepsilon \right) = 2\varepsilon,$$

which is arbitrarily small. The result follows.

Now prove necessity. Given $\varepsilon > 0$, assume there are step functions ϕ and ψ with $\phi \leq f \leq \psi$ so $\int \psi - \int \phi = \int (\psi - \phi) < \varepsilon$, which algebraically implies $\int \phi > \int \psi - \varepsilon$. It is always the case that $\underline{\int_a^b} f \leq \overline{\int_a^b} f$, which means (1) $\overline{\int_a^b} f$ is an upper bound for the set $S = \{\int \phi : \phi \text{ is a step function and } \phi \leq f\}$. But its property as an infimum means $\overline{\int_a^b} f$ also satisfies $\int \psi \geq \overline{\int_a^b} f$. Therefore (2) $\int \phi > \int \psi - \varepsilon \geq \overline{\int_a^b} f - \varepsilon$. Hence $\overline{\int_a^b} f$ satisfies both conditions for the alternate characterization of a supremum, and so $\overline{\int_a^b} f = \sup S$. By definition, that fact means $\overline{\int_a^b} f = \underline{\int_a^b} f$, and f is Riemann integrable. ∎

Darboux's theorem will help us directly calculate the Riemann integral of many functions, including polynomial functions. We will use it repeatedly in this way. It will also be instrumental in showing a function's Riemann integral value (whenever it exists) equals its corresponding Lebesgue integral. Finally, it will lead to a simple characterization of when a function f is Riemann integrable. Darboux's theorem changes the way we are able to think about the calculation of a Riemann integral: instead of having to calculate very complicated infimum and supremum values (which define the upper and lower Riemann integrals in Definition 2.1.3), we can calculate the Riemann integral as a limit. Suprema and infima of complicated sets are often complicated to determine. Limits are often much simpler to find.

Theorem 2.1.5 Darboux's Theorem: *Suppose $\{\phi_n\}$ is a nondecreasing sequence of step functions and $\{\psi_n\}$ is a nonincreasing sequence of step functions, where $\phi_n \leq f \leq \psi_n$ for every n and where $\lim\limits_{n \to \infty} \int_a^b \psi_n - \phi_n = 0$. Then*

1. *The sequences $\int_a^b \phi_n$ and $\int_a^b \psi_n$ converge to the same limit.*
2. *The function f is Riemann integrable.*
3. *We may calculate the Riemann integral of f in terms of the integrals of the step-function sequences:*

$$\lim_{n \to \infty} \int_a^b \psi_n(x)\, dx = R\text{-}\int_a^b f(x)\, dx = \lim_{n \to \infty} \int_a^b \phi_n(x)\, dx.$$

Proof. 1. Theorem 1.3.3 says, since $\{\phi_n\}_{n=1}^\infty$ is nondecreasing, so is $\{\int_a^b \phi_n\}_{n=1}^\infty$. Also, every term of the sequence $\{\int_a^b \phi_n\}_{n=1}^\infty$ is bounded above (for example, by $\int_a^b \psi_1$). Hence, by the axiom of completeness (see Subsection 1.4.1), the limit $l \equiv \lim\limits_{n \to \infty} \int_a^b \phi_n$ exists. Similarly, $\{\int_a^b \psi_n\}_{n=1}^\infty$ is nonincreasing and bounded below, for example by $\int_a^b \phi_1$. Hence the limit $L \equiv \lim\limits_{n \to \infty} \int_a^b \psi_n$ also exists. Putting these facts together, $L - l = \lim\limits_{n \to \infty} \int_a^b \psi_n - \lim\limits_{n \to \infty} \int_a^b \phi_n = \lim\limits_{n \to \infty} \int_a^b (\psi_n - \phi_n) = 0$ by the assumption. The result for this first part follows.

2. The sequence $\{\int_a^b (\psi_n - \phi_n)\}_{n=1}^\infty$ approaches 0 in the limit. Therefore, given $\varepsilon > 0$, there exists $N > 0$ such that, for $n \geq N$, $\int_a^b (\psi_n - \phi_n) = |\int_a^b (\psi_n - \phi_n) - 0| < \varepsilon$. Theorem 2.1.4 then proves f is Riemann integrable.

3. For any of the given step functions ϕ_n and ψ_n, since $\phi_n \leq f \leq \psi_n$ and by the definition of the infimum and supremum,

$$\int_a^b \phi_n \leq \sup\left\{\int_a^b \phi : \phi \text{ is a step function and } \phi \leq f\right\}$$

$$\leq \inf\left\{\int_a^b \psi : \psi \text{ is a step function and } f \leq \psi\right\} \leq \int_a^b \psi_n.$$

2.1 The Riemann Integral

Taking limits and applying Definition 2.1.3 (along with the fact from part 2 that f is Riemann integrable),

$$l \equiv \lim_{n \to \infty} \int_a^b \phi_n \leq R\text{-}\int_a^b f(x)\,dx \leq \lim_{n \to \infty} \int_a^b \psi_n \equiv L.$$

By the sandwich theorem and part 1, all the terms are equal. ∎

Darboux's theorem is useful because the standard construction and its dual sequence often provide just the right kind of sequences of step functions—they often satisfy the conditions in Darboux's theorem. The sequences $\{\phi_n\}$ and $\{\psi_n\}$ from Definition 2.1.2 are, respectively, nondecreasing and nonincreasing, $\phi_n \leq f \leq \psi_n$ for every n, and we often have $\lim_{n \to \infty} \int_a^b \psi_n - \phi_n = 0$. When that happens, part 2 of Darboux's theorem says f is integrable, and part 3 says we can take the limit of either of these sequences to determine the Riemann integral of f. In calculus, when you took limits of Riemann sums either from inscribed or circumscribed rectangles, you were using Darboux's theorem to find the given (continuous) function's Riemann integral as a limit (rather than as a supremum or infimum).

We will soon see any continuous function on an interval $[a, b]$ is integrable on $[a, b]$, and Darboux's theorem will therefore apply. The next example illustrates this. We've chosen it because it includes many complicated nuances, making the calculations a bit untidy, but allowing us to display details that can arise in these types of problems. You'll want to be graciously patient as you work through it.

Example 2.1.3 Evaluate $R\text{-}\int_{-1}^1 x^2\,dx$, using Darboux's theorem with the standard construction of Definition 2.1.2 and its dual sequence. Along the way, show part 3 of Darboux's theorem holds, obtaining $\lim_{n \to \infty} \int_0^1 \psi_n(x)\,dx = R\text{-}\int_0^1 x^2\,dx = \lim_{n \to \infty} \int_0^1 \phi_n(x)\,dx$. The main point of this example is to illustrate the power of Darboux's theorem: we do not have to consider upper or lower Riemann integrals to evaluate the integral, and we obtained its value from limits of step-function integrals (i.e., as limits of Riemann sums).

Solution. The standard construction forms sequences $\{\phi_n\}$ and $\{\psi_n\}$ piecewise defined over 2^n subintervals of the form $I_k = [-1 + 2(k-1)/2^n, -1 + 2k/2^n)$, $k = 1, 2, 3, \ldots, 2^n$. Then $\phi_n = \sum_{k=1}^{2^n} m_k \cdot \mathcal{X}_{I_k}$, where $m_k = \inf\{x^2 : x \in I_k\}$. The function $f(x) = x^2$ is decreasing over those subintervals I_k that consist of negative values, and therefore the infimum of x^2 over the subinterval is found by evaluating x^2 at the right endpoint. Since these I_k are the first half of the 2^n subintervals in the list, $m_k = [-1 + 2k/2^n]^2$ if $k = 1, 2, 3, \ldots, 2^{n-1}$. In the same way, analyzing the I_k subintervals that consist of nonnegative values (and over which x^2 is increasing), we have $m_k = [-1 + 2(k-1)/2^n]^2$ if $k = 2^{n-1} + 1, 2^{n-1} + 2, \ldots, 2^n$. Hence

$$\phi_n(x) = \sum_{k=1}^{2^{n-1}} [-1 + 2k/2^n]^2 \cdot \mathcal{X}_{I_k}(x) + \sum_{k=2^{n-1}+1}^{2^n} [-1 + 2(k-1)/2^n]^2 \cdot \mathcal{X}_{I_k}(x).$$

Using a similar analysis but with the supremum, the dual sequence has step functions

$$\psi_n(x) = \sum_{k=1}^{2^{n-1}} [-1 + 2(k-1)/2^n]^2 \cdot \mathcal{X}_{I_k}(x) + \sum_{k=2^{n-1}+1}^{2^n} [-1 + 2k/2^n]^2 \cdot \mathcal{X}_{I_k}(x).$$

We use the fact that $m(I_k) = 1/2^{n-1}$ to evaluate the integrals of the step functions. For example,

$$\int \phi_n = \sum_{k=1}^{2^{n-1}} \left(-1 + \frac{k}{2^{n-1}}\right)^2 \cdot \frac{1}{2^{n-1}} + \sum_{k=2^{n-1}+1}^{2^n} \left(-1 + \frac{k-1}{2^{n-1}}\right)^2 \cdot \frac{1}{2^{n-1}}.$$

After obtaining a rather messy closed-form expression in n (using techniques similar to the examples from Section 1.5), the sum simplifies to $\int \phi_n = -2 + \frac{4(2^n+1)(2 \cdot 2^n+1)}{3 \cdot 2^{2n}} - \frac{6}{2^n}$. Taking the limit (the second term's limit is found as the ratio of its bold-faced coefficients) and using part 3 of Darboux's theorem, we obtain the desired result: $R\text{-}\int_{-1}^{1} x^2\,dx = \lim_{n\to\infty} \int \phi_n = -2 + 8/3 - 0 = 2/3$.

We can evaluate $\int \psi_n$ in a similar way.

$$\int \psi_n = \sum_{k=1}^{2^{n-1}} \left(-1 + \frac{k-1}{2^{n-1}}\right)^2 \cdot \frac{1}{2^{n-1}} + \sum_{k=2^{n-1}+1}^{2^n} \left(-1 + \frac{k}{2^{n-1}}\right)^2 \cdot \frac{1}{2^{n-1}}.$$

As a closed-form expression in n, $\int \psi_n = -2 + \frac{4(2^n+1)(2 \cdot 2^n+1)}{3 \cdot 2^{2n}} - \frac{2}{2^n} - \frac{3}{2^{2n}}$. As predicted by Darboux's theorem, we get the same value as before when taking the limit: $\lim_{n\to\infty} \int \psi_n = -2 + 8/3 = 2/3$. We could have used either limit to determine $R\text{-}\int_{-1}^{1} x^2\,dx = 2/3$. ∎

Question 2.1.3 Find the step-function sequences $\{\phi_n\}$ and $\{\psi_n\}$ from the standard construction and its dual sequence for the function $f(x) = 3x + 1$ on the interval $[0, 2]$. Note f is continuous and therefore (as we will soon prove) Riemann integrable. Evaluate $\int \phi_n$ and $\int \psi_n$ as closed-form expressions in n, and show $\lim_{n\to\infty} \int \phi_n = \lim_{n\to\infty} \int \psi_n$. Then use Darboux's theorem to find the value of $R\text{-}\int_0^2 3x + 1\,dx$. ∎

The next theorem will be useful in the next subsection to characterize which functions are Riemann integrable. The proof is algebraic, working with inequalities to compare different integrals of step functions to show a certain limit holds. Such proofs often seem like bookkeeping, and they can be dry. We present the result here because of our need for it in the upcoming subsection, and because it formulates integral calculations of the now-familiar standard construction's functions.

Theorem 2.1.6 *For a bounded function f on an interval $[a, b]$, the standard construction sequence $\{\phi_n\}$ and its dual sequence $\{\psi_n\}$ of Definition 2.1.2 satisfy*

$$\lim_{n\to\infty} \int \phi_n = \underline{\int_a^b} f(x)\,dx \quad \text{and} \quad \lim_{n\to\infty} \int \psi_n = \overline{\int_a^b} f(x)\,dx.$$

Proof. We prove the first equality. By the alternative characterization of the supremum (using $\varepsilon/2 > 0$) applied to the definition of the lower Riemann integral, there is a step function $\phi = \sum_{i=1}^{m} c_i \cdot J_i$ with $\phi \leq f$ on $[a, b]$ and $\int \phi > \underline{\int_a^b} f(x)\,dx - \varepsilon/2$. Using the same J_i intervals, define the corresponding step function $\Phi = \sum_{i=1}^{m} d_i \cdot J_i$, where $d_i = \inf\{f(t) : t \in J_i\}$. On any one of the m subintervals J_i, both ϕ and Φ are constant (equal to c_i and d_i, respectively). Since $\phi(t) \leq f(t)$

2.1 The Riemann Integral

for all $t \in J_i$, we have, for any $x \in J_i$, $\phi(x) = \inf\{\phi(t) : t \in J_i\} \leq \inf\{f(t) : t \in J_i\} = \Phi(x)$. In summary, $\phi \leq \Phi \leq f$.

How does Φ compare with the standard construction's ϕ_n? Take any of ϕ_n's subintervals I_k and examine two cases:

Case 1: Suppose $I_k \subseteq J_i$ for some one of Φ's subintervals J_i. Theorem 1.3.3, together with the fact from Theorem 2.1.1 that $\inf\{f(t) : t \in J_i\} \leq \inf\{f(t) : t \in I_k\}$, then implies $\int_{I_k} \Phi \leq \int_{I_k} \phi_n$.

Case 2: Suppose I_k overlaps with more than one of the m disjoint J_i subintervals so that it contains points from perhaps several of them. In this case, $\int_{I_k} \Phi \leq \int_{I_k} \phi_n$ will not always be true. What's the worst that can happen to violate this inequality? It would be when Φ is as big as possible over as many of these J_i as possible, while ϕ_n is as small as possible on I_k. If the bound on f is expressed as $|f| \leq K$, this worst case happens when $\phi_n = -K$ on I_k (so that $\int_{I_k} \phi_n = -K \cdot m(I_k) = -K \cdot (1/2^n)$) and $\Phi = K$ throughout as many of the J_i subintervals overlapping with I_k as possible, a situation that would at least result in $\int_{I_k} \Phi \leq K \cdot m(I_k) = K \cdot (1/2^n)$. Putting the two inequalities together, $\int_{I_k} \Phi - 2K \cdot (1/2^n) \leq -K \cdot (1/2^n) = \int_{I_k} \phi_n$. In summary, $\int_{I_k} \Phi - 2K \cdot (1/2^n) \leq \int_{I_k} \phi_n$, and this could happen at most $m - 1$ times, since there are only m of the J_i subintervals available to overlap with any I_k.

In either case, $\int \phi_n = \sum_{k=1}^{2^n} \int_{I_k} \phi_n \geq \sum_{k=1}^{2^n} \int_{I_k} \Phi - \sum_{i=1}^{m-1} 2K \cdot (1/2^n) = \int \Phi - (m-1)2K/2^n > \int \Phi - \varepsilon/2$ whenever $(m-1)2K/2^n < \varepsilon/2$. More simply put, $\int \phi_n > \int \Phi - \varepsilon/2$ whenever $n > N$, where $N \equiv \log_2(4(m-1)K/\varepsilon)$.

Putting it together, given $\varepsilon > 0$, whenever $n > N$ we have $\int \phi_n > \int \Phi - \varepsilon/2 > \int \phi - \varepsilon/2 > (\int_a^b f(x)\,dx - \varepsilon/2) - \varepsilon/2 = \int_a^b f(x)\,dx - \varepsilon$, which implies $-\varepsilon < \int \phi_n - \int_a^b f(x)\,dx < \varepsilon$, or $|\int \phi_n - \int_a^b f(x)\,dx| < \varepsilon$. By Definition 1.4.1, $\lim_{n \to \infty} \int \phi_n = \int_a^b f(x)\,dx$. The first limit is established.

The second limit follows similarly from a nearly identical set of constructions and calculations. ∎

2.1.4 The Riemann-Lebesgue Theorem

The Riemann-Lebesgue theorem precisely characterizes which functions are Riemann integrable. Proved by Lebesgue in his 1902 dissertation, it is a beautiful fact that immediately stunned the mathematical community by its remarkable straightforwardness: a bounded function is Riemann integrable on $[a, b]$ if and only if it is continuous almost everywhere on $[a, b]$. That makes the theorem useful: with it, you can know almost immediately (for most functions f you'll run across in practice) if f is Riemann integrable. It also tells you how to think about constructing a function f that is not—you'd better be looking for a function discontinuous on more than a set of measure zero. We formally state and prove the result in the next theorem. It will not surprise you to know the standard construction and its dual sequence (from Definition 2.1.2), along with Darboux's theorem, provide the keys to proving the result.

Theorem 2.1.7 The Riemann-Lebesgue Theorem: *A bounded function f is Riemann integrable over an interval $[a, b]$ if and only if the points of discontinuity of f on $[a, b]$ form a set of measure zero. In this case, the Riemann integral equals Lebesgue's: $R\text{-}\int_a^b f\,dx = \int_a^b f\,dx$.*

Proof. Start by assuming the points of discontinuity form a set of measure zero and set out to prove f is Riemann integrable. Use the standard construction to define the sequence with step functions $\phi_n(x) = \sum_{k=1}^{2^n} m_k \cdot \mathcal{X}_{I_k}(x)$, where $m_k = \inf\{f(t) : t \in I_k\}$ and $I_k = [a + (b-a)(k-1)/2^n, a + (b-a)k/2^n)$ for $k = 1, 2, \ldots, 2^n$. Similarly define the dual sequence with $\psi_n(x) = \sum_{k=1}^{2^n} M_k \cdot \mathcal{X}_{I_k}$, where $M_k = \sup\{f(t) : t \in I_k\}$. As usual, $\phi_n \leq f \leq \psi_n$. Also, Theorem 2.1.2 says $\{\phi_n\}$ is nondecreasing and $\{\psi_n(x)\}$ is nonincreasing. By the axiom of completeness, both limits $\lim_{n\to\infty} \phi_n$ and $\lim_{n\to\infty} \psi_n$ exist. Furthermore, $\lim_{n\to\infty} \int \phi_n$ and $\lim_{n\to\infty} \int \psi_n$ exist, since they are also nondecreasing and nonincreasing, respectively, and since they are bounded, respectively, from above and below. (For example, since f is bounded, we know $\int \phi_n$ is bounded from above by $M \cdot (b-a)$, where $M = \sup\{f(x) : x \in [a, b]\}$.)

Now examine any value $x \in [a, b]$ where f is continuous at x. By Definition 1.4.3, given $\varepsilon > 0$, there exists $\delta > 0$ such that $|f(t) - f(x)| < \varepsilon$ whenever $|t - x| < \delta$. Choose a natural number N so large that $(b-a)/2^N < \delta$. Then one of ϕ_N's 2^N subintervals, say I_j, contains x and has all its points t within a distance δ of x, since the total width of any subinterval is $(b-a)/2^N$. That means $|f(t) - f(x)| < \varepsilon$ for all $t \in I_j$.

In general, for real A and $B > 0$, the inequality $|A| < B$ means the distance from A to 0 is less than B, which is the same as the equivalent expression $-B < A < B$. Applying this, $-\varepsilon < f(t) - f(x) < \varepsilon$, or $f(x) - \varepsilon < f(t) < f(x) + \varepsilon$ for $t \in I_j$. Since $\phi_N(x) = \inf\{f(x) : x \in I_j\}$, $f(x) - \varepsilon \leq \phi_N(x) \leq f(x) + \varepsilon$. But the sequence $\{\phi_n(x)\}$ is nondecreasing up to $f(x)$, and so, for $n \geq N$, $f(x) - \varepsilon \leq \phi_N(x) \leq \phi_n(x) \leq f(x) \leq f(x) + \varepsilon$.

Therefore $f(x) - \varepsilon \leq \phi_n(x) \leq f(x) + \varepsilon$, which is $-\varepsilon \leq \phi_n(x) - f(x) \leq \varepsilon$, whenever $n \geq N$. This says $|\phi_n(x) - f(x)| \leq \varepsilon$ whenever $n \geq N$. By Definition 1.4.4, $\lim_{n\to\infty} \phi_n(x) = f(x)$. An equivalent analysis applied to ψ_n similarly implies $\lim_{n\to\infty} \psi_n(x) = f(x)$ at any point x where f is continuous.

A first conclusion is that Definition 1.5.1 is satisfied: $\{\phi_n\}$ is a nondecreasing sequence of step functions that converges pointwise almost everywhere to f, and hence $f \in L^0$ is Lebesgue integrable (if it is in L^0, then it is certainly in L^1). Furthermore, Definition 1.5.1 says the Lebesgue integral of f on $[a, b]$ is $\int_a^b f = \lim_{n\to\infty} \int \phi_n$.

A second conclusion is almost immediate if we apply the analysis for the dual sequence to the negation of f. The sequence $\{-\psi_n\}$ is nondecreasing, $\lim_{n\to\infty} -\psi_n(x) = -f(x)$ a.e. (it holds wherever f is continuous), and (by Definition 1.5.1) $\int_a^b (-f) = \lim_{n\to\infty} \int (-\psi_n)$. The linearity of L^1 then implies $\int_a^b f = \lim_{n\to\infty} \int \psi_n$.

We have shown $\lim_{n\to\infty} \int_a^b \psi_n = \int_a^b f = \lim_{n\to\infty} \int_a^b \phi_n$, and so $\lim_{n\to\infty} \int_a^b (\psi_n - \phi_n) = 0$. By part 2 of Darboux's theorem, f is Riemann integrable. Furthermore, part 3 of Darboux's theorem proves the Riemann integral of f is equal to the Lebesgue integral, since we have $R\text{-}\int_a^b f = \lim_{n\to\infty} \int_a^b \phi_n = \int_a^b f$. This establishes sufficiency.

Now we prove necessity. For f bounded and Riemann integrable over $[a, b]$, $\int_a^b f(x)\,dx = \int_a^b f(x)\,dx$. For $x \in [a, b]$ and $n = 1, 2, 3, \ldots$, let J_n be the I_k interval, as defined for the standard construction's step function ϕ_n, that contains x. Then $J_1 \supseteq J_2 \supseteq \ldots$ and $\lim_{n\to\infty} m(J_n) = 0$. If x is not one of the (countably many) endpoints of any J_n, then define

2.1 The Riemann Integral

$g(x) = \lim_{n\to\infty} (\inf\{f(t) : t \in J_n\})$ and $h(x) = \lim_{n\to\infty} (\sup\{f(t) : t \in J_n\})$. If x is one of the endpoints of J_n, then define $g(x) = h(x) = f(x)$.[4] Theorem 2.1.3 says g and h are well-defined, and $g \leq f \leq h$ on $[a, b]$. Furthermore, the standard construction defines $\phi_n(x) = \inf\{f(t) : t \in J_n\}$, and the dual sequence defines $\psi_n(x) = \sup\{f(t) : t \in J_n\}$. Therefore $g(x) = \lim_{n\to\infty} \phi_n(x)$ and $h(x) = \lim_{n\to\infty} \psi_n(x)$ for any x not an endpoint of any J_n, which is almost everywhere on $[a, b]$. By Definition 1.5.1, $g \in L^0$ and $\int g = \lim_{n\to\infty} \int \phi_n$. Applying Definition 1.5.1 to the negation of h, and noting the linearity of L^1, we see $h \in L^1$ with $\int h = \lim_{n\to\infty} \int \psi_n$.

Now apply Theorem 2.1.6 to obtain $\int g = \lim_{n\to\infty} \int \phi_n = \underline{\int_a^b} f(x)\, dx = \overline{\int_a^b} f(x)\, dx = \lim_{n\to\infty} \int \psi_n = \int h$. We have $(h - g)(x) \geq 0$ almost everywhere and $\int (h - g) = \int h - \int g = 0$. That turns out to mean, as we will easily and independently be able to show in Corollary 2.2.1, that $h - g = 0$. With that fact graciously and kindly granted (you may glance ahead if you wish), we have $h(x) = f(x) = g(x)$ for almost all x.

That means f is continuous almost everywhere. For, given $\varepsilon > 0$, the definitions of g and h as sequence limits imply there exists $N > 0$ such that $|\inf\{f(t) : t \in J_n\} - g(x)| < \varepsilon$ and $|\sup\{f(t) : t \in J_n\} - h(x)| < \varepsilon$ whenever $n \geq N$ and x is not an endpoint of any J_n. The equivalent expressions for the absolute value inequalities imply $g(x) - \varepsilon < \inf\{f(t) : t \in J_n\}$ and $\sup\{f(t) : t \in J_n\} < h(x) + \varepsilon$. Substituting $f(x)$ for its equal values $g(x)$ and $h(x)$, they imply $f(x) - \varepsilon < \inf\{f(t) : t \in J_n\} \leq f(s) \leq \sup\{f(t) : t \in J_n\} < f(x) + \varepsilon$ for $s \in J_n$ and $n > N$. Since x is not an endpoint of any J_n, there is an interval of the form $(x - \delta, x + \delta) \subseteq J_n$ for some $\delta > 0$. Hence $f(x) - \varepsilon < f(s) < f(x) + \varepsilon$ whenever $s \in (x - \delta, x + \delta)$. Stated equivalently, $|f(s) - f(x)| < \varepsilon$ whenever $|s - x| < \delta$. By Definition 1.4.3 of Section 1.4, f is continuous at any of these x values, which is almost everywhere on $[a, b]$. ∎

The Riemann-Lebesgue theorem's characterization of Riemann integrability in terms of f being continuous almost everywhere is fairly simple to determine. Any bounded function that is continuous a.e. has a Riemann integral that exists, and then Theorem 2.1.6 implies, for example, that $R\text{-}\int_a^b f(x)\, dx = \lim_{n\to\infty} \int \phi_n$, where ϕ_n comes from the standard construction. The next example illustrates these concepts in action. Since every Riemann integrable function is also Lebesgue integrable and is in L^0, you'll recognize many of the steps in the next example. You've already seen them used in Section 1.5 to calculate the Lebesgue integral of an L^0 function.

Example 2.1.4 Use the Riemann-Lebesgue theorem to determine if
$f(x) = \begin{cases} (x^2 - 1)/(x - 1) & \text{if } x \neq 1 \\ 0 & \text{if } x = 1 \end{cases}$ is Riemann integrable on $[0, 2]$. If so, find the value of $R\text{-}\int_0^2 f(x)\, dx$.

Solution. The function is continuous at every point except for $x = 1$. In fact, $f(x) = x + 1$ when $x \neq 1$. Also, f is bounded on $[0, 2]$. By the Riemann-Lebesgue theorem, the function is then Riemann integrable. The standard construction determines $\int_0^2 f(x)\, dx = \int_0^2 x + 1\, dx$. Here, $\phi_n = \sum_{k=1}^{2^n} [2(k-1)/2^n + 1] \cdot I_k$, which satisfy $\int_0^2 \phi_n = \sum_{k=1}^{2^n} [2(k-1)/2^n + 1] \cdot (2/2^n) =$

---
[4] This choice, for the J_n endpoints, of $g(x)$ and $h(x)$ is made for the sake of convenience, since it works well in the remaining statements of the proof.

$(4/2^{2n})[\sum_{k=1}^{2^n}(k-1)] + 2 = 2(2^n - 1)/2^n + 2$. Take the limit to produce the well-known result:
$R\text{-}\int_0^2 f(x)\,dx = \lim_{n\to\infty} \int \phi_n = \lim_{n\to\infty} 2(2^n - 1)/2^n + 2 = 2 + 2 = 4$. ∎

The next question gives you practice on a similar problem—think of it as a warm-up for the exercises.

Question 2.1.4 Determine points of discontinuity for $f(x) = (x-1)(x^2 + 5x + 4)/(x^2 - 1)$. What does the Riemann-Lebesgue theorem say about f? What is $R\text{-}\int_{-10}^{10} f(x)\,dx$? ∎

Solutions to Questions

2.1.1 (a) inf $A = -40$ and sup $A = 12$. The supremum is in A but the infimum is not. (b) inf $B = 0$ and sup $B = 5$. Neither the supremum nor the infimum is in B. (c) inf $C = 0$ and sup $C = 5$. Both are in C. (d) inf $D = 0$ and sup $D = 1$. The infimum is in D but the supremum is not. (e) inf $E = 1$ and sup $E = 9$. The infimum is in E but the supremum is not. (f) inf $F = 0$ and sup $F = 1$. The infimum is in F but the supremum is not.

2.1.2 (1) $R\text{-}\int_0^5 f(x)\,dx = 1$. (2) $R\text{-}\int_0^5 f(x)\,dx = 1/2 + 2/3 + 3/4 + 4/5 = 163/60$ (3) $R\text{-}\int_0^5 f(x)\,dx$ does not exist. The upper Riemann integral is 5 but the lower Riemann integral is 0. (4) $R\text{-}\int_0^5 f(x)\,dx = 1 \cdot (4 - (0)) = 4$.

2.1.3 Define ϕ_n and ψ_n on 2^n equal-sized subintervals of $[0, 2)$ as $\phi_n(x) = 6(k-1)/2^n + 1$ and $\psi_n(x) = 6k/2^n + 1$ for x in the kth subinterval, where $k = 1, 2, \ldots, 2^n$. Then $\lim_{n\to\infty} \int \phi_n = \lim_{n\to\infty} \sum_{k=1}^{2^n}(6(k-1)/2^n + 1) \cdot (2/2^n) = \lim_{n\to\infty} \frac{12 \cdot 2^n(2^n+1)}{2 \cdot 2^{2n}} - \frac{12}{2^n} + 2 = 8$. Similarly, $\lim_{n\to\infty} \int \psi_n = 8$. By Darboux's theorem, $R\text{-}\int_0^2 3x + 1\,dx = 8$.

2.1.4 $S = \{1, -1\}$ contains the points of discontinuity. Since $m(S) = 0$, f is Riemann integrable. $f(x) = x + 4$ a.e. on $\mathbb{R}$. $R\text{-}\int_{-10}^{10} f(x)\,dx = R\text{-}\int_{-10}^{10} x + 4\,dx = \lim_{n\to\infty} \sum_{k=1}^{2^n}[(-10 + 20(k-1)/2^n) + 4] \cdot 20/2^n = \lim_{n\to\infty}(20/2^{2n})20 \cdot \frac{2^n(2^n-1)}{2} - 6 \cdot 20 = 80$.

Reading Questions for Section 2.1

1. For a real-valued set A with an upper bound, what two properties make $s \in \mathbb{R}$ the supremum of A?

2. Answer the last question using the alternative characterization of the supremum of a set A.

3. For a given bounded function f on $[a, b]$, what is the definition of the lower Riemann integral? The upper Riemann integral?

4. When is a bounded function f Riemann integrable on $[a, b]$?

5. Give an example of a function not Riemann integrable on $[0, 1]$.

6. What does Darboux's theorem say?

7. What does the Riemann-Lebesgue theorem say?

2.1 The Riemann Integral

Exercises for Section 2.1

In Exercises 1–5, give examples of sets with the stated feature.

1. A set not bounded above.

2. An uncountable set that is bounded above and contains its supremum of 4.

3. An uncountable set that is bounded above but does not contain its supremum of 5.

4. A countable set that is bounded above and contains its supremum of 6.

5. A countable set that is bounded above but does not contain its supremum of 7.

In Exercises 6–14, find the supremum and infimum of the set, and determine if the supremum and/or the infimum is a member of the set.

6. $A = \{-3, 1, 4, 6, 9, 10, 14\}$

7. $B = [0, 2) \cup [3, 19)$

8. $C = [0, 4) \cap [3, 19)$

9. $D = \{0, 2, 2\frac{1}{2}, 2\frac{2}{3}, 2\frac{3}{4}, 2\frac{4}{5}, \ldots\}$

10. $E = \{0, 1, 2\frac{1}{2}, 3\frac{2}{3}, 4\frac{3}{4}, 5\frac{4}{5}, \ldots\}$

11. $F = \{x : x = 5/n + 3, \text{ where } n = 1, 2, 3, \ldots\}$

12. $G = \{g(x) = x^3 : x \in [3/2^5, 4/2^5)\}$

13. $H = \{h(x) = 3x^2 + 2x + 4 : x \in [3/11, 4/11)\}$

14. $I = \{f(x) = 3x + 5 : x \in [5/17, 6/17)\}$

For each function f in Exercises 15–20, use the standard construction to determine a nondecreasing sequence $\{\phi_n\}$ that converges to f on the given integral $[a, b]$. Also find the dual sequence $\{\psi_n\}$. Do **not** use your answers to calculate the integral of f.

15. $f(x) = x^2 + 3x + 5$ on $[4, 10]$

16. $f(x) = -3x + 7$ on $[-2, 3]$

17. $f(x) = -3x^2 + 7$ on $[-1, 1]$

18. $f(x) = e^x$ on $[0, b]$, for b a fixed positive real number

19. $f(x) = \sin x$ on $[-\pi/2, \pi/2]$

20. $f(x) = \cos x - x$ on $[0, \pi]$

The Riemann-Lebesgue theorem guarantees the Riemann integral of the functions in Exercises 21–27 exists. Use Theorem 2.1.6 with the standard construction to find them.

21. $R\text{-}\int_0^2 x \, dx$

22. $R\text{-}\int_1^3 x \, dx$

23. $R\text{-}\int_0^1 2x^2 \, dx$

24. $R\text{-}\int_0^1 (3x^2 - 1) \, dx$

25. $R\text{-}\int_1^2 3x^2\,dx$

27. $R\text{-}\int_{-1}^3 (x^3 - 4)\,dx$

26. $R\text{-}\int_1^2 (2x^2 - x + 1)\,dx$

In Exercises 28–34, use the Riemann-Lebesgue theorem to determine if the Riemann integral exists. Say why or why not, including identification of the function's points of discontinuity.

28. $R\text{-}\int_2^7 (4x^5 - 2x^3 - 8x + 7)\,dx$

29. $R\text{-}\int_0^{2\pi} 4\sin(1/x)\,dx$

30. $R\text{-}\int_0^1 f(x)\,dx$, where $f(x) = \begin{cases} 1 & \text{if } x \in [0, .5) \\ 0 & \text{if } x \in [.5, 1) \end{cases}$

31. $R\text{-}\int_0^4 f(x)\,dx$, where $f(x) = \begin{cases} 1 & \text{if } x \in [0, 1) \\ x & \text{if } x \in [1, 2) \\ x^2 & \text{if } x \in [2, 3) \\ \ln(x) & \text{if } x \in [3, 4] \end{cases}$

32. $R\text{-}\int_0^1 f(x)\,dx$, where $f(x) = \begin{cases} 1 & \text{if } x = \frac{m}{2n} \text{ for } m, n \in \mathbb{Z} \\ 0 & \text{otherwise} \end{cases}$

33. $R\text{-}\int_0^1 f(x)\,dx$, where $f(x) = \begin{cases} x & \text{if } x = \frac{1}{2^n} \text{ for } n \in \mathbb{N} \\ x+1 & \text{otherwise} \end{cases}$

34. $R\text{-}\int_0^1 f(x)\,dx$, where $f(x) = \begin{cases} 1 & \text{if } x \text{ is in the Cantor set } C \\ 0 & \text{otherwise} \end{cases}$

Exercises 35 and 36 explore variations on the Dirichlet function.

35. Named after the German mathematician Johannes Karl Thomae, *Thomae's function*, like the Dirichlet function, has nonzero range only on the rationals in [0, 1]. The function is:

$$f(x) = \begin{cases} \frac{1}{n} & \text{if } x = \frac{m}{n} \text{ in lowest terms is a rational number,} \\ 0 & \text{if } x \text{ is irrational.} \end{cases}$$

(a) Determine the set of discontinuities of f, showing that f is continuous on the irrationals. (*Hint*: you may use the fact that x may be chosen close enough to a given irrational y so that either x is irrational or x has a large denominator.)

(b) What, then, does the Riemann-Lebesgue theorem conclude about the integrability of f?

(c) What does the Riemann-Lebesgue theorem then say is the Riemann integral $R\text{-}\int_0^1 f(x)\,dx$?

36. Enumerate the rationals as $\mathbb{Q} = \{a_1, a_2, a_3, \ldots\}$ and define $f_1 = \mathcal{X}_{\{a_1\}}$, $f_2 = \mathcal{X}_{\{a_1, a_2\}}$, ..., so that $f_n(x) = \mathcal{X}_{\{a_1, a_2, \ldots, a_n\}}(x)$ for $n = 1, 2, 3, \ldots$.

(a) Identify the points of discontinuity for each function $f_n(x)$ in the sequence. What does the Riemann-Lebesgue theorem say about the Riemann integrability of f_n?

(b) For $n = 1, 2, 3, \ldots$, find $R\text{-}\int_0^1 f_n(x)\,dx$.

(c) What function $f(x)$ is $\lim_{n\to\infty} f_n(x)$?

(d) Use your result from part (c) to show the set of Riemann integrable functions is not closed, in the sense that some limit functions are not contained in the set.

Exercises 37 and 38 deal with the alternative characterization of the infimum.

37. Show the alternative characterization of the infimum is equivalent to the definition of the infimum in Definition 2.1.1. Feel free to mimic the argument found in Subsection 2.1.3 for the supremum.

38. Use the alternative characterization of the infimum and supremum to prove the following facts about real sets A and B:
    (a) If $A \subseteq B$, then $\inf A \geq \inf B$. (This is stated but not proved in Theorem 2.1.1.)
    (b) If we define the set $A + B = \{a + b : a \in A \text{ and } b \in B\}$, then $\sup(A + B) = \sup A + \sup B$ and $\inf(A + B) = \inf A + \inf B$.

2.2 Properties of the Lebesgue Integral

This section presents and proves several theorems that express properties of the Lebesgue integral. Some results are familiar, such as the fundamental theorem of calculus. But others are completely new. The monotone convergence theorem goes well beyond any property of the Riemann integral. We've seen that the Lebesgue integral is more powerful than Riemann's in the sense that it integrates many functions the Riemann integral cannot. The monotone convergence theorem points to a further power: the Lebesgue integral behaves better when taking limits. It lets us evaluate a limit of integrals by taking the integral of a limit. Commuting the operations of taking limits and taking integrals forms a category of "integral convergence theorems." They will be important as we study spaces of functions, just as limit theorems are important as we study spaces of points. The theorems will also be valuable tools to calculate many types of integrals. Much of the next two sections will describe Lebesgue convergence theorems that will empower us in such settings. The sections will also mix in other results that help us use the convergence theorems effectively.

2.2.1 The Monotone Convergence Theorem

The monotone convergence theorem is not too difficult to prove, but it relies on a slightly simpler result to which we give the (admittedly ungraceful) name the nondecreasing convergence theorem for L^0, because it applies to a nondecreasing sequence of functions in L^0. The theorem is really just a stepping stone to the stronger monotone convergence theorem result, and so it is a lemma. The proof follows a presentation by Riesz (see, e.g., [98, p. 30] and [121, p. 94]).

Lemma 2.2.1 Nondecreasing Convergence for L^0: *Assume $\{f_n\}$ is an increasing sequence of functions in L^0, where the integrals collectively satisfy $\int f_n \leq M$ for some finite upper bound M. Then the sequence $\{f_n\}$ converges a.e. to a function f in L^0, and*

$$\int \lim_{n \to \infty} f_n = \int f = \lim_{n \to \infty} \int f_n.$$

Proof. By Definition 1.5.1, we know for each individual function f_m in the sequence (where m is fixed), there is an increasing sequence of step functions $\phi_{m1}, \phi_{m2}, \phi_{m3}, \ldots$ that converges a.e. to f_m and has $\lim_{k \to \infty} \int \phi_{mk} = \int f_m$. Now for each $n = 1, 2, 3 \ldots$, define the function $\phi_n = \max\{\phi_{mk} : 1 \leq m, k \leq n\}$. By Exercise 29 of Section 1.3, $\{\phi_n\}$ is a sequence of step functions, and it is nondecreasing since

$$\{\phi_{mk} : 1 \leq m, k \leq n\} \subseteq \{\phi_{mk} : 1 \leq m, k \leq n+1\}.$$

Furthermore, $\phi_n \leq f_n$, since the nondecreasing nature of each sequence $\{\phi_{mk}\}$ implies

$$\max\{\phi_{mk} : 1 \leq m, k \leq n\} \leq \max\{f_m : 1 \leq m \leq n\}.$$

Then by Theorem 1.5.3, $\int \phi_n \leq \int f_n \leq M$.

Summarizing, $\{\phi_n\}$ is a nondecreasing sequence of step functions whose integrals are bounded from above by a constant M. That fact turns out to mean (from a dry and detailed analysis that we quickly present here) that the sequence $\{\phi_n\}$ converges a.e. to some function f, which serves as the function f in the conclusion of the theorem. For we can set S equal to the set of values for which the limit $\lim_{n \to \infty} \phi_n(x) \equiv f(x)$ does not exist, and then the axiom of choice implies S is the set of x values for which $\{\phi_n(x)\}$ grows without bound. Then, for $\varepsilon > 0$ and $n = 1, 2, 3, \ldots$, we can let S_n be the set of points x for which $\phi_n(x) > M/\varepsilon$. By the disjoint interval representation of ϕ_n, guaranteed by Lemma 1.3.1, S_n may be written as a union of a finite number of disjoint intervals, say $S_n = I_1 \cup I_2 \cup \ldots \cup I_k$. Then, by considering the sequence $\{\phi_n - \phi_1\}$, without any loss of generality we may always assume that $\{\phi_n\}$ is a sequence of nonnegative step functions. But that means $\phi_n \geq (M/\varepsilon) \cdot \mathcal{X}_{S_n}$. Since $m(S_n) = m(I_1) + \cdots + m(I_k)$, these facts imply $\int \phi_n \geq (M/\varepsilon) \sum_{j=1}^{k} m(I_j)$. Since $M \geq \int \phi_n$, this last fact says $m(S_n) \leq \varepsilon$, for any $n = 1, 2, 3, \ldots$. Then $S \subseteq \bigcup_{n=1}^{\infty} S_n$, because, if $\{\phi_n(x)\}$ grows without bound, then for some N, we would have $\phi_N(x) > M/\varepsilon$. And since $\phi_n \leq \phi_{n+1}$, we know $S_n \subseteq S_{n+1}$. Then $S \subseteq \bigcup_{k=1}^{\infty} I_k$, where the I_k are disjoint intervals for $k = 1, 2, 3, \ldots$, and where the bounded total measure $\sum_{k=1}^{\infty} m(I_k)$ is at most ε. Finally, an application of Definition 1.2.2 (and the fact that, since both S_n and S_{n+1} are a finite union of disjoint intervals, $S_{n+1} \setminus S_n = \{x : x \in S_{n+1} \text{ but } x \notin S_n\}$ must also be a finite union of disjoint intervals) allows us to conclude $m(S) = 0$.

By Definition 1.5.1, $f \in L^0$ and $\lim_{n \to \infty} \int \phi_n = \int f$. Furthermore, we know $\phi_{mn} \leq \phi_n$ for each $n = 1, 2, \ldots$ and $1 \leq m \leq n$, and so $\lim_{n \to \infty} \phi_{mn} \leq \lim_{n \to \infty} \phi_n$ for any m. Evaluating the limits, we get $f_m \leq f$ a.e. for each $m = 1, 2, 3, \ldots$, which means, by the countable union theorem, that the set of points for which any member of the entire collection of functions f_m is greater than f must be a set of measure zero. In short, $\phi_n \leq f_n \leq f$ a.e. Taking a limit as n approaches infinity, the sandwich theorem (and the fact that $\phi_n \to f$ a.e.) implies $\lim_{n \to \infty} f_n = f$ a.e. We now use the same sandwich theorem technique on the inequality $\int \phi_n \leq \int f_n \leq \int f$, obtaining $\lim_{n \to \infty} \int f_n = \int f$, and the result follows. ∎

The monotone convergence theorem was first proved in 1906 at the University of Piacenza, near Turin, Italy, by Beppo Levi, who was 30 years old at the time. Levi's lifetime work was mostly in an area of mathematics that studies algebraic curves, and therefore his successful foray into the newly established area of measure theory was an indication of his impressive abilities. One important type of algebraic curve is an elliptic curve, which in 1995 turned out to be the mathematical construction fundamental to Andrew Wiles's method of proof of Fermat's last theorem.

2.2 Properties of the Lebesgue Integral

Along with 90 other Jewish mathematicians, in 1938 Levi was expelled from his academic post (his was then at the University of Bologna) as a result of Mussolini's rise to power. The ruling Italian fascist government forced these scholars to find other countries of refuge. Levi subsequently left his native Italy and emigrated to Rosario, Argentina, where he spent the rest of his life. A testament to his positive will, Beppo Levi taught at the Universitad del Litoral in Rosario until he was 84 years old. During this period he founded and regularly contributed (in Spanish) to the first mathematics journal in Argentina, the *Mathematicæ Notæ* (see, e.g., [108]). Levi lived until 1961 and proved a broad and impressive set of mathematical results throughout his life.

Figure 2.3. Beppo Levi around 1930.

Here's how Levi's theorem on Lebesgue integration works. (Similar to the nondecreasing convergence theorem for L^0, the proof presented here follows an approach that traces back to Riesz; see [98, pp. 33–35] or [121, p. 96].) It involves a monotone sequence (one that is either nondecreasing or nonincreasing) of integrable functions. You can therefore see that the monotone convergence theorem's setting is more general than for the nondecreasing convergence theorem. Instead of nondecreasing, the sequence simply needs to be monotone, and instead of being elements of L^0, the functions simply need to be in L^1. Both theorems point to an important fact: such sequences of functions can play the same role that step functions did in the definition of the Lebesgue integral. A nondecreasing (or nonincreasing) sequence of integrable functions, having bounded integrals, will converge a.e. to a function $f \in L^1$, and the Lebesgue integral of f is determined as the limit of the corresponding sequence of integrals. The full upshot, expressed in terms of the monotone convergence theorem's conclusion, says we can always pass the limit through the integral sign when examining a bounded monotone sequence of integrable functions. The exact statement of the theorem follows.

Theorem 2.2.1   The Monotone Convergence Theorem (Beppo Levi): *Assume $\{f_n\}$ is a monotone sequence of functions in L^1, where the integrals collectively satisfy $|\int f_n| \leq M$ for some finite upper bound M. Then the sequence $\{f_n\}$ converges a.e. to a function f in L^1, and*

$$\int \lim_{n \to \infty} f_n = \int f = \lim_{n \to \infty} \int f_n.$$

Proof. The proof is constructive. We may assume, without any loss of generality, the monotone sequence is nondecreasing (for if $\{f_n\}$ is nonincreasing, we could consider $\{-f_n\}$) and nonnegative (otherwise we could then simply consider $\{f_n - f_1\}$). With those facts in mind, we now define a sequence of functions, $s_1 = f_1$ and $s_n = f_n - f_{n-1}$ for $n = 2, 3, \ldots$, which allows us to think of the sequence elements f_n as partial sums: $f_n = s_1 + s_2 + \cdots + s_n$. Each $s_n \geq 0$, since $\{f_n\}$ is nondecreasing. Also, each $s_n \in L^1$ (as each is the difference of two L^1 functions). Definition 1.5.2 then says $s_n = g_n - h_n$, where g_n and h_n are in L^0, for each $n = 1, 2, 3, \ldots$.

By Definition 1.5.1, for any one of these functions h_n, there is a nondecreasing sequence of step functions $\{\phi_k\}$ converging a.e. to h_n and with $\lim_{k \to \infty} \int \phi_k = \int h_n$. By Definition 1.4.1, for $\varepsilon = 1/2^n > 0$, we may choose N such that $0 < \int h_n - \int \phi_N < 1/2^n$. At a point x where $\{\phi_k\}$ converges to h_n, we define $u_n(x) = h_n(x) - \phi_N(x)$, and we set $u_n = 0$ at all other points. Then $0 < \int u_n < 1/2^n$, $u_n \in L^0$, and $u_n \geq 0$.[5] We also define $v_n = g_n - \phi_N$. Then, similarly, $v_n \in L^0$. By construction, $s_n = g_n - h_n = (g_n - \phi_N) - (h_n - \phi_N) = v_n - u_n$. Therefore $v_n \geq 0$, since $v_n = s_n + u_n$ is the sum of two nonnegative functions.

Now we define functions $G_n = v_1 + v_2 + \cdots + v_n$ and $H_n = u_1 + u_2 + \cdots + u_n$, obtaining $f_n = G_n - H_n$ for $n = 1, 2, 3, \cdots$. As a sum of functions in L^0, G_n and H_n are in L^0. Also, $\{G_n\}$ and $\{H_n\}$ are nondecreasing, since the additional terms v_n and u_n are nonnegative. We have

$$\int H_n = \int u_1 + \int u_2 + \cdots + \int u_n < 1/2 + 1/2^2 + \cdots + 1/2^n,$$

which is bounded above by the infinite geometric series $\sum_{n=1}^{\infty}(1/2^n) = 1$. Since $\int G_n = \int f_n + \int H_n < M + 1$, we also have $\int G_n$ bounded above. By the nondecreasing convergence theorem for L^0, there exist functions g and h in L^0 for which $\lim_{n \to \infty} G_n = g$, $\lim_{n \to \infty} H_n = h$, $\lim_{n \to \infty} \int G_n = \int g$, and $\lim_{n \to \infty} \int H_n = \int h$.

Since $f_n = G_n - H_n$, these facts imply that $\{f_n\}$ converges almost everywhere to $f \equiv g - h$, which is in L^1 by Definition 1.5.2. Furthermore, $\lim_{n \to \infty} \int f_n = \lim_{n \to \infty} \int(G_n - H_n) = \int g - \int h = \int f = \int \lim_{n \to \infty} f_n$. ∎

We have just proved that the monotone convergence theorem is always true when using the Lebesgue integral setting. But the conclusion of the theorem sometimes fails for the Riemann integral, as the next example shows. It's an important realization: the monotone convergence theorem (and other integral convergence theorems yet to come) show forcefully why the Lebesgue integral is much more powerful than Riemann's. Many descriptions of how Lebesgue integration is more powerful will appear throughout the remainder of the book.

Example 2.2.1 This example displays a sequence of functions whose elements are both Riemann integrable and Lebesgue integrable, but for which the monotone convergence theorem holds only in the Lebesgue integral sense. Toward that result, define a sequence $\{f_n\}$ as follows: let $f_1(x) = 1$ if $x = 0$ or 1, and $f_1(x) = 0$ otherwise. Let $f_2(x) = 1$ if $x = 0, 1/2$, or 1, and $f_2(x) = 0$ otherwise. Let $f_3(x) = 1$ if $x = 0, 1/2, 1/3, 2/3$, or 1, and $f_3(x) = 0$ otherwise. In general, let f_n be the function that is identically 0 except at any fraction in $[0, 1]$ in lowest terms whose denominator is less than or equal to n:

$$f_n(x) = \begin{cases} 1 & \text{if } x = i/k, \text{ where } k = 1, 2, \ldots, n \text{ and (indexed for each } k\text{) } i = 0, 1, \ldots, k \\ 0 & \text{otherwise.} \end{cases}$$

f_n is continuous except at the finite number of x values where $f_n(x) = 1$, and is therefore, by the Riemann-Lebesgue theorem, Riemann integrable with $R\text{-}\int f_n(x)\, dx = 0$ for each n. The

---

[5] We know $u_n \in L^0$ because ϕ_N is a step function, and so $\{\phi_k - \phi_N\}$ is a sequence of step functions that serves as the nondecreasing sequence for u_n called for in Definition 1.5.1.

2.2 Properties of the Lebesgue Integral

Riemann-Lebesgue theorem then guarantees that each function's Lebesgue integral also exists and equals the Riemann integral: $\int f_n = 0$ for each n.

As n approaches infinity, more and more rationals x in $[0, 1]$ have function values $f_n(x) = 1$, and each successive function in the sequence equals 1 wherever a previous function in the sequence did. So the sequence is monotone nondecreasing. Also, by the way the functions are defined, every rational in $[0, 1]$ eventually has $f_n(x) = 1$ for some n. Hence $f(x) = \lim_{n \to \infty} f_n(x) = 1$ if $x \in [0, 1]$ is rational, and $f(x) = 0$ otherwise. The limit function $f(x)$ is the Dirichlet function $D(x)$ described in Section 2.1.2. The monotone convergence theorem checks out in the sense of Lebesgue: $\lim_{n \to \infty} \int f_n = \lim_{n \to \infty} 0 = 0 = \int D(x)\, dx = \int f(x)\, dx = \int \lim_{n \to \infty} f_n$. But the monotone convergence theorem fails for the Riemann integral: $\lim_{n \to \infty} R\text{-}\int f_n(x)\, dx = 0$, but $R\text{-}\int f(x)\, dx$ does not even exist, and so they are not equal. ∎

We can't emphasize enough that the sequence in the last example is another great illustration of why the Lebesgue integral is so much better than Riemann's. Important limiting processes work under the Lebesgue integral when they fail in Riemann's. We will see in Chapter 3 how they allow us to form spaces of functions with the right limiting properties—it's similar to being able to form complete spaces of numbers. The Riemann integrable functions are analogous to rational numbers, in that we can take convergent limits of Riemann integrable functions, as we did in the last example, and fail to get a Riemann integrable function in the limit—just as we can take convergent limits of rational numbers, such as 3, 3.1, 3.14, 3.141, 3.1415, ..., and fail to get a rational number in the limit (obtaining π instead). Chapter 3 will show, using a famous theorem due to Riesz, that the Lebesgue integrable functions in L^1 are analogous to the real numbers, in that any appropriately defined and convergent sequence of L^1 functions always produces a functional limit in L^1—just as an appropriately defined and convergent sequence of real numbers always produces a finite, real-number limit. We'll need the Lebesgue integral to get that result.

As we've mentioned, Beppo Levi's monotone convergence theorem allows a monotone sequence of L^1 functions (whose integrals are bounded by a common value M) to play the same role that step functions played in Section 1.5. That can be useful to determine easily the integral of a function over an infinite range of integration. The next example illustrates this powerful use. Note how different the approach is from the method that must be employed using the Riemann integral, which is defined only over a finite interval $[a, b]$. Over an infinite range of integration, the Riemann integral therefore demands, as you recall from your calculus class, a limiting process that evaluates the integral as an improper integral. In contrast, the Lebesgue integral in the next example is defined according to the definition put in place in Section 1.5. The monotone convergence theorem makes the calculation of the integral easy.

You've already worked with the Riemann integral extensively in your calculus courses, utilizing several of the integral's properties. For example, the fundamental theorem of calculus (see [78, p. 282]) says $\int_a^b f(x)\, dx = F(b) - F(a)$ when f is continuous on $[a, b]$ and $F'(x) = f(x)$ on $[a, b]$. The notation $F'(x)$ here stands, as usual, for the derivative of the function F, so that $F'(x) = \lim_{h \to 0} \frac{F(x+h) - F(x)}{h}$, and F is identified as an antiderivative of f. Mathematicians use the fundamental theorem to avoid the tedious and often impractical calculations involved with the standard construction of Section 1.5 or 2.1, where we had to take the limit of a nondecreasing sequence of step-function integrals to find the integral of f. Instead, the fundamental theorem of calculus simply requires us to find the antiderivative F and evaluate it at the points a and b. What a reduction in degree of difficulty!

We've seen the Lebesgue integral of a function f exists whenever its Riemann integral does, and in this case they are equal. Therefore, any theorem that holds for the Riemann integral will also automatically hold for the Lebesgue integral (applied to any function that satisfies the theorem's assumptions). The fundamental theorem of calculus is a good example. Other examples include the theorems that describe integration techniques such as integration by parts or by u-substitution. Therefore, as usual from calculus, when we integrate continuous functions, we'll hope to be able to find an antiderivative. For instance, when $f(x) = e^{-x}$, a simple u-substitution obtains the antiderivative $F(x) = -e^{-x}$. That fact will help us calculate the Lebesgue integral in the next example. As is common, we write $F(x)|_a^b$ for $F(b) - F(a)$.

Example 2.2.2 Use the monotone convergence theorem to find $\int_0^\infty e^{-x}\, dx$.

Solution. Thinking of it as defined on all of $\mathbb{R}$, the function we are integrating is $f(x) = \begin{cases} e^{-x} & \text{if } x \geq 0 \\ 0 & \text{otherwise.} \end{cases}$ To use the monotone convergence theorem, we need to find a sequence of integrable functions f_n with bounded integrals that converge almost everywhere to f. A common and effective choice in a situation such as this is to truncate f so that the nonzero range values for f_n do not extend beyond n. To this end, we set $f_n(x) = \begin{cases} e^{-x} & \text{if } 0 \leq x \leq n \\ 0 & \text{otherwise.} \end{cases}$

f_n truncates f at $x = n$. Figure 2.4 illustrates the effect, displaying the functions f, f_4, and f_{10}. You can also see from the graph several of the functions' properties: (i) the sequence is nondecreasing ($f_n \leq f_{n+1}$, since the two graphs are identical up to $x = n$ and then $f_n = 0$ while f_{n+1} continues to have positive values until $x = n+1$); (ii) each function f_n is continuous almost everywhere (except at $x = 0$ and $x = n$), and so, by the Riemann-Lebesgue theorem, each f_n is in L^1; and (iii) each function's integral (which is the area of the region under the function's curve) is contained in the area under the curve of f (suggesting that each integral $\int f_n$ may be bounded above by a finite value M). In fact, we can readily find a value for M: on $[0, n]$, since an antiderivative is $F_n(x) = -e^{-x}$, the fundamental theorem of calculus implies

$$0 \leq \int f_n = \int_0^n f_n(x)\, dx = \int_0^n e^{-x}\, dx = -e^{-x}\big|_0^n = -e^{-n} - (-e^{-0}) = 1 - e^{-n} < 1.$$

In summary, $\{f_n\}$ satisfies all of the conditions of the monotone convergence theorem: it is a monotone (nondecreasing) sequence of functions in L^1 whose integrals satisfy $|\int f_n| \leq 1$. As

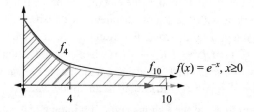

Figure 2.4. The functions $f = e^{-x}$, $x \geq 0$ along with f_4 and f_{10}. The areas of the shaded regions in dark grey and light grey represent the integrals of f_4 and f_{10}, respectively.

2.2 Properties of the Lebesgue Integral

n approaches infinity, f_n approaches f (the two functions are identical up to $x = n$), and so the monotone convergence theorem says we can pass the limit through the integral sign, obtaining the desired integral:

$$\int_0^\infty e^{-x}\, dx = \int f = \int \lim_{n\to\infty} f_n(x)\, dx = \lim_{n\to\infty} \int f_n(x)\, dx = \lim_{n\to\infty} \int_0^n e^{-x}\, dx$$
$$= \lim_{n\to\infty}(1 - e^{-n}) = 1. \quad \blacksquare$$

Now's your chance to practice putting the monotone convergence theorem into action. Take time to go through each part of the next question, so that you fully appreciate the way the theorem plays out.

Question 2.2.1 Find each integral $\int f$ by finding a monotone sequence of functions $f_n \in L^1$ that approach f almost everywhere and whose integrals are bounded by some real value M, and then by applying the monotone convergence theorem. The sequence $\{f_n\}$ can be formed by truncating f appropriately, as in the last example. As you use the theorem for each function f, passing the limit through the integral to determine $\int f$, make sure that you explicitly state why each of the monotone convergence theorem's assumptions holds.

1. $\int_1^\infty \frac{1}{x^2}\, dx$ (Hint: $f(x) = \begin{cases} x^{-2} & \text{if } x \geq 1 \\ 0 & \text{otherwise} \end{cases}$ is the function we are integrating.)
2. $\int_{-\infty}^\infty e^{-|x|}\, dx$ (Hint: $\int_{-\infty}^\infty f(x)\, dx = \int_{-\infty}^0 f(x)\, dx + \int_0^\infty f(x)\, dx$, and $e^{-|x|} = e^x$ when $x < 0$.)
3. $\int_0^1 \frac{1}{\sqrt{x}}\, dx$ (To form f_n, truncate f at $x = \frac{1}{n}$ for $n = 2, 3, \ldots$. Then $f_n(x) = 0$ when $x \notin [\frac{1}{n}, 1]$.) $\blacksquare$

An important corollary results from the monotone convergence theorem. It was used in the proof of the Riemann-Lebesgue Theorem, and so its proof here is the payment of a long-awaited debt. You'll see the result is so immediate and easily proven from the monotone convergence theorem that it would have been foolish to have presented it earlier.

Corollary 2.2.1 *For $f \in L^1$ with $f \geq 0$, if $\int f = 0$, then $f = 0$ almost everywhere.*

Proof. Define a nondecreasing sequence according to $f_n(x) = nf(x)$, $n = 1, 2, 3, \ldots$. Since $\lim_{n\to\infty} \int f_n = \lim_{n\to\infty} n \int f = n \cdot 0 = 0$, the monotone convergence theorem implies $\{f_n\}$ converges to some function almost everywhere. But $\lim_{n\to\infty} f_n = n \cdot f$ cannot exist almost everywhere unless $f = 0$ almost everywhere. $\blacksquare$

Question 2.2.2 Use Corollary 2.2.1 to prove $f(x) = 0$ a.e., where $f(x) = \lim_{n\to\infty}(1/n^2) \cdot \mathcal{X}_{[0,n]}(x)$. $\blacksquare$

2.2.2 Additional Integration Theorems

As mentioned, the fundamental theorem of calculus is automatically in play as it applies to the Riemann integral. Perhaps not surprisingly, the Lebesgue integral generates a slightly more general version of the fundamental theorem of calculus in terms of L^1. The generalization can be especially important when proving other theorems. We will apply it, for example, to prove several results in Section 2.4.

Theorem 2.2.2 The L^1 Fundamental Theorem of Calculus: *For f in $L^1[a,b]$, define the function $F(x) = \int_a^x f(t)\,dt$ for $x \in [a,b]$, and $F(x) = 0$ elsewhere.*
1. Then $F'(x) = f(x)$ at a point $x \in [a,b]$ where f is continuous.[6]
2. If f is continuous on $[a,b]$ and F is now any antiderivative of f (i.e., $F'(x) = f(x)$ on $[a,b]$), then

$$\int_a^b f(x)\,dx = F(b) - F(a).$$

Proof. 1. Assume f is continuous at a point $x \in (a,b)$. (The cases $x = a$ and $x = b$ will be considered in the exercises.) Then, given $\varepsilon > 0$, there exists δ such that the open interval $(x - \delta, x + \delta)$ is contained in $[a,b]$ and, by continuity, for which $-\varepsilon < f(t) - f(x) < \varepsilon$ whenever $0 \le |t - x| < \delta$. For h with $0 < |h| < \delta$,

$$\frac{F(x+h) - F(x)}{h} = \frac{1}{h}\left[\int_a^{x+h} f(t)\,dt - \int_a^x f(t)\,dt\right] = \frac{1}{h}\left[\int_x^{x+h} f(t)\,dt\right].[7]$$

Since $f(x) - \varepsilon < f(t) < f(x) + \varepsilon$ whenever $0 \le |t - x| < \delta$, we get $f(x) - \varepsilon < \frac{\int_x^{x+h} f(t)\,dt}{h} < f(x) + \varepsilon$. Hence $-\varepsilon < \frac{\int_x^{x+h} f(t)\,dt}{h} - f(x) < \varepsilon$, and so $-\varepsilon < \frac{F(x+h)-F(x)}{h} - f(x) < \varepsilon$. Thus $|\frac{F(x+h)-F(x)}{h} - f(x)| < \varepsilon$ for $0 \le |h| < \delta$. By Definition 1.4.2, $F'(x) = \lim_{h \to 0} \frac{F(x+h)-F(x)}{h} = f(x)$.

2. A standard calculus course uses the mean value theorem to show any two antiderivatives differ at most by an arbitrary constant, which we label C. Therefore, using the result from part 1, $F(x) = \int_a^x f(t)\,dt + C$,

$$F(b) - F(a) = \left(\int_a^b f(t)\,dt + C\right) - \left(\int_a^a f(t)\,dt + C\right) = \int_a^b f(t)\,dt,$$

and the result follows. ∎

The next example uses the fundamental theorem of calculus to examine familiar types of integrals.

Example 2.2.3 Part 1: Determine the definite integrals

$$\text{(A)} \int_{-\pi/2}^{\pi} 6\sin x\,dx \quad \text{and} \quad \text{(B)} \int_1^4 (2e^x + 3x^2 - 2)\,dx$$

---

[6] We clarify both the derivative and continuity at the interval's endpoints a and b. They are understood to be one-sided concepts. For example, $F'(a)$ refers to the right-hand derivative, where the limit of the difference quotient $(F(a+h) - F(a))/h$ is taken by considering only values of h that are positive and tending toward 0. Similarly, continuity of f at b means, given $\varepsilon > 0$, there exists δ such that $|f(x) - f(b)| < \varepsilon$ whenever $0 \le b - x < \delta$ (in other words, we consider only x values to the left of b).

[7] Here we use a standard convention used in a calculus course. We define, for $f \in L^1[a,b]$, $\int_a^b f(t)\,dt = -\int_b^a f(t)\,dt$. The convention allows for consideration of negative values of h.

2.2 Properties of the Lebesgue Integral

Solution: (A) An antiderivative of $f(x) = 6\sin x$ is $F(x) = -6\cos x$. By the second part of the fundamental theorem, $\int_{-\pi/2}^{\pi} 6\sin x \, dx = F(\pi) - F(-\pi/2) = -6\cos x|_{-\pi/2}^{\pi} = -6\cos \pi - (-6\cos(-\pi/2)) = 6 - 0 = 6$. (B) An antiderivative of $f(x) = 2e^x + 3x^2 - 2$ is $F(x) = 2e^x + x^3 - 2x$. Therefore $\int_1^4 (2e^x + 3x^2 - 2) \, dx = 2e^x + x^3 - 2x|_1^4 = (2e^4 + 4^3 - 2 \cdot 4) - (2e^1 + 1^3 - 2 \cdot 1) = 2e^4 - 2e + 57$.

Part 2: Find the derivative of $F(x) = \int_0^x e^{-t^2} dt$.

Solution. The first part of the fundamental theorem of calculus identifies $F'(x)$ as the function f that is being integrated, which means $F'(x) = f(x) = e^{-x^2}$. ■

Question 2.2.3 Use the fundamental theorem of calculus to find the following values.

1. $\int_{-2}^{2} (4x^5 - 3e^x + 7x - 1) \, dx$
2. $\int_{-\pi}^{\pi} 7\cos x \, dx$
3. $\int_{3\pi/4}^{\pi} (3\tan t - 2t^3) \, dt$
4. $F'(x)$, where $F(x) = \int_0^x (4\sin t^2 - 5e^{t^3}) \, dt$
5. $F'(x)$, where $F(x) = \int_0^x (4\cos(e^t) - 5e^t \sin t^2) \, dt$ ■

Other integration techniques besides the fundamental theorem of calculus can be shown to work in a more general setting than could be expressed with the Riemann integral. We will soon prove such is the case with integration by parts, describing it generally in Theorem 2.2.3. Furthermore, the situation for u-substitution is addressed in Exercises 36–40. It is important to realize, however, that in practical terms the integration techniques are still best to be considered as connected to the fundamental theorem of calculus. They work best when thought of as methods that help obtain an antiderivative for an integrand, and therefore as procedures that help us apply the fundamental theorem of calculus. The next example and question illustrate this fact for the familiar u-substitution technique.

Example 2.2.4 Using a standard calculus u-substitution technique, determine the value of $\int_0^{\pi} 5\sin^2(3x) \, dx$.

Solution. The double-angle formula for the sine function implies $\sin^2(t) = (1/2)(1 - \cos(2t))$, and so $\int_0^{\pi} 5\sin^2(3x) \, dx = \int_0^{\pi} (5/2)(1 - \cos(6x)) \, dx$. Letting $u = 6x$, $du = 6 \, dx$, $u = 0$ when $x = 0$, and $u = 6\pi$ when $x = \pi$. Applying the fundamental theorem of calculus, the integral is $(5/12) \int_0^{6\pi} (1 - \cos u) \, du = (5/12)(u - \sin u)|_0^{6\pi} = 5\pi/2$. ■

Question 2.2.4 Apply the fundamental theorem of calculus, using the u-substitution technique (or any technique that helps you find the antiderivative), to evaluate

1. $\int_0^{\pi} 4\sin^3(2x) \, dx$
2. $\int_0^{\pi} 10x \cdot e^{2x^2} \, dx$
3. $\int_1^{\pi} x^2 \cdot \ln(4x^3) \, dx$ ■

You'll recognize the format of the next theorem from calculus (though you may have used different notation, such as "du" where we will use "u," and "dv" where we will use "v"). In practice, we would typically use the fundamental theorem of calculus to integrate the functions as the theorem requires, and so assumptions about continuity would be in play. But you will note there is no assumption about the continuity of any function involved. Instead the theorem characterizes the result in terms of general L^1 functions.

Theorem 2.2.3 Integration by Parts: *Assume the functions u and v are in $L^1[a,b]$ and the functions U and V are defined, for any $x \in [a,b]$, as*

$$U(x) = \int_a^x u(t)\,dt + C_1 \text{ and } V(x) = \int_a^x v(t)\,dt + C_2,$$

where C_1 and C_2 are arbitrary constants.[8] *Then*

$$\int_a^b U\,v = U \cdot V\Big|_a^b - \int_a^b V\,u.$$

Proof. Without any loss of generality, it turns out we may assume the arbitrary constants C_1 and C_2 are both 0. Otherwise the calculations are messier but work out exactly the same. The proof proceeds by considering increasingly sophisticated types of functions u and v, progressing from constant functions to step functions to functions in L^0 that are nonnegative, and then to general L^1 functions. We consider each type as a separate case:

Case 1: If $u = c$ and $v = d$ are both constants, then $U(x) = cx - ca$ and $V(x) = dx - da$. Hence

$$\int_a^b U v = \int_a^b (cx - ca)d\,dx = \left(cd\frac{x^2}{2} - cadx\right)\Big|_a^b = cd\frac{b^2 - a^2}{2} - cad(b-a) = cd(b-a)^2/2,$$

and

$$U \cdot V\Big|_a^b - \int_a^b V u = (cx-ca)(dx-da)\Big|_a^b - \int_a^b (dx-da)c\,dx$$

$$= [cd(b-a)^2 - 0] - \left(cd\frac{x^2}{2} - cdax\right)\Big|_a^b = cd(b-a)^2/2.$$

The result therefore holds in this case.

Case 2: Assume u and v are both step functions. Then the real line (and hence the interval $[a,b]$) can be partitioned into a finite number of bounded intervals, say $I_1, I_2, \ldots, I_n$, over which both u and v are constant. Now, for any integrable function f, we have $\int_a^b f = \int_{I_1} f + \int_{I_2} f + \cdots + \int_{I_n} f$. Hence the result follows from a straightforward calculation: we break up the integrals $\int_a^b U v$ and $\int_a^b V u$ into such sums of integrals (over whose domains of integration both u and v are constant), and then we apply case 1 to the resulting integrals.

Case 3: Assume u and v are both nonnegative functions in L^0. Then there exist nondecreasing sequences of step functions $\{\phi_n\}$ and $\{\psi_n\}$ converging almost everywhere to u and v, respectively. By Case 2,

$$\int_a^b \Phi_n\,\psi_n = \Phi_n \cdot \Psi_n\Big|_a^b - \int_a^b \Psi_n\,\phi_n,\,{}^9$$

---

[8] We will see neither C_1 nor C_2 play any predominant role in this theorem; we mention them to make the definitions of U and V as general as possible.

[9] It should be clear we are using the upper-case notation Φ_n for $\Phi_n = \int_a^x \phi_n(t)\,dt$, and similarly for Ψ_n. This notational convenience for the relationship between an upper-case function and its lower-case relative is also used in case 4.

2.2 Properties of the Lebesgue Integral

for each $n = 1, 2, \ldots$ (where $\Phi_n = \int \phi_n$ and $\Psi_n = \int \psi_n$). The result follows as we take limits on both sides of the equation and apply the monotone convergence theorem to evaluate the limits.

Case 4: Assume u and v are both in L^1. Then $u = g - h$ and $v = s - t$, where g, h, s, and t are nonnegative functions in L^0. Hence $U = G - H$ and $V = S - T$ (where each upper-case function is the integral of the lower-case function). The result follows directly from the linearity of the integral and by applying case 3. ∎

Example 2.2.5 We evaluate $\int_1^2 x \ln x \, dx$.

Solution: Set $U(x) = \ln x$ and $v(x) = x$. Then $V(x) = \int_1^x t \, dt + C = x^2/2 - 1/2 + C$. For simplicity's sake, we set $C = 1/2$, so $V(x) = x^2/2$. Since $\ln x = U(x) = \int_1^x u(t) \, dt$, the fundamental theorem of calculus identifies $u(x) = U'(x) = 1/x$. Using integration by parts,

$$\int_1^2 x \ln x \, dx = \int_1^2 U v$$

$$= U \cdot V \Big|_1^2 - \int_1^2 V u = \ln x \cdot \frac{x^2}{2}\Big|_1^2 - \int_1^2 \frac{x^2}{2} \cdot \frac{1}{x} \, dx = 2\ln 2 - \frac{3}{4}. \quad \blacksquare$$

The next question involves integration by parts. These integrals likely seem familiar—you've surely worked with them in a calculus course. Further practice with integration by parts is in the section's exercises.

Question 2.2.5 Evaluate each of the following integrals, using integration by parts.

1. $\int_0^\pi x \cos x \, dx$
2. $\int_0^5 x \, e^x \, dx$
3. $\int_0^\infty x \, e^{-x} \, dx$ (*Hint:* you can also apply the monotone convergence theorem.) ∎

The last theorem this section presents is an innocent-looking but impressive property of the Lebesgue integral. Using a bit of slang language to put some fun into the result, we can say "We can *crash through* the integral sign with absolute values." When doing so we might not get equality, but we certainly won't get something smaller. Stating the theorem's main use, "If f is integrable, then so is $|f|$."

Theorem 2.2.4 *If $f \in L^1$, then so is $|f|$, and $|\int f| \leq \int |f|$.*

Proof. $f = g - h$, where $g, h \in L^0$, and so there are nondecreasing sequences of step functions ϕ_n and φ_n that converge a.e. to g and h, respectively. For any two values, say A and B, $|A - B| = \max\{A, B\} - \min\{A, B\}$. Hence $|f| = \max\{g, h\} - \min\{g, h\}$, and so $|f| = \lim_{n \to \infty} \max\{\phi_n, \varphi_n\} - \min\{\phi_n, \varphi_n\}$. Furthermore, both $\max\{\phi_n, \varphi_n\}$ and $\min\{\phi_n, \varphi_n\}$ are step functions for each n, by Exercise 1.3.29, converging a.e. and nondecreasingly to $\max\{g, h\}$ and $\min\{g, h\}$, respectively. By Definition 1.5.2, $|f| \in L^1$ with $\int |f| = \int (\max\{g, h\} - \min\{g, h\}) = \lim_{n \to \infty} \int (\max\{\phi_n, \varphi_n\} - \min\{\phi_n, \varphi_n\})$. Since $|f| - f \geq 0$, $\int |f| - \int f \geq 0$. Hence $\int |f| \geq \int f$. Similarly, $\int |f| \geq -\int f$. Putting the two facts together implies $\int |f| \geq |\int f|$. ∎

In full disclosure, we note there are examples of peculiar functions f that are not Lebesgue integrable but have a Riemann integral that exists over an infinite range. Such examples are limited to the situation where the Riemann integral is defined as improper, where, for example, $R\text{-}\int_0^\infty f(x)\,dx \equiv \lim_{T\to\infty} R\text{-}\int_0^T f(x)\,dx$. One such function is $f(x) = \sin x/x$. It turns out (see Exercises 16–18 in Section 2.3) that $R\text{-}\int_0^\infty \sin x/x\,dx = \pi/2$. But the last theorem indicates why $f \notin L^1(0, \infty)$. For if it were, then so would $|\sin x/x|$ be in $L^1(0, \infty)$. Examining the portion of the region of integration that stretches, for $k = 1, 2, \ldots$, from $(k-1)\pi$ to $k\pi$, we have $\int_{(k-1)\pi}^{k\pi} |\sin x/x|\,dx \geq 2/(k\pi)$.[10] Hence, for $n = 1, 2, 3, \ldots$,

$$\int_0^{n\pi} |\sin x/x|\,dx = \sum_{k=1}^n \int_{(k-1)\pi}^{k\pi} |\sin x/x|\,dx \geq (2/\pi)(1 + 1/2 + 1/3 + \cdots + 1/n).$$

Truncating the function $|\sin x/x|$ at $x = n$, the monotone convergence theorem would then imply

$$\int_0^\infty |\sin x/x|\,dx = \lim_{n\to\infty} \int_0^n |\sin x/x|\,dx \geq \lim_{n\to\infty} (2/\pi)(1 + 1/2 + 1/3 + \cdots + 1/n).$$

But the harmonic series limit is not a finite value, and so the Lebesgue integral cannot exist as a finite value.

From this discussion, we can conclude that there are rare cases where the Riemann integral, defined as an improper type, does sometimes manage to evaluate integrals of functions over infinite ranges as finite when the Lebesgue integral does not exist as a finite value. But don't place too much importance on this fact—it doesn't mean the Riemann integral has any real advantage over Lebesgue's. The examples manifest themselves as instances of finite integrals in Riemann's setting but not in Lebesgue's. You should come away from this section understanding the Lebesgue integral as much more powerful than Riemann's in the sense of convergence. For a monotone sequence of appropriately bounded Lebesgue integrable functions, the limit function is guaranteed to exist almost everywhere and be integrable in the Lebesgue sense but not necessarily in the Riemann sense. The guarantee comes from useful tools such as the monotone convergence theorem, and it will have crucial implications in many applications. The next section will present more of these powerful convergence theorems, and subsequent sections will then show their applications and importance.

Solutions to Questions

2.2.1 (1) Set $f_n(x) = x^{-2}$ if $1 \leq x \leq n$ (and $f_n(x) = 0$ otherwise) for $n = 2, 3, \ldots$. Then $\int f_n = -x^{-1}\big|_1^n = 1 - 1/n < 1$. By the monotone convergence theorem, $\int f = \lim_{n\to\infty} \int f_n = 1$.
(2) Let $f_n(x) = e^{-x}$ if $0 \leq x \leq n$ (and $f_n(x) = 0$ otherwise) for $n = 1, 2, 3, \ldots$. Then $\int_0^\infty e^{-x}\,dx = \lim_{n\to\infty} \int_0^\infty f_n(x)\,dx = \lim_{n\to\infty} -e^{-x}\big|_0^n = \lim_{n\to\infty} 1 - e^{-n} = 1$. Similarly, $\int_{-\infty}^0 e^x\,dx = 1$. Adding, $\int_{-\infty}^\infty e^{-|x|}\,dx = 2$. (3) Set $f_n(x) = x^{-1/2}$ if $1/n \leq x \leq 1$ (and $f_n(x) = 0$ otherwise) for $n = 2, 3, \ldots$. Then $\int f_n = 2x^{1/2}\big|_{1/n}^1 = 2(1 - 1/\sqrt{n}) < 2$. Hence $\int_0^1 1/\sqrt{x}\,dx = \lim_{n\to\infty} \int f_n = 2$.

---

[10] This fact follows from the inequality, valid on $(k-1)\pi \leq x \leq k\pi$, that says $|\sin x/x| \geq |(\sin x)/(k\pi)|$, and then from the evaluation of the resulting integral: $\dfrac{1}{k\pi}\int_{(k-1)\pi}^{k\pi} |\sin x|\,dx = 2/(k\pi)$.

2.2.2 By the monotone convergence theorem, $\int \lim_{n\to\infty} [(1/n^2) \cdot \mathcal{X}_{[0,n]}(x)] = \lim_{n\to\infty} \int (1/n^2) \cdot \mathcal{X}_{[0,n]}(x)\, dx = \lim_{n\to\infty} (1/n^2) \cdot n = \lim_{n\to\infty} 1/n = 0$. By Corollary 2.2.1, $0 = \lim_{n\to\infty} (1/n^2) \cdot \mathcal{X}_{[0,n]}(x)$.

2.2.3 (1) $(2x^6/3 - 3e^x + 7x^2/2 - x)|_{-2}^{2} = -4 - 3(e^2 - e^{-2})$. (2) $7\sin x|_{-\pi}^{\pi} = 0$.
(3) $-3\ln|\cos t| - t^4/2|_{3\pi/4}^{\pi} = -3\ln(\sqrt{2}/2) - 175\pi^4/512$. (4) $4\sin x^2 - 5e^{x^3}$.
(5) $4\cos(e^x) - 5e^x \sin x^2$.

2.2.4 (1) $-2(\cos 2x - (1/3)\cos^3 2x)|_0^{\pi} = 0$. (2) $(5/2)e^{2x^2}|_0^{\pi} = (5/2)(e^{2\pi^2} - 1)$.
(3) $(1/12)(4x^3)[\ln(4x^3) - 1]|_1^{\pi} = (1/12)[(4\pi^3)(\ln(4\pi^3) - 1) - 4\ln 4 - 4]$.

2.2.5 (1) $x\sin x|_0^{\pi} - \int_0^{\pi} \sin x\, dx = -2$. (2) $(xe^x - e^x)|_0^5 = 4e^5 + 1$.
(3) $\lim_{n\to\infty} -xe^{-x} - e^{-x}|_0^n = 1$.

Reading Questions for Section 2.2

1. State the fundamental theorem of calculus as presented in this section. Give an example of a function $f \in L^0$ whose definite integral is easily evaluated with the fundamental theorem of calculus but is difficult to evaluate using Definition 1.5.1.

2. Write at least one integral that can easily be evaluated using the u-substitution technique along with the fundamental theorem of calculus.

3. State the nondecreasing convergence theorem for L^0.

4. State Beppo Levi's monotone convergence theorem and give at least one reason why it is important.

5. Provide an example of an integral evaluated using the monotone convergence theorem.

6. State the integration by parts formula in its generality presented in this section. Provide an example of a definite integral of a function $f \in L^1$ easily evaluated using integration by parts.

Exercises for Section 2.2

In Exercises 1–16, use the fundamental theorem of calculus, along with a u-substitution technique, integration by parts, or some other integration technique from your calculus course to evaluate each integral.

1. $\int_0^1 (3x+5)^4\, dx$
2. $\int_{-10}^{10} x^4 (7x^5 - 12)^{10}\, dx$
3. $\int_0^{\pi} \sin(3x)\, dx$
4. $\int_{-\pi}^{\pi} \sin(nx)\, dx$, where n is any natural number $n = 1, 2, 3, \ldots$
5. $\int_{-\pi}^{\pi} \sin(nx)\cos(nx)\, dx$, where n is any natural number $n = 1, 2, 3, \ldots$
6. $\int_{-\pi}^{\pi} \sin(nx)\cos(mx)\, dx$, where n and m are any natural numbers $1, 2, 3, \ldots$
7. $\int_0^{\pi/2} \sin^2(5x)\, dx$
8. $\int_0^{\pi/2} \sin^3(2x)\, dx$
9. $\int_1^e x^2 \ln(x^3)\, dx$
10. $\int_1^e x^3 \ln(x^3)\, dx$
11. $\int_1^e x^4 \ln(x^3)\, dx$
12. $\int_0^{\pi} e^x \sin x\, dx$

13. $\int_e^{e^2} \frac{dx}{x \ln x}$

14. $\int_0^\pi x^3 \sin(2x^3)\, dx$

15. $\int_0^\pi e^{ax} \sin(bx)\, dx$, where $a, b \in \mathbb{R}$

16. $\int_0^{1/2} \frac{dx}{\sqrt{1-x^2}}$

Exercises 17–26 involve integrals evaluated using the monotone convergence theorem. For each integral $\int f$, truncate f to construct a sequence f_n of integrable functions that converge a.e. to f, and then apply the monotone convergence theorem to find the integral. As you pass the limit through the integral to determine $\int f$, make sure you state why each of the monotone convergence theorem's assumptions holds.

17. $\int_0^1 1/\sqrt{3x}\, dx$

18. $\int_0^5 (5x+2)^{-1/3}\, dx$

19. $\int_I^\infty a \cdot s^a \cdot x^{-(a+1)}\, dx$, where $a, s > 0$ and $I \geq a$ is a real variable. (In probability theory, the integrand is often labeled the Pareto distribution for I. The integral has many applications, including in economics as a mathematical model that describes the proportion of a population whose annual income exceeds I.)

20. $\int_e^\infty \frac{\ln x - 1}{x^2}\, dx$ (Hint: set $u = \ln x$ or, equivalently, examine $F(x) = -x^{-1} \ln x$.)

21. $\int_2^\infty x^{-2}\, dx$

22. $\int_1^\infty \frac{x}{(6x^2+3)^3}\, dx$

23. $\int_\pi^\infty \frac{|\sin x|}{x^2}\, dx$ (Hint: use Taylor series.)

24. $\int_{-\infty}^{-1} x^{-5}\, dx$

25. $\int_{-\infty}^0 \frac{x}{(1+x^2)^2}\, dx$

26. $\int_0^\infty \frac{dx}{1+x^2}$ (Hint: examine $F(x) = \arctan x$.)

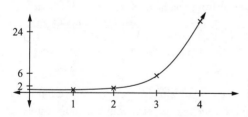

Figure 2.5. The graph of the gamma function.

Exercises 27–31 deal with the gamma function $\Gamma(x)$, which is defined, for $x > 0$, as $\Gamma(x) = \int_0^\infty t^{x-1} e^{-t}\, dt$. (More generally, the gamma function may be similarly defined for any complex value $z = x + iy$ where $x > 0$, but we restrict our attention here to a real domain.) The gamma function provides a way to generalize the factorial $n!$, defined as $n! = n \cdot (n-1) \cdot \ldots \cdot 2 \cdot 1$ to any nonnegative value, in the sense it produces a (continuous) differentiable curve that passes through each point (x, y) related by $y = (x-1)!$ for each (discrete) natural number x. Figure 2.5 illustrates this effect.

27. Use the monotone convergence theorem to prove $\Gamma(x)$ is finite for all $x > 0$. Consider the integral separately over $(0, 1)$ and $(1, \infty)$.

28. Use integration by parts to show $\Gamma(x+1) = x\Gamma(x)$ for all $x > 0$.

29. Show why $\Gamma(1) = 1$.

30. Using the last two exercises, prove $\Gamma(n) = (n-1)!$ for all values $n \in \mathbb{N}$. (Remember, 0! is defined to equal 1, so as to make such formulas work.)

2.2 Properties of the Lebesgue Integral

31. Show $\Gamma(1/2) = \sqrt{\pi}$ by making the substitution $u = \sqrt{t}$ in the integral, and using the fact (which we will prove in Chapter 3) that $\int_{-\infty}^{\infty} e^{-u^2} \, du = \sqrt{\pi}$.

Exercises 32–36 ask you to prove a property about the Lebesgue integral.

32. Prove the following theorem by considering the sequence of functions $\{f_n\}$ defined as $f_n = n \cdot f$ and applying the monotone convergence theorem. Theorem: If f is a nonnegative function in L^1 and $\int f = 0$, then $f = 0$ almost everywhere.

33. Prove the following theorem by using the last exercise. Theorem: Define the function M, whose domain consists of sets S having $\mathcal{X}_S \in L^1$, by $M(S) = \int \mathcal{X}_S$. Then $M(S) = 0$ implies S is a set that has measure zero.

34. Prove the converse of the theorem stated in the last exercise. In other words, prove the following result, where the function M is defined in the last exercise. Theorem: $M(S) = 0$ whenever S is a set that has measure zero.

35. Prove $\max\{f, g\} = (f+g)/2 + |f-g|/2$ and $\min\{f, g\} = (f+g)/2 - |f-g|/2$. Apply Theorem 2.2.4 to conclude $\max\{f, g\}$ and $\min\{f, g\}$ are both in L^1 whenever f and g are.

36. An inequality associated with the famous arithmetic and geometric means implies $\sqrt[n]{a_1 a_2 \cdots a_n} \leq (a_1 + a_2 + \cdots + a_n)/n$.
    (a) Use this to prove $(1 - \frac{x}{n})^n \leq (1 - \frac{x}{n+1})^{n+1}$ for any x such that $|x|/n < 1$ with $n = 1, 2, 3, \ldots$.
    (b) Show $\frac{x-1}{x} \leq \int_1^x \frac{dt}{t} \leq x - 1$ whenever $x > 0$.
    (c) Apply part (b), along with the monotone convergence theorem, to show $\lim_{n \to \infty} \int_0^n (1 - \frac{x}{n})^n x^{a-1} = \Gamma(a)$, where the function Γ is defined as in Exercises 27–31. (*Hint*: use the fact, first given by Jacob Bernoulli in 1683, that $e^x = \lim_{n \to \infty} (1 + \frac{x}{n})^n$.)

Advanced Exercises. Exercises 37–41 outline a proof of a general form of the u-substitution integration technique, which we can express in terms of the following theorem. *Integration by u-Substitution:* For x in $[a, b]$, suppose $u(x) = \int_a^x u'(t) \, dt + C$, where u' is a nonnegative function in $L^1(a, b)$. (We are not stipulating that u' is necessarily the derivative of u.) Then for $f \in L^1(u(a), u(b))$, $\int_a^b f(u(x)) \cdot u'(x) \, dx = \int_{u(a)}^{u(b)} f(t) \, dt$.

37. In the next section we will show u is continuous on $[a, b]$. Show u is nondecreasing, and therefore (by the intermediate value theorem) that it assumes each value in $[u(a), u(b)]$ at least once.

38. Assume, as a first case, that f is a characteristic function $f = \mathcal{X}_I$, where I is an interval with endpoints c and d contained in $[u(a), u(b)]$. Show the result holds in this case, calculating the right-hand side and the left-hand side of the theorem's concluding equation. (*Hint*: use Exercise 37.)

39. Use Exercise 38 to prove the result is true, as a second case, for any step function f.

40. Prove the result is true for any function in L^0.
    (a) What definition implies the existence of a sequence of step functions $\{\phi_n\}$ converging to f except possibly on a set S of measure zero?
    (b) Define the set $A = \{x : x \in [a, b], u'(x) \neq 0, \text{ and } u(x) \in S\}$. Use the fact $m(A) = 0$ (an admittedly nontrivial fact) to show $\{\phi_n(u(x)) \cdot u'(x)\}$ is a nondecreasing sequence

of functions in L^1 (after all, $\mathcal{X}_I \cdot u' \in L^1$ for any interval I) that converges almost everywhere to $f(u(x)) \cdot u'(x)$.

(c) Apply part (b), Exercise 39, and the monotone convergence theorem to prove the result in the case of $f \in L^0$.

41. Use Exercise 40 to show the result holds for any function $f \in L^1$.

2.3 Dominated Convergence and Further Properties of the Integral

This section explores four more properties of the Lebesgue integral, but it mainly focuses on the famous Lebesgue dominated convergence theorem (LDCT, for short), and it presents that result first. Proved by Lebesgue in a published paper of 1908, and with a more detailed explanation of the proof provided in 1910,[11] the dominated convergence theorem deals with the same issue as did the monotone convergence theorem: for a given sequence of functions, when can you switch the order of taking the Lebesgue integral and of taking the limit almost everywhere? For the monotone convergence theorem, the sequence of functions had to be nondecreasing or nonincreasing. For the dominated convergence theorem, any convergent (almost everywhere) sequence will work, as long as its functions f_n are collectively bounded (or "dominated") by an integrable function g. Neither theorem is more general than the other, and you might want to think of them as a complementary pair of screwdrivers, one a flathead and one a Phillips-head, that can drive a limit past an integral sign. The last section showed the usefulness of the monotone convergence theorem when integrating certain functions over an infinite range. The dominated convergence theorem requires a function g that dominates, and then it can be similarly applied in many useful ways.

The dominated convergence theorem has many theoretical applications, which we will eventually see, especially as we apply it to investigate limits within function spaces. Indeed, we will discuss many of its applications throughout the rest of the text—the Lebesgue dominated convergence theorem has a usefulness that goes well beyond finding an individual function's integral value. We will see a few of those applications almost immediately, in this section.

Theorem 2.3.1 The Lebesgue Dominated Convergence Theorem: *Suppose $\{f_n\}$ is a sequence of functions in L^1 that converges almost everywhere to a function f. Also suppose the sequence functions are dominated by an integrable function g, in the sense that $|f_n| \leq g$ for all n, where $g \in L^1$. Then $f \in L^1$ and its Lebesgue integral can be evaluated by passing the limit through the integral:*

$$\lim_{n \to \infty} \int f_n = \int \lim_{n \to \infty} f_n = \int f.$$

---

[11] The timeline history of the proofs of the four main integral convergence theorems, all of them eventually described by the end of this section (the monotone convergence, dominated convergence, bounded convergence, and Fatou's) differs from the order of presentation that this text provides. Lebesgue proved the bounded convergence theorem in his 1902 dissertation. As mentioned in Section 2.2, Levi proved the monotone convergence theorem in 1906. Almost simultaneously, Fatou independently proved his result (Fatou's lemma). Then came Lebesgue's dominated convergence theorem two years later (see [55]).

2.3 Dominated Convergence and Further Properties of the Integral

Proof. If $\lim_{n\to\infty} f_n(x) = f(x)$, define two new sequences $\{i_n(x)\}$ and $\{s_n(x)\}$:

$$i_n(x) = \inf\{f_n(x), f_{n+1}(x), f_{n+2}(x), \ldots\} \text{ and } s_n(x) = \sup\{f_n(x), f_{n+1}(x), f_{n+2}(x), \ldots\},$$

for $n = 1, 2, \ldots$.

Since $\{f_{n+1}(x), f_{n+2}(x), \ldots\} \subseteq \{f_n(x), f_{n+1}(x), \ldots\}$, Theorem 2.1.2 implies $i_n(x) \leq i_{n+1}(x)$ and $s_n(x) \geq s_{n+1}(x)$, and so the sequences are monotone.

For other x, those in the measure-zero set of values where f_n does not converge to f, we define $i_n(x) = 0 = s_n(x)$, which (trivially) provides monotonicity for the sequences at these points, too.

In this way, we have two monotone sequences of functions $\{i_n\}$ and $\{s_n\}$ that satisfy, due to their construction as infima and suprema, $i_n \leq f_n \leq s_n$ almost everywhere. Both converge to f almost everywhere. For if x is a value where $\lim_{n\to\infty} f_n(x) = f(x)$, then, given $\varepsilon > 0$, we can find N so that $|f_n(x) - f(x)| < \varepsilon$ whenever $n \geq N$. In its equivalent expression, $f(x) - \varepsilon < f_n(x) < f(x) + \varepsilon$ for $n \geq N$. But that means $f(x) - \varepsilon$ and $f(x) + \varepsilon$ are lower and upper bounds, respectively, for $\{f_n(x), f_{n+1}(x), f_{n+2}(x), \ldots\}$, and so $f(x) - \varepsilon \leq i_n(x) \leq s_n(x) \leq f(x) + \varepsilon$ for $n \geq N$. Taking limits, $f(x) - \varepsilon \leq \lim_{n\to\infty} i_n(x) \leq \lim_{n\to\infty} s_n(x) \leq f(x) + \varepsilon$. Here, ε is arbitrarily small, and so it must be the case (letting ε go to zero) that $f(x) \leq \lim_{n\to\infty} i_n(x) \leq \lim_{n\to\infty} s_n(x) \leq f(x)$. Hence $\lim_{n\to\infty} i_n(x) = f(x) = \lim_{n\to\infty} s_n(x)$.

In summary, $\{i_n\}$ and $\{s_n\}$ are two monotone convergent sequences that each converge almost everywhere to f, and for which $i_n \leq f_n \leq s_n$. If i_n and s_n are in L^1 for all n, then $\int i_n \leq \int f_n \leq \int s_n$, and, taking limits, $\lim_{n\to\infty} \int i_n \leq \lim_{n\to\infty} \int f_n \leq \lim_{n\to\infty} \int s_n$. We can define an upper bound $M = \int g$ for the integrals from the facts that $|\int i_n|$ and $|\int s_n|$ are both bounded by $\int g$. We can therefore apply the monotone convergence theorem to pass the limit through the integral sign for the two outside integrals, which then implies $\int f \leq \lim_{n\to\infty} \int f_n \leq \int f$ and proves the result.

But how do we know i_n and s_n are in L^1? To show that, we're going to have to construct two more subsequences, one that converges (for any fixed n) to i_n and one that converges to s_n. To that end, fix n and define, for $k = 1, 2, \ldots$,

$$m_k(x) = \min\{f_n(x), f_{n+1}(x), \ldots, f_{n+k}(x)\} \text{ and } M_k(x) = \max\{f_n(x), f_{n+1}(x), \ldots, f_{n+k}(x)\}.$$

By construction $\lim_{k\to\infty} m_k = i_n$ and $\lim_{k\to\infty} M_k = s_n$. Furthermore, for any A and B, $\min\{A, B\} = (1/2)(A + B - |A - B|)$ and $\max\{A, B\} = (1/2)(A + B + |A - B|)$ are linear combinations of A, B, and $|A - B|$. Both m_k and M_k are therefore linear combinations of the terms $f_n, f_{n+1}, \ldots, f_{n+k}$ and the absolute values of their pairwise differences. By assumption and by Theorem 2.2.4, the terms are integrable. The linearity of L^1 implies m_k and M_k are integrable, too. Finally, the domination of g implies $|\int m_k| \leq \int g$ and $|\int M_k| \leq \int g$, and so all the necessary conditions to apply the monotone convergence theorem are satisfied. The conclusion is that i_n and s_n are integrable for each n, with $\lim_{k\to\infty} \int m_k = \int i_n$ and $\lim_{k\to\infty} \int M_k = \int s_n$. That completes the proof. ∎

Just as for the monotone convergence theorem, the dominated convergence theorem always holds in the framework of the Lebesgue integral. But, just as there were for the monotone convergence theorem, there are counterexamples that show the Lebesgue dominated convergence theorem is not always true for the Riemann integral. The next example is a simple illustration.

Example 2.3.1 This example demonstrates a dominated (but non-monotone) sequence of functions that satisfies the Lebesgue dominated convergence theorem. However, for this sequence, the conclusion of the LDCT fails in the setting of the Riemann integral. Toward that result, define a sequence $\{f_n\}$ whose graph is a step function $\sum_j \mathcal{X}_{I_j}$ where, for each n, each of the finite number of intervals I_j is centered at a rational number in $[0, 1]$ and has width equal to $2/4^n$ (except for the first and last, which have width $1/4^n$). See Figure 2.6 for the graphs of the first three functions in the sequence. More exactly, define the sequence so that (similar to the functions in Example 2.2.1) each $f_n(x)$ equals 1 at $x = i/k$ for $k = 0, 1, \ldots, n$ and (indexed for each k) $i = 0, 1, \ldots, k$ (some of the indexed values may be repeats). But here (as opposed to the sequence in Example 2.2.1), on $[0, 1]$ each $f_n(x)$ also equals 1 in the interval $[i/k - 1/4^n, i/k + 1/4^n]$. For all other values, $f_n(x) = 0$.

Comparing function values, it's easy to see the sequence $\{f_n\}$ is not monotonic. For example, it is not monotonically nondecreasing, since, for instance, $f_1(1/4) = 1$, which is strictly greater than $f_2(1/4) = 0$. Similarly, it is not monotonically nonincreasing, since, for instance, $f_1(1/2) = 0$, while $f_2(1/2) = 1$. In conclusion, the monotone convergence theorem does not apply.

But the sequence *is* dominated by the characteristic function $g(x) = \mathcal{X}_{[0,1]}$, which is in L^1 with $\int g = 1$. Furthermore, each f_n is in L^1 (since it is a step function). Also, $\int f_n \leq (n+1)(n+2)/4^n$. This follows from the observation that each rectangle has area $2/4^n$ except for the first and last, which have combined area $2/4^n$. The index system of the center of each interval I_n shows there are at most $(n+1)(n+2)/2$ rectangles, and so the total rectangular areas, which equals $\int f_n$, must be at most $(n+1)(n+2)/4^n$. Finally, the sequence $\{f_n\}$ converges everywhere to the Dirichlet function $D(x)$, where $D(x) = 1$ if $x \in [0, 1]$ is rational, and $D(x) = 0$ otherwise.

We may view this example as a verification of the Lebesgue dominated convergence theorem, since all the theorem's conditions are satisfied and the LDCT's conclusion follows:

$$\lim_{n \to \infty} \int f_n \leq \lim_{n \to \infty} (n+1)(n+2)/4^n = 0 = \int D(x)\, dx.$$

In comparison, however, the limit function's Riemann integral does not exist: $\lim_{n \to \infty} \int f_n = 0 \neq R\text{-}\int D(x)\, dx$, since D is discontinuous at every $x \in [0, 1]$. The dominated convergence theorem's powerfully successful conclusion does not hold for the sequence $\{f_n\}$ when using the Riemann integral. ∎

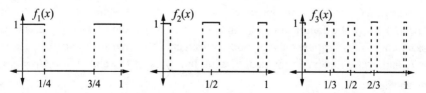

Figure 2.6. The graphs of the first three sequence functions in Example 2.3.1. All rectangular areas have width $2/4^n$, except for the first and last (formed using the key values $0/n = 0$ and $n/n = 1$). They form rectangles of width $1/4^n$, since the other half of their intervals would extend outside $[0, 1]$.

2.3 Dominated Convergence and Further Properties of the Integral

The sequence in the last example (and the one in Example 2.1.1) is a good one to keep in mind as an illustration of the power of the Lebesgue integral when we also consider additional theoretical results. The next example and question show how the LDCT can calculate many integrals of unbounded functions or integrals over an infinite range.

Example 2.3.2 Use the Lebesgue dominated convergence theorem to write $\lim_{n\to\infty} \int_1^n \cos(x^{-1})/x^2 \, dx$ as a Lebesgue integral of an integrable function, without a limit sign. Then find the integral.

Solution. Since the integrand function is continuous over the interval $[1, n]$, the function defined as $f_n(x) = \cos(x^{-1})/x^2 \cdot \mathcal{X}_{[1,n]}(x)$ is in L^1 for $n = 1, 2, \ldots$. f_n is dominated by the integrable function $g(x) = x^{-2} \cdot \mathcal{X}_{[0,\infty)}(x)$. Also, $\lim_{n\to\infty} f_n = \cos(x^{-1})/x^2 \cdot \mathcal{X}_{[1,\infty)}(x)$. By the LDCT, $\lim_{n\to\infty} \int_1^n \cos(x^{-1})/x^2 \, dx = \int_1^\infty \cos(x^{-1})/x^2 \, dx$. Substituting $u = x^{-1}$, the integral is therefore $\lim_{n\to\infty} \int_{1/n}^1 \cos u \, du = \lim_{n\to\infty} \sin u \big|_{1/n}^1 = \sin 1 - \sin 0 = \sin 1$. ∎

Question 2.3.1 Use the Lebesgue dominated convergence theorem to write $\lim_{n\to\infty} \int_0^n x e^{-x^2} \, dx$ as a Lebesgue integral of an integrable function, without a limit sign. Then find the integral. ∎

The last example also illustrates the next fact (the third in the quartet of convergence results). It is a corollary that follows immediately from the LDCT. The functions f_n in the last example will satisfy the conditions and the conclusion of the corollary, but once again the Riemann integral fails to produce the result.

Corollary 2.3.1 The Bounded Convergence Theorem:[12] *Suppose $\{f_n\}$ is a sequence of functions in $L^1(I)$ for a finite interval I that converges almost everywhere on I to a function f. Also suppose the sequence functions satisfy $|f_n(x)| \leq M$ for some real number M and all $x \in I$. Then $f \in L^1(I)$, and its Lebesgue integral satisfies $\lim_{n\to\infty} \int_I f_n = \int \lim_{n\to\infty} f_n = \int_I f$.*

Proof. The function $g = M \cdot \mathcal{X}_I$ dominates the sequence. Now apply the LDCT. ∎

Question 2.3.2 For a constant $K > 1$, use the bounded convergence theorem to write $\lim_{n\to\infty} \int_0^K n x e^{-nx} \, dx$ as a Lebesgue integral of an integrable function, without a limit sign. Then find the integral. (*Hint*: the functions nxe^{-nx} are bounded on $[0, K)$. Find their maxima.) ∎

The next theorem is another famous tool in the Lebesgue convergence theory toolbox. In a course based on a full treatment of measure theory, it is often presented first, and then used (as a lemma) to prove the Lebesgue dominated convergence theorem. Hence its name, Fatou's lemma. Our presentation reverses this order, but the result is still useful, as an example that follows its proof will show.

---

[12] The theorem is sometimes known as Arzelà's theorem, because a slightly weaker version of it is true for the Riemann integral and was proven by Cesare Arzelà in 1885 (see [83]).

Theorem 2.3.2 Fatou's Lemma: *Suppose $\{f_n\}$ is a sequence of nonnegative L^1 functions that converges almost everywhere to a function f. Also suppose $\int f_n \leq M$ for some real number M. Then $f \in L^1$, and its Lebesgue integral satisfies*

$$\int f \leq M.$$

An appropriate choice for M is always $M = \liminf \int f_n$, which is

$$\lim_{n \to \infty} \inf \{\int f_m : m = n, n+1, \ldots\}.$$

Proof. Just as in the proof of the Lebesgue dominated convergence theorem, we can define the functions

$$m_k(x) = \min\{f_n(x), f_{n+1}(x), \ldots, f_{n+k}(x)\}, \text{ for any } n \text{ and } k = 1, 2, \ldots.$$

Also as shown in the proof of the LDCT, the functions are in L^1, and for any n, the sequence $\{m_k\}$ is nonincreasing, its functions are in L^1, and it converges almost everywhere to the function $i_n = \inf\{f_n, f_{n+1}, f_{n+2}, \ldots\}$. Also, since $0 \leq \int f_n \leq M$, it must be the case, for any of these functions, that $0 \leq \int m_k \leq M$. All the hypotheses of the monotone convergence theorem are satisfied, and so each i_n is integrable with $0 \leq \int i_n \leq M$. Applying the monotone convergence theorem, again, to the nondecreasing sequence $\{i_n\}$, which converges a.e. to f, we see f is integrable with $\int f \leq M$. An appropriate choice of M is always $M = \lim_{n \to \infty} \inf \{\int f_n\}$, whenever that value is finite (if it is infinite, the result follows trivially). For we have $0 \leq \int m_k \leq \min\{\int f_n, \ldots, \int f_{n+k}\}$, and so, letting k approach infinity, $0 \leq \int i_n \leq \inf\{\int f_n, \int f_{n+1}, \ldots\}$. Taking the limit as n approaches infinity (and since $\lim_{n \to \infty} \int i_n = \int f$), the result follows. ∎

The next example demonstrates the use of Fatou's lemma. It shows how, for various sequences, we can obtain both a strict inequality and an equality in the lemma's conclusion.

Example 2.3.3 Show how Fatou's lemma applies to the nonnegative function sequences:

(1) $f_n(x) = \begin{cases} 1/n & \text{if } x \in (0, n] \\ 0 & \text{otherwise} \end{cases}$ and (2) $g_n(x) = \begin{cases} e^{-x} & \text{if } 0 \leq x \leq n \\ 0 & \text{otherwise,} \end{cases}$

where $n = 1, 2, 3, \ldots$

Solution. (1) The sequence $\{f_n\}$ converges almost everywhere (in fact, everywhere) on $\mathbb{R}$ to $f(x) = 0$. Also, $\int f_n = \int_0^n 1/n \, dx = 1$, which means $\{\int f_m : m = n, n+1, \ldots\} = \{1, 1, \ldots\}$. Taking the infimum for each n and the resulting limit of the constant sequence of 1s, we obtain $\liminf \int f_n = 1$. A strict inequality therefore exists in the conclusion of Fatou's lemma: $\int f = \int 0 = 0 < 1 = \liminf \int f_n$.

(2) Example 2.2.2 examined this sequence, showing $g(x) = e^{-x}$ satisfies $\lim_{n \to \infty} g_n = g$. The example also proved $\int g(x) = \lim_{n \to \infty} \int g_n(x) = 1$, which is the same as $\lim_{n \to \infty} \inf \{\int g_m : m = n, n+1, \ldots\} = \lim_{n \to \infty} \inf \{\int_0^m e^{-x} dx\}_{m=n}^\infty = \lim_{n \to \infty} \inf \{1 - e^{-m}\}_{m=n}^\infty = \lim_{n \to \infty} 1 - e^{-n} = 1$. ∎

Question 2.3.3 Show how Fatou's lemma applies to:

$$f_n(x) = \begin{cases} 1 + 1/n & \text{if } x \in [\frac{n}{n+1}, 1] \\ 0 & \text{otherwise.} \end{cases}$$

∎

2.3 Dominated Convergence and Further Properties of the Integral

The last powerful result of this section gives a condition on when the operation of taking a derivative can be passed through the integral. Because a derivative is a limit of a difference quotient, we would expect the ability to interchange the processes of taking a derivative and taking an integral to be related to the integral convergence theorems. The strength of the theorem statement speaks for itself.

> **Theorem 2.3.3** *Assume $f(x, t)$ is an integrable function of x for each value of t and differentiable as a function of t for each value of x. Also assume $\left|\frac{\partial f(x,t)}{\partial t}\right| \leq g(x)$ for all x, t where $g \in L^1$. Then*
>
> $$\frac{\partial}{\partial t} \int f(x,t)\, dx = \int \frac{\partial f(x,t)}{\partial t}\, dx.$$

Proof. Recall the mean value theorem (MVT) from calculus (see [78, p. 174]): for f continuous on a closed interval $[t, t+h]$ and differentiable on the open interval $(t, t+h)$, there exists a number c in $(0, 1)$ such that $\frac{f(t+h)-f(t)}{h} = f'(t+ch)$. We restate the MVT in the case of a two-dimensional function $f(x, t)$, but where x is being held constant. Then the derivative of f, taken with respect to t, is a *partial derivative* $\partial f/\partial t$, where $\frac{\partial f(x,t)}{\partial t} = \lim_{h \to 0} \frac{f(x,t+h)-f(x,t)}{h}$. As we hold x constant and assume that, with respect to its second variable, f is continuous on a closed interval $[t, t+h]$ and differentiable on the open interval $(t, t+h)$, the MVT then implies there exists a number c in $(0, 1)$ such that $\frac{f(x,t+h)-f(x,t)}{h} = \frac{\partial f(x,t+ch)}{\partial t}$. Therefore,

$$\begin{aligned}
\frac{\partial}{\partial t} \int f(x,t)\, dx &= \lim_{h \to 0} \frac{\int f(x,t+h)\, dx - \int f(x,t)\, dx}{h} \quad \text{(by definition of the partial derivative)} \\
&= \lim_{h \to 0} \int \frac{f(x,t+h) - f(x,t)}{h}\, dx \quad \text{(by linearity of the integral)} \\
&= \lim_{h \to 0} \int \frac{\partial}{\partial t} f(x, t+ch)\, dx \quad \text{(by the MVT)} \\
&= \lim_{n \to \infty} \int \frac{\partial}{\partial t} f\left(x, t+c \cdot \frac{1}{n}\right) dx \quad \text{(letting } h = 1/n\text{).}
\end{aligned}$$

By the assumption of dominance $\left|\frac{\partial f(x,t)}{\partial t}\right| \leq g(x)$, we can apply the dominated convergence theorem to attain $\frac{\partial}{\partial t} \int f(x,t)\, dx = \int \lim_{n \to \infty} \frac{\partial}{\partial t} f(x, t + c \cdot \frac{1}{n})\, dx = \int \frac{\partial f(x,t)}{\partial t}\, dx$. ∎

Example 2.3.4 Let $F(t) = \int_0^\infty e^{-tx} \frac{\sin x}{x}\, dx$, where $t > 0$. Find $\frac{\partial}{\partial t} F(t)$.

Solution: We can pass the derivative through the integral sign because: (i) $f(x, t) = e^{-tx} \frac{\sin x}{x}$, where $x > 0$, is an integrable function of x for each t (realizing t is treated as a constant and $|\sin x/x| \leq 1$, which was essentially proven in Example 2.2.2); (ii) $f(x, t)$ is also differentiable as a function of t for each value of x—the derivative is $-e^{-tx} \sin x$; and (iii) the absolute value of the derivative, $e^{-tx}|\sin x|$, which is $\left|\frac{\partial f(x,t)}{\partial t}\right|$, is bounded above by $g(x) = e^{-tx}$ for all x and t, and this function g is in L^1 (by Example 2.2.2, as we treat t as a constant). In short, $f(x, t)$ satisfies all the conditions of Theorem 2.3.3.

Then (remembering to hold x fixed and therefore to treat it as a constant),

$$\frac{\partial}{\partial t} F(t) = \frac{\partial}{\partial t} \int_0^\infty e^{-tx} \frac{\sin x}{x} dx = \int_0^\infty \frac{\partial}{\partial t} \left(e^{-tx} \frac{\sin x}{x} \right) dx = \int_0^\infty -e^{-tx} \sin x \, dx.$$

Using the Lebesgue dominated convergence theorem (with $g(x) = e^{-tx}$, where $x \geq 0$, and the truncated function $f_n = -e^{-tx} \sin x$, $0 \leq x \leq n$) along with integration by parts twice, we obtain

$$\frac{\partial}{\partial t} F(t) = \int_0^\infty -e^{-tx} \sin x \, dx = -(1+t^2)^{-1}.$$

∎

In short: we are guaranteed we can always pass the derivative through the integral sign under the conditions of Theorem 2.3.3. The Riemann integral isn't powerful enough to do that, but the Lebesgue integral can.

Question 2.3.4 Find $\frac{\partial F(t)}{\partial t}$, when $F(t) = \int_0^\infty \frac{e^{-tx+1} \sin x}{x} dx$. ∎

Solutions to Questions

2.3.1 Define $f_n(x) = xe^{-x^2} \mathcal{X}_{[0,n]}(x)$. Because f_n is continuous a.e., it is in L^1. Furthermore, f_n is dominated by $g(x) = e^{-x} \mathcal{X}_{[0,\infty]}(x) \in L^1$. By the LDCT, $\int_0^\infty xe^{-x^2} dx = \lim_{n \to \infty} \int_0^n xe^{-x^2} dx = 1/2$.

2.3.2 From its critical values, we see the maximum of $f_n(x)$ over the interval $I = [0, K]$ equals $f(1/n) = 1/e$. By the bounded convergence theorem, $0 = \int_0^K 0 \, dx = \int_0^K \lim_{n \to \infty} nxe^{-nx} dx = \lim_{n \to \infty} \int_0^K nxe^{-nx} dx$.

2.3.3 The sequence $\{f_n(x)\}$ converges a.e. to $f(x) = 0$. Also, $\int f_n = (1 + 1/n)(1 - n/(n+1)) = 1/n$, which means $\inf\{\int f_m : m = n, n+1, n+2, \ldots\} = \inf\{1/n, 1/(n+1), 1/(n+2), \ldots\} = 0$. Fatou's lemma confirms the value of the integral: $0 = \int 0 = \int f \leq 0$.

2.3.4 By Theorem 2.3.3 and the LDCT, $\frac{\partial}{\partial t} F(t) = \int_0^\infty \frac{\partial}{\partial t} \left(\frac{e^{-tx+1} \sin x}{x} \right) dx = \int_0^\infty -e^{-tx+1} \sin x \, dx = \lim_{n \to \infty} \int_0^n -e^{-tx+1} \sin x \, dx$. Using integration by parts, this limit equals $\lim_{n \to \infty} (e - e^{1-nt}(\cos n + t \sin n))/(1 + t^2) = e/(1 + t^2)$.

Reading Questions for Section 2.3

1. Correctly state the Lebesgue dominated convergence theorem and provide an example of a sequence of functions to which it applies.

2. State the bounded convergence theorem and provide its proof, based on the LDCT.

3. State Fatou's lemma. Give an example of a sequence for which strict inequality holds in its conclusion, and one for which equality holds.

4. Is it always true that $\frac{\partial}{\partial t} \int f(x, t) \, dx = \int \frac{\partial f(x,t)}{\partial t} dx$? Explain your answer.

Exercises for Section 2.3

In Exercises 1–6, use the Lebesgue dominated convergence theorem to write the limit of integrals as a Lebesgue integral of an integrable function, without a limit sign.

1. $\lim_{n \to \infty} \int_1^n \sin x / x^2 \, dx$

2. $\lim_{n \to \infty} \int_0^n e^{-x} \cos x \, dx$

3. $\lim_{n \to \infty} \int_0^\infty (1 + \frac{x}{n})^{-n} \sin(\frac{x}{n}) \, dx$

4. $\lim_{n \to \infty} \int_0^1 x/(1 + n^2 x^2) \, dx$

5. $\lim_{n \to \infty} \int_0^1 nx/(1 + n^2 x^2) \, dx$

6. $\lim_{n \to \infty} \int_1^n \cos(\frac{x}{n})/x^2 \, dx$

In Exercises 7–10, evaluate.

7. $\lim_{n \to \infty} \int_0^n e^{-x} \sin x \, dx$

8. $\lim_{n \to \infty} \int_0^\infty (1 + nx^2)(1 + x^2)^{-n} \, dx$

9. $\lim_{n \to \infty} \int_0^1 (1 + \frac{x}{n})^{-n} \cos(\frac{x}{n}) \, dx$

10. $\lim_{n \to \infty} \int_0^1 n/(1 + n^2 x^2) \, dx$

In Exercises 11–15, use a convergence theorem to evaluate the integral as a limit of a sequence of integrable functions. Clearly state which convergence theorem you are applying, and show why the theorem's conditions are satisfied.

11. $\int_0^\infty x e^{-x} \, dx$

12. $\int_{-\infty}^\infty x e^{-|x|} \, dx$

13. $\int_0^\infty e^{-x} \cos x \, dx$

14. $\int_0^\infty f(x) \, dx$, where $f(x) = (-1)^n / n^2$ if $n - 1 < x \leq n$ for $n = 1, 2, 3, \ldots$

15. $\int_1^\infty 1/x^n \, dx$, where $n = 2, 3, 4, \ldots$

Exercises 16–18 show why $R\text{-}\int_0^\infty \sin x / x \, dx = \pi/2$.

16. Define, for any x, $F(t) = \int_0^\infty e^{-tx} \sin x / x \, dx$. Use the result of Example 2.3.4 (which showed $F'(t) = -(1 + t^2)^{-1}$ for $t > 0$) to prove $F(t) = C - \arctan t$ for $t > 0$.

17. Show $C = \pi/2$. (*Hint:* consider the sequence $\{F(n)\}$, $n = 1, 2, 3, \ldots$.)

18. Use the result of the last exercise to prove $\lim_{T \to \infty} \int_0^T \sin x / x \, dx = \pi/2$.

Exercises 19–24 examine a sequence $\{f_n\}_{n=1}^\infty$ where $\lim_{n \to \infty} \int f_n$ converges but $\lim_{n \to \infty} f_n(x)$ does not converge a.e. To that end, let each f_n function's graph be diagonal-line segments connecting discrete x points having $f(x) = 0$ with discrete x points having $f(x) = 1$. This creates a jagged sawtooth function curve between $x = 0$ and $x = 1$ ($f_n(x) = 0$ off of $[0, 1]$). Figure 2.7 shows the first three functions' graphs. More exactly, define the sequence so that (similar to those in Example 2.2.1) $f_n(x)$ equals 1 at $x = i/k$ for $k = 0, 1, \ldots, n$ and, indexed for each k, $i = 0, 1, \ldots, k$. But here (as opposed to those in Example 2.2.1), on $[0, 1]$ each $f_n(x)$ equals 0 only at "halfway" x values, so-named because they are distributed halfway between the x values. For example, $f_1(x) = 0$ when $x = 1/2$, because $1/2$ is halfway between $x = 0$ and $x = 1$. Similarly, $f_2(x) = 0$ at $x = 1/4$ and $x = 3/4$, because they are halfway between $x = 0$ and $x = 1/2$, and $x = 1/2$ and $x = 1$, respectively. Other f_n's have halfway values defined similarly. Finally, each graph of f_n is formed in straight-line segments that connect x-value points (having $y = 1$) to its two neighboring halfway x-value points (having $y = 0$).

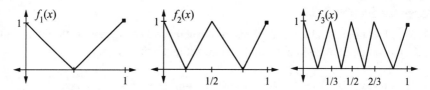

Figure 2.7. The graphs of the first three functions for Exercises 19–24.

19. Show the sequence $\{f_n\}$ is not monotonic, by comparing the values $f_1(1/4)$ and $f_2(1/4)$ to show it is not monotonically nondecreasing. Then compare $f_1(1/2)$ with $f_2(1/2)$ to similarly show it is not monotonically nonincreasing. Conclude that the monotone convergence theorem does not apply.

20. Show the sequence $\{f_n\}$ is dominated by the characteristic function $g(x) = \mathcal{X}_{[0,1]}$. Also, state why $g \in L^1$ with $\int g = 1$.

21. Give a reason supporting why f_n is in L^1. Also, say why $\int f_n = 1/2$ for each $n = 1, 2, \ldots$.

22. Determine $\lim_{n \to \infty} \int f_n$.

23. Evaluate $\lim_{n \to \infty} f_n(x)$ when x is rational. Does $\lim_{n \to \infty} f_n(x)$ exist for almost every x value? Why not? (*Hint:* examine the behavior of $\{f_n(x)\}$ for irrational x.)

24. Explain this sequence's behavior in the setting of the Lebesgue dominated convergence theorem: why doesn't $\lim_{n \to \infty} \int f_n = \int \lim_{n \to \infty} f_n$?

Advanced Exercises. We have indicated the function sequences in each of the LDCT, the nondecreasing convergence theorem for L^0, and the monotone convergence theorem play the role of step functions in Definition 1.5.1 for L^0 functions. The fact that the integral convergence theorems may identify an integrand function f as a member of L^0 means, by Definition 1.5.1, that there is indeed a nondecreasing sequence of step functions that approach f almost everywhere. In Exercises 25–28, find a nondecreasing sequence of step functions that converges a.e. to the integrand over the range of integration. Then write (but do not evaluate) the integral of each step function in your sequence as a finite sum.

25. $\int_0^\infty e^{-x}\, dx$
26. $\int_1^\infty x^{-2}\, dx$
27. $\int_0^\infty x e^{-x^2}\, dx$
28. $\int_0^1 1/\sqrt{x}\, dx$

2.4 Application: Fourier Series

A French mathematician who lived from 1768 to 1830, Joseph Fourier managed to find a functional representation of the way heat flowed through different structures. The representation is in terms of infinite series that involves sines and cosines, which is a Fourier series (or a trigonometric series). Fourier's investigations began in 1807, and he published his results in 1822 in a groundbreaking and much heralded work *Théorie Analytique de la Chaleur* [43].

To obtain the mathematical representation of heat flow, Fourier described Newton's law of cooling in terms of a differential equation involving both time t and position x. Such an equation

2.4 Application: Fourier Series

is a partial differential equation because it involves partial derivatives. The law of cooling, which describes the flow of heat between two nearby points as proportional to the distance between them, is $k^2 \partial^2 f/\partial x^2 = \partial f/\partial t$, where f is the temperature and k^2 is a proportionality constant. This is the "heat equation."

Fourier studied under the brilliant Joseph Louis Lagrange, whose work you likely saw in multivariable calculus when investigating Lagrange multipliers. By the mid-1700s, Lagrange had investigated the properties of several partial differential equations, publishing his results in the proceedings' records of a society he helped found (the Turin Academy of Sciences). One equation involved the position of a vibrating string; it turned out to be a precursor of Fourier's series representation of heat. Many mathematicians had earlier published partial results describing the solution, dating back to Jean le Rond d'Alembert in 1747 (as he produced a format that the solution must follow) and Leonhard Euler in 1748 (as he independently studied the position of the string in terms of coefficients in a representative series), and including Daniel Bernoulli (who applied physical properties to claim any solution could be expressed as a sum of what he called "modal" functions) and

Figure 2.8. Jean Baptiste Joseph Fourier.

Lagrange. As the string vibrates, its form can be thought of as a functional curve on the xy plane. For example, under basic physical assumptions d'Alembert proved the solution $f(x, t)$ could be expressed as $f(x, t) = (g(x + t) + g(x - t))/2$, where t is time. In other words and in a physical sense, f must be the average of two waves—the two expressions in g—that travel in opposite directions. Describing this, physicists say f is the "superposition" of the two waves. Furthermore, d'Alembert knew the waves (and f) were periodic when the ends of the string were tied down to $y = 0$, and the waves then also had to be tied down periodically. Under basic assumptions, Lagrange's solution found an underlying formula (Bernoulli's modal function) for the curve: $y = c \cdot \sin mx \cdot \sin nt$, where m and n are positive integers and c is a constant (determined from some conditions imposed on its original position and vibration). In fact, Lagrange went beyond this fact to derive a complete solution [25, p. 2]) in the form of a trigonometric series. These results, which were precursors for Fourier's analysis, were only small achievements for Lagrange in the total scope of his work. The extent and sophistication of the full body of his discovery of mathematical theory goes well beyond what anyone would expect to find from a single mathematician. For instance, his 1811 text on mechanics [76] derives, from the physical law of virtual work, what can essentially be considered a holistic treatment of solid and fluid mechanics. He published more than 200 papers in a single 20-year productive period. He is credited with proving scores of fundamental and important theorems across many fields. For example, in number theory, he showed any positive integer can be written as the sum of four integers squared, such as $11 = 3^2 + 1^2 + 1^2 + 0^2$. He made highly significant contributions to the three-body problem, which asks how to predict the motion and position of three celestial objects (such as the Sun, Earth, and Moon) as they interact with one another due to their gravity. Surely his insights on the solution to the string equation taught and guided his student Fourier and enabled his accomplishments. Unfortunately though, in the later 1700s Lagrange convinced himself that series of trigonometric functions were not appropriate as representations of vibrating strings. His reasoning, misplaced, was that strings having corners (discontinuous derivatives) were examples of functions not representable as trigonometric series. As a result, Lagrange actually opposed some acclaim of Fourier's methods. He claimed they were too nonrigorous a use of trigonometric series. That concern led to his denial (he was an appointed judge) of an 1812 award to Fourier from the Paris Academy for work on Fourier series [25, p. 6].

Fourier's method to solve the heat equation is extremely powerful. For example, it produces the complete solution for the vibrating string, as well as for other partial differential equations that are discussed in a typical undergraduate course such as Mathematical Physics. Fourier's theory of trigonometric series forms a cornerstone of the mathematical subfield known as harmonic analysis, which studies functions that arise (quite often in our physical world) as a superposition of basic waves.

Fourier's trigonometric series are just one form of a more general mathematical development of Fourier series, which can use other collections of functions (besides sines and cosines) to obtain series representations of functions. The functions in the collection can be thought of as forming an underlying basis of functions in a function space, and so we will examine this general development after we discuss function spaces in more detail, in Chapter 3. There we will also give several examples of collections (whose functions have a so-called "orthogonal" relationship to each other). We will also provide more extensive convergence theorems, possible only with the Lebesgue integral and not with Riemann's. For now and in this section, we will keep things simple, introducing Fourier series in a classical format in terms of sines and cosines.

2.4.1 Basic Formulation

Because sines and cosines are periodic, we formulate the Fourier series of a function f over the interval $(-\pi, \pi]$, assuming f is defined on not only that interval but also off of it—so that f is periodic. (If f is not defined off of $(-\pi, \pi]$, then we simply extend it across other intervals periodically.) So the shape of f's curve is repeated over each subsequent interval of length 2π. We have chosen the interval $(-\pi, \pi]$ to be half-open because of a convenience that will occur when we investigate convergence issues (and one that will allow for convenient notation whenever we wish to expand the domain of the function to all of $\mathbb{R}$). But the choice of $(-\pi, \pi]$ is a convenience that will not be necessary to derive these convergence theorems. The construction of the Fourier series and the subsequent conversion theory depend on the choice only in specific entries in the formulas. We also assume f is integrable over the open interval $(-\pi, \pi)$. This is not the only set of assumptions that may be used—different ones turn out to have marvelous consequences, such as the assumption that f is "square-integrable," which means $\int_{-\pi}^{\pi} |f(t)|^2 \, dt$ is finite. Fourier didn't examine the last assumption, incorrectly believing that his series representation would converge properly for any function whose coefficients were properly defined. We know today that convergence is not automatic, even for continuous functions.

Assuming $f \in L^1(-\pi, \pi)$ is periodic with period 2π, the classic Fourier series for f is

$$a_0 + \sum_{n=1}^{\infty} a_n \cos nx + b_n \sin nx,$$

where the coefficients are in terms of f:

$$a_0 = \frac{1}{2\pi} \int_{-\pi}^{\pi} f(t) \, dt, \ a_n = \frac{1}{\pi} \int_{-\pi}^{\pi} f(t) \cos nt \, dt, \ \text{and} \ b_n = \frac{1}{\pi} \int_{-\pi}^{\pi} f(t) \sin nt \, dt, \ n = 1, 2, 3, \ldots.$$

2.4 Application: Fourier Series

An alternative representation of the Fourier series develops from a famous formula of Euler's: $e^{it} = \cos t + i \sin t$, where $i = \sqrt{-1}$ and $t \in \mathbb{R}$. Then $\cos t = \frac{1}{2}(e^{it} + e^{-it})$ and $\sin t = \frac{1}{2i}(e^{it} - e^{-it})$ (see Exercises 1 and 2). Substituting into the series produces (see Exercises 3–5) the alternative format for the Fourier series:

$$\sum_{n=-\infty}^{\infty} c_n e^{inx},$$

where $c_0 = a_0$, $c_n = \frac{1}{2}(a_n - ib_n)$, and $c_{-n} = \frac{1}{2}(a_n + ib_n)$, $n = 1, 2, 3, \ldots$. In a simple description,

$$c_n = \frac{1}{2\pi} \int_{-\pi}^{\pi} f(t) e^{-int} dt, \quad n \in \mathbb{Z}.$$

You can see the complex formulation is more compact—there is only one coefficient format. In most calculations, we'll stick to the sine and cosine representation to avoid confusion possibly resulting from integration of complex terms. (This text does not assume knowledge of complex function theory.)

The next example illustrates the Fourier series representation of a basic linear function over $(-\pi, \pi]$. Periodically repeating this function over intervals of length 2π produces a shape that resembles the teeth on a saw, and so the function is known as the sawtooth function.

Example 2.4.1 The Sawtooth Function. We find the Fourier series for $f(x) = x$, $\pi < x \leq \pi$ (and where $f(x)$ is repeated periodically over other intervals of length 2π). Figure 2.9 presents the graph of f. We calculate the Fourier series coefficients, using integration by parts to find them when $n > 0$:

$$a_0 = \frac{1}{2\pi} \int_{-\pi}^{\pi} t \, dt = \frac{1}{2\pi} \frac{t^2}{2} \Big|_{-\pi}^{\pi} = 0, \, a_n = \frac{1}{\pi} \int_{-\pi}^{\pi} t \cos nt \, dt = \frac{1}{n\pi} \left(t \sin nt + \frac{1}{n} \cos nt \right) \Big|_{-\pi}^{\pi} = 0,$$

and

$$b_n = \frac{1}{\pi} \int_{-\pi}^{\pi} t \sin nt \, dt = \frac{1}{n\pi} \left(-t \cos nt + \frac{1}{n} \sin nt \right) \Big|_{-\pi}^{\pi} = \frac{2(-1)^{n+1}}{n} \text{ for } n = 1, 2, 3, \ldots.$$

Substituting into the series formula, we obtain the Fourier series for $f(x) = x, x \in (-\pi, \pi]$:

$$2 \sin x - \sin 2x + \frac{2}{3} \sin 3x - \ldots = 2 \sum_{n=1}^{\infty} \frac{(-1)^{n+1}}{n} \sin nx.$$

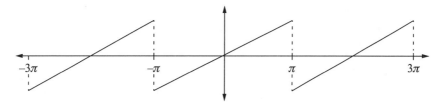

Figure 2.9. The sawtooth function $f(x) = x$, where $x \in (-\pi, \pi]$ is extended periodically across other intervals of length 2π.

Later in this section we will show how this infinite series converges at each x to the sawtooth function on $(-\pi, \pi)$ and each subsequent periodic open interval of length 2π. In fact, those types of convergence issues are the focus of the next subsection. ∎

The next questions lead you through derivations of Fourier series for a couple of straightforward examples.

Question 2.4.1 What is the Fourier series for $f(x) = x^2, x \in (-\pi, \pi]$? To answer this question, calculate[13] the coefficients a_0, a_n, and b_n for $n = 1, 2, \ldots$. Then substitute them into the formula for the Fourier series. ∎

Question 2.4.2 Show the Fourier series for the function $f + g$ is the sum of the Fourier series for f and the Fourier series for g. Then, for $c \in \mathbb{R}$, find the Fourier series for the functions $g(x) = x + c, x \in (-\pi, \pi], h(x) = x + x^2, x \in (-\pi, \pi]$, and $g(x) + h(x)$. ∎

2.4.2 Convergence Issues

After deriving a Fourier series for f, it is natural to wonder when the series converges to it. Such investigations are familiar to you from Calculus II, where you found the Taylor series for f (a power series), and then wondered when the series converged to f. The ratio test determined the radius of convergence and the corresponding interval of convergence. Convergence issues for Fourier series are more difficult, but there are many results we can derive. The Fourier series representation is also more broadly applicable than the power series representation. The only functions that have convergent power series are analytic (differentiable) ones. Many more functions have convergent Fourier series.

One very natural question is when the series converges to f pointwise, where we wonder if the series (in terms of its partial sums) converges to f at each point x in $(-\pi, \pi]$. One powerful result in this area is the Dirichlet-Jordan theorem, where the periodic function f is assumed integrable and "of bounded variation" over the open interval $(-\pi, \pi)$. It gives a necessary condition for the Fourier series at x to converge to $f(x)$ in the pointwise sense of Definition 1.4.4. The theorem originated in 1829 with a proof by Gustav Lejeune Dirichlet of a slightly simpler result [36]. Then in 1881 Camille Jordan generalized the result [67], basing it on the concept of bounded variation (which we will define and explain as we proceed with the proof). You can see from the dates of discovery that the result will not depend on the Lebesgue integral, and you might expect other more powerful results, which use the Lebesgue integral, have superseded the Dirichlet-Jordan theorem. If so, then you'd be right, but the modern results (some of which are described in this chapter's notes) are often difficult to prove, and so we start with the classical theorem of Dirichlet and Jordan. The proof presented here, since it will use the Lebesgue integral, will be fairly direct and straightforward. To proceed, we must state several theorems and definitions that build up to it.

The first is the Riemann-Lebesgue lemma, which Lebesgue proved in 1903. It is important in its own right, because it implies (using the function $f \cdot \mathcal{X}_{(-\pi, \pi]}$) that the Fourier coefficients

---

[13] You'll notice Example 2.4.1 produced $a_n = 0$. That happened because $t \cdot \cos nt$ is an odd function. (Odd functions g have the property that $g(-t) = -g(t)$.) The integral of an odd function over any symmetric bounds, such as from $-\pi$ to π, is always 0. In a helpful way, you may wish to note $t^2 \cdot \sin nt$ is odd for $n \in \mathbb{N}$.

2.4 Application: Fourier Series

a_n and b_n exist for any Lebesgue integrable function and converge to 0 as n gets large. And the theorem is even more general than that, because n can be any real value, not just $n = 1, 2, 3, \ldots$. Furthermore, we will soon use the Riemann-Lebesgue lemma in the proof of the Dirichlet-Jordan theorem.

Theorem 2.4.1 The Riemann-Lebesgue Lemma: *If $f \in L^1$, then for $n \in \mathbb{R}$,*

$$\int_{-\infty}^{\infty} f(x) \cos nx \, dx \quad \text{and} \quad \int_{-\infty}^{\infty} f(x) \sin nx \, dx$$

exist and converge to 0 as n approaches infinity.

Proof. This proof of the result is for the first integral (with $\cos nx$). The proof for the second integral works the same way. Proceed by examining the following four cases: (i) $f = \mathcal{X}_I$ for an interval $I = (a, b)$; (ii) f is a step function; (iii) $f \in L^0$; and (iv) $f \in L^1$. For case (i), if $f = \mathcal{X}_I$, where $I = (a, b)$ is a finite interval, then $\int_{-\infty}^{\infty} f(x) \cos nx \, dx = \int_a^b \cos nx \, dx = (\sin nb - \sin na)/n$, where a and b are the endpoints of I. Since $|\sin x| \leq 1$ for all x, this integral's value as a function in n converges to 0 as $n \to \infty$. Case (ii), where f is a step function, now follows immediately by linearity of the integral.

For case (iii), when $f \in L^0$, Definition 1.5.1 provides a nondecreasing sequence $\{\phi_k\}$ of step functions that converges to f almost everywhere (hence $\lim_{k \to \infty} \phi_k(x) \cos x = f(x) \cos x$ a.e.) and for which $\lim_{k \to \infty} \int \phi_k = \int f$. Since f is integrable, so is $|f|$ by Theorem 2.2.4, and since $|f(x) \cos x| \leq |f|$, so is $f(x) \cos x$.

From the definition of limit, given $\varepsilon > 0$, there exists a real K such that $\int |f - \phi_K| = \int f - \phi_K < \varepsilon$. For this K, ϕ_K is a step function, and so by case (ii) there exists a value N such that $|\int \phi_K(x) \cos nx \, dx| < \varepsilon$ whenever $n \geq N$. Therefore, subtracting and adding the term $\phi_K(x) \cos nx$ inside the integral and then using the triangle inequality,

$$\left| \int f(x) \cos nx \, dx \right| \leq \left| \int (f(x) - \phi_K(x)) \cos nx \, dx \right| + \left| \int \phi_K(x) \cos nx \, dx \right|$$

$$< \int |f(x) - \phi_K(x)| \, dx + \varepsilon < 2\varepsilon$$

whenever $n \geq N$. Since we have made the expression's absolute value arbitrarily small, it must go to zero in the limit. That proves case (iii). Case (iv) now follows immediately by linearity of the integral. ∎

We now define two objects used in the proof of the Dirichlet-Jordan theorem.[14]

Definition 2.4.1 *The Dirichlet kernel function for any $n = 1, 2, \ldots$ is $D_n(x) = 1/2 + \sum_{k=1}^{n} \cos kx$, $x \in \mathbb{R}$, and the Fourier series partial sum for a function f at x is $s_n(x) = a_0/2 + \sum_{k=1}^{n} (a_k \cos kx + b_k \sin kx)$.*

Question 2.4.3 Find the values for $D_1(\pi/2)$, $D_2(0)$, and the derivative value $D_2'(\pi/2)$. ∎

---

[14] The outline used here follows a similar presentation in [121].

It is natural to ask, "Where in the world does the formula for the Dirichlet kernel come from, and why should I be interested in it?" The answer is that D_n shows up in formulas as soon as we start investigating representations of s_n (the partial sums for the Fourier series) that come from the Fourier coefficient formulas. The next theorem lists several properties of D_n, and the second one, in particular, shows how D_n arises naturally in connection to s_n. More than any other fact, the statement and proof of the second property motivates the formulation and study of the Dirichlet kernel D_n.

Theorem 2.4.2 Dirichlet Kernel Properties: *For* $n = 1, 2, \ldots$,

(i) $\int_0^\pi D_n(t)\, dt = \pi/2$.

(ii) $s_n(x) = \frac{1}{\pi} \int_{-\pi}^\pi f(t) \cdot D_n(t - x)\, dt$.

(iii) $s_n(x) = \frac{1}{\pi} \int_0^\pi ((f(x + t) + f(x - t)) \cdot D_n(t)\, dt$.

(iv) For $x \neq 2k\pi$ for $k \in \mathbb{Z}$, $D_n(x) = \dfrac{\sin(nx + \frac{x}{2})}{2 \sin \frac{x}{2}}$.

(v) $|\int_a^b D_n(t)\, dt| < 2$ for any $0 \leq a, b \leq \pi$.

Proof. (i) When $k = 1, 2, \ldots$, $\int_0^\pi \cos kx\, dx = (\sin kx)/k|_0^\pi = 0$. Hence $\int_0^\pi D_n(x)\, dx = \int_0^\pi 1/2\, dx = \pi/2$.

(ii) We have defined the partial sum of the Fourier series as $s_n(x) = a_0 + \sum_{k=1}^n a_k \cos kx + b_k \sin kx$. Substituting the integral expressions for a_0, a_n, and b_n and simplifying produces

$$s_n(x) = \frac{1}{\pi} \int_{-\pi}^\pi f(t) \left(\frac{1}{2} + \sum_{k=1}^n \cos kt \cos kx + \sin kt \sin kx \right) dt.$$

Apply the identity $\cos u \cos v + \sin u \sin v = \cos(u - v)$ and the first item of Definition 2.4.1 to prove the result.

(iii) A change of variables from $t - x$ to t (see Section 2.2's Exercises 37–41) in the equation in part (ii) produces $s_n(x) = \frac{1}{\pi} \int_{x-\pi}^{x+\pi} f(x + t) \cdot D_n(t)\, dt$. We are assuming f is periodic with period 2π, and so is D_n (as it is built up from cosine functions), and so $s_n(x) = \frac{1}{\pi} \int_{-\pi}^\pi f(x + t) \cdot D_n(t)\, dt$. Another simple change of variables (from t to $-t$) shows $\int_{-\pi}^0 f(x + t) \cdot D_n(t)\, dt = \int_0^\pi f(x - t) \cdot D_n(-t)\, dt$. But $\cos x = \cos(-x)$, and so $D_n(x) = D_n(-x)$. Hence

$$s_n(x) = \frac{1}{\pi} \left\{ \int_{-\pi}^0 f(x + t) \cdot D_n(t)\, dt + \int_0^\pi f(x + t) \cdot D_n(t)\, dt \right\}$$

$$= \frac{1}{\pi} \left\{ \int_0^\pi f(x - t) \cdot D_n(t)\, dt + \int_0^\pi f(x + t) \cdot D_n(t)\, dt \right\}$$

$$= \frac{1}{\pi} \int_0^\pi \{f(x - t) + f(x + t)\} \cdot D_n(t)\, dt.$$

(iv) Since $2 \sin u \cos v = \sin(u + v) - \sin(v - u)$,

$$2 \sin(x/2) \sum_{k=1}^n \cos kx = \sum_{k=1}^n \{\sin(k + 1/2)x - \sin(k - 1/2)x\}.$$

2.4 Application: Fourier Series

The series telescopes. (All the terms cancel except for the first and last.) Hence $2\sin(x/2)\sum_{k=1}^{n}\cos kx = \sin(n+1/2)x - \sin(1/2)x$. That implies $2\sin(x/2)D_n(x) = \sin(nx + x/2)$, which is equivalent to the result.

(v) This result follows from several strict calculations. The substitution $u = 2t$ and the trigonometric identities $\sin(u+v) = \sin u \cos v + \cos u \sin v$ and $\cos^2 u = \frac{1}{2}(1+\cos 2u)$ imply

$$\int_0^{2\pi/3} D_1(t)\,dt = \frac{1}{2}\int_0^{2\pi/3} \sin(3t/2)/\sin(t/2)\,dt = \int_0^{\pi/3} \sin 3u/\sin u\,du$$

$$= \int_0^{\pi/3} \cos 2u + 2\cos^2 u\,du = \frac{1}{2}(\sin 2u + 2u + \sin 2u)\big|_0^{\pi/3} = \sqrt{3}/2 + \pi/3.$$

The absolute maximum value of the function $F(x) = \int_0^x D_n(t)\,dt$ over the integral $[0, \pi]$ is found by calculus methods. The integrand D_n is continuous, and so the derivative always exists and equals $F'(x) = D_n(x)$ by the fundamental theorem of calculus. Setting the derivative equal to zero produces the critical values[15] $x = k\pi/(n+1/2)$ for $k = 1, 2, \ldots, n$. Comparing the value of F at each critical point, it is not hard to show the maximum occurs at the first one, and hence equals $\int_0^{\pi/(n+1/2)} D_n(t)\,dt$. The values form a sequence of maximum values in n, and they turn out to be strictly decreasing as n increases. Hence, for $0 \le x \le \pi$, $F(x)$ is less than the first element in this sequence, the one that has $n = 1$. In fact, because the magnitude of areas between successive critical values decreases as k gets larger, $|\int_a^b D_n(t)\,dt| \le F(3\pi/2)$ for $a, b \in [0, \pi]$. In short, $|F(b) - F(a)| \le \int_0^{\pi/(1+1/2)} D_1(t)\,dt$. But that integral is the one just evaluated as $\sqrt{3}/2 + \pi/3$, which is a number smaller than 2. The result follows. ∎

Theorem 2.4.3 *Suppose $f \in L^1(-\pi, \pi)$ is periodic with period 2π. Then, for h with $0 < h \le \pi$,*

$$\lim_{n\to\infty} \int_0^h \{f(x+t) + f(x-t) - 2s(x)\} \cdot D_n(t)\,dt = 0 \text{ for some function } s$$

if and only if $\lim_{n\to\infty} s_n(x)$ (the limit of the Fourier series' partial sums) exists and equals the function $s(x)$.

Proof. Apply part (i) of Theorem 2.4.2 to obtain an intermediate result. That theorem's statement is algebraically equivalent to $1 = \frac{1}{\pi}\int_0^\pi 2D_n(t)\,dt$. Multiply both sides by $s(x)$ and then subtract it from the equation in Theorem 2.4.2 part (iii) to get $s_n(x) - s(x) = \frac{1}{\pi}\int_0^\pi \{f(x+t) + f(x-t) - 2s(x)\}D_n(t)\,dt$. Taking the limit, $\lim_{n\to\infty}\int_0^\pi \{f(x+t) + f(x-t) - 2s(x)\} \cdot D_n(t)\,dt = 0$ if and only if $\lim_{n\to\infty} s_n$ exists and equals $s(x)$.

On the interval $[h, \pi]$, the function $(\sin\frac{t}{2})^{-1}$ is continuous (as it is the composition of two continuous functions) and is therefore bounded on $[h, \pi]$. Therefore, as an integral in t and by

---

[15] Each critical value turns out to be a local maximum or minimum for F. They alternate as local maxima and minima at successive critical points.

part (iv) of Theorem 2.4.2,

$$\lim_{n \to \infty} \int_h^\pi \{f(x+t) + f(x-t) - 2s(x)\} \cdot D_n(t)\, dt$$

$$= \lim_{n \to \infty} \int_h^\pi \{f(x+t) + f(x-t) - 2s(x)\} \frac{\sin(nt + \frac{t}{2})}{2 \sin \frac{t}{2}}\, dt$$

$$= \lim_{n \to \infty} \int_h^\pi f(x+t) \frac{\sin(nt + \frac{t}{2})}{2 \sin \frac{t}{2}}\, dt + \int_h^\pi f(x-t) \frac{\sin(nt + \frac{t}{2})}{2 \sin \frac{t}{2}}\, dt - \int_h^\pi 2s(x) \frac{\sin(nt + \frac{t}{2})}{2 \sin \frac{t}{2}}\, dt.$$

Applying the boundedness of $(\sin \frac{t}{2})^{-1}$, each of the integrals behave nicely in the limit: the Riemann-Lebesgue lemma (Theorem 2.4.1) implies they converge to zero. (The first two follow from Theorem 2.4.1 applied to the functions $f(x+t) \cdot \mathcal{X}_{[h,\pi]}$ and $f(x-t) \cdot \mathcal{X}_{[h,\pi]}(t)$, respectively, and the third from Theorem 2.4.1 applied to $s(x) \cdot \mathcal{X}_{[h,\pi]}(t)$, where $s(x)$ is treated as a constant in t.) The result follows:

$$\lim_{n \to \infty} \int_0^\pi \{f(x+t) + f(x-t) - 2s(x)\} \cdot D_n(t)\, dt$$

$$= \lim_{n \to \infty} \int_0^h \{f(x+t) + f(x-t) - 2s(x)\} \cdot D_n(t)\, dt$$

$$+ \lim_{n \to \infty} \int_h^\pi \{f(x+t) + f(x-t) - 2s(x)\} \cdot D_n(t)\, dt,$$

which is zero if and only if (by the intermediate result in the preceding paragraph) $\lim_{n \to \infty} s_n$ exists and equals $s(x)$. ∎

Definition 2.4.2 *We say $f \in L^1(-\pi, \pi)$ is of bounded variation when we can express f as the difference $g - h$ of two nondecreasing functions.*[16]

Proved by Dirichlet and Camille Jordan, the next theorem is this subsection's main result. Dirichlet's work was first; he proved only the case when f was monotonically nondecreasing or nonincreasing. Jordan's result was essentially as presented here. In fact, Jordan's 1881 paper was the first to use the concept of bounded variation. Many other function-theoretic issues are described for functions of bounded variation, as it has proven to be a natural setting—functions with bounded variation are increasingly understood to be special in analysis, especially when examining the integral. As might be familiar from calculus, the limit symbol used in the theorem's statement, "$\lim_{t \to 0^+}$," indicates the limit is being taken from the right, with only positive values of t approaching 0.

---

[16] The term "bounded variation," for example on $[-\pi, \pi]$, comes from a more fundamental and intricate definition, which says, for $f \in L^1[-\pi, \pi]$ and for any finite partition of x values formed as $-\pi = x_0 < x_1 < \ldots < x_m = \pi$, the sums $\sum_{j=1}^{m} |f(x_j) - f(x_{j-1})|$ are bounded. The supremum of all such sums is the total variation of f on $[-\pi, \pi]$. It is not difficult to show this definition is equivalent to Definition 2.4.2. See Exercises 31–34.

2.4 Application: Fourier Series

Theorem 2.4.4 **The Dirichlet-Jordan Theorem:** *Suppose $f \in L^1(-\pi, \pi)$ is periodic with period 2π and, for $x \in \mathbb{R}$, has bounded variation on an interval $[x-h, x+h]$, where $0 < h \leq \pi$. Then the Fourier series of f converges at x to $\lim_{t \to 0^+}(f(x+t) + f(x-t))/2$.*

Proof. On the interval $[x-h, x+h]$, we write $f = g - h$, where g and h are nondecreasing functions on that interval. A standard fact from calculus (see Exercise 36) implies $\lim_{t \to 0^+} g(x+t)$, $\lim_{t \to 0^+} g(x-t)$, $\lim_{t \to 0^+} h(x+t)$, and $\lim_{t \to 0^+} h(x-t)$ exist. By linearity of the limit, so do $\lim_{t \to 0^+} f(x+t)$ and $\lim_{t \to 0^+} f(x-t)$.

At x, define $u(t) = f(x+t) + f(x-t) - \lim_{t \to 0^+} f(x+t) - \lim_{t \to 0^+} f(x-t)$, $t \in [0, h]$. Then $\lim_{t \to 0^+} u(t) = 0$. Also, u is of bounded variation on $[0, h]$, hence $u = v - w$ on that interval, where v and w are nondecreasing and both tend to 0 as $t \to 0^+$. Applying the limit definition to both functions, given $\varepsilon > 0$, there exists $\delta > 0$ (and smaller than h) such that $0 \leq v(t), w(t) < \varepsilon$ whenever $0 < t < \delta$.

A version of the mean value theorem (see [121, p. 134]) guarantees, for some value d with $0 < d < \delta$, that $|\int_0^\delta v(t) D_n(t)\, dt| = \lim_{r \to 0^+} v(\delta - r) |\int_d^\delta D_n(t)\, dt|$ and $|\int_0^\delta w(t) D_n(t)\, dt| = \lim_{r \to 0^+} w(\delta - r) |\int_d^\delta D_n(t)\, dt|$. By part (v) of Theorem 2.4.2, the two terms are less than $\lim_{r \to 0^+} v(\delta - r) \cdot 2$ and $\lim_{r \to 0^+} w(\delta - r) \cdot 2$, respectively, and so they are bounded by 2ε. Then, since $u = v - w$, the triangle inequality implies

$$\left| \int_0^\delta u(t) D_n(t)\, dt \right| \leq \left| \int_0^\delta v(t) D_n(t)\, dt \right| + \left| \int_0^\delta w(t) D_n(t)\, dt \right| < 2\varepsilon + 2\varepsilon = 4\varepsilon.$$

By the Riemann-Lebesgue lemma, there exists N such that $|\int_\delta^h u(t) D_n(t)\, dt| < \varepsilon$ whenever $n \geq N$. Therefore

$$\left| \int_0^h u(t) D_n(t)\, dt \right| \leq \left| \int_0^\delta u(t) D_n(t)\, dt \right| + \left| \int_\delta^h w(t) D_n(t)\, dt \right| < 4\varepsilon + \varepsilon = 5\varepsilon \text{ for } n \geq N.$$

Since u is of the form $u(t) = f(x+t) + f(x-t) - \lim_{t \to 0^+} f(x+t) - \lim_{t \to 0^+} f(x-t)$, Theorem 2.4.3 guarantees the existence of the limit of the Fourier series partial sums and identifies that limit as $s(x)$, where

$$2s(x) = \lim_{t \to 0^+} f(x+t) + \lim_{t \to 0^+} f(x-t). \qquad \blacksquare$$

Example 2.4.2 **The Castle Wave Function:** Set $f(x) = \mathcal{X}_{(0,\pi]}(x) - \mathcal{X}_{(-\pi,0]}(x)$, and then extend f periodically over the rest of $\mathbb{R}$ in intervals of length 2π.

Since f is an odd function over $(-\pi, \pi]$, it is easy to see $a_n = 0, n = 1, 2, \ldots$. For $n = 1, 2, \ldots$, the sine coefficient is $b_n = \frac{1}{\pi}(\int_0^\pi \sin nt\, dt - \int_{-\pi}^0 \sin nt\, dt) = \frac{2(1 - \cos n\pi)}{n\pi} = \frac{4}{n\pi}$ when n is odd (and is 0 when n is even). The Fourier series is $\frac{4}{\pi}\left(\sin x + \frac{1}{3}\sin 3x + \ldots\right) = \frac{4}{\pi} \sum_{k=1}^\infty \frac{1}{2k-1} \sin(2k-1)x$.

We note several interesting facts from the Dirichlet-Jordan theorem. First, this example's f is of bounded variation (it is, in fact, a nondecreasing function on $(-\pi, \pi]$). Hence the Fourier series converges to the value $\lim_{t \to 0^+}(f(x+t) + f(x-t))/2$. This value equals $f(x)$ at any point

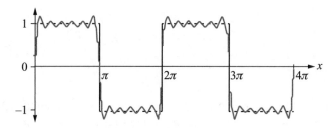

Figure 2.10. The castle wave function on $[0, 4\pi]$, along with the approximating partial sum Fourier series. Convergence issues exist at multiples of π; the approximation overshoots its mark there. That characteristic is the "Gibbs' phenomenon." (Josiah Willard Gibbs is considered to be the founder of chemical thermodynamics.) It is standard at jump discontinuities of f, essentially caused by the form of the Dirichlet kernel D_n.

of continuity, which is at any point except $x = -\pi, 0$, and π. There the Fourier series converges to 0, since at each one we have $\lim_{t \to 0^+} f(x+t) = -\lim_{t \to 0^+} f(x-t)$ (at $x = -\pi, 0$, or π, one of these values is always 1 while the other is -1). The convergence at any other point is always to $f(x)$. To illustrate how the convergence works, Figure 2.10 shows a graph of f with the partial sum at $n = 51$. One of the most interesting x values that we could choose is $x = \pi/2$, since $\sin(2k-1)\pi/2 = (-1)^{k+1}$. At that point, the Dirichlet-Jordan theorem says the Fourier series $\frac{4}{\pi}\left(1 - \frac{1}{3} + \frac{1}{5} - \ldots\right)$ converges to $f(\pi/2) = 1$. Multiplying both sides by $\pi/4$, we obtain a closed-form expression for the alternating odd-termed harmonic series in an expression for π: $\frac{\pi}{4} = 1 - \frac{1}{3} + \frac{1}{5} - \ldots$. ∎

Question 2.4.4 Examine Example 2.4.1's sawtooth function $f(x) = x$, where $x \in (-\pi, \pi]$ and f is extended periodically across $\mathbb{R}$: determine the limit of f's Fourier series, $\lim_{n \to \infty} s_n(x)$, at any point x. ∎

Solutions to Questions

2.4.1 $a_0 = \frac{\pi^2}{3}$, $a_n = \frac{4 \cdot (-1)^n}{n^2}$, and $b_n = 0$. The series is $\frac{\pi^2}{3} + \sum_{n=1}^{\infty} \frac{4 \cdot (-1)^n}{n^2} \cos nx$.

2.4.2 The cosine coefficient of the series for $f + g$ is $a_n = \frac{1}{\pi}\int_{-\pi}^{\pi}[f(t) + g(t)]\cos nt \, dt = \frac{1}{\pi}\int_{-\pi}^{\pi} f(t)\cos nt \, dt + \frac{1}{\pi}\int_{-\pi}^{\pi} g(t)\cos nt \, dt$, which is the sum of the cosine coefficients for f and g, respectively. The other coefficients work similarly. Since an infinite series can be split apart linearly, we see the Fourier series for $f + g$ is the sum of the series for f and g. The given function $g(x) = x + c$ has Fourier coefficients $a_0 = c$, $a_n = 0$, and $b_n = \frac{2(-1)^{n+1}}{n}$. The function $h(x) = x + x^2$ has coefficients $a_0 = \frac{\pi^2}{3}$, $a_n = \frac{4 \cdot (-1)^n}{n^2}$, and $b_n = \frac{2(-1)^{n+1}}{n}$. Adding, the function $g + h$ has Fourier coefficients $a_0 = \frac{\pi^2}{3} + c$, $a_n = \frac{4 \cdot (-1)^n}{n^2}$, and $b_n = \frac{4(-1)^{n+1}}{n}$. They construct the Fourier series in a straightforward way; for example, the Fourier series for $h + g$ is $\frac{\pi^2}{3} + c + \sum_{n=1}^{\infty} \frac{4 \cdot (-1)^n}{n^2} \cos nx + \frac{4(-1)^{n+1}}{n} \sin nx$.

2.4.3 $D_1(\pi/2) = 1/2 + \sum_{k=1}^{1} \cos k\pi/2 = 1/2 + \cos \pi/2 = 1/2$. $D_2(0) = 1/2 + \sum_{k=1}^{2} \cos(k \cdot 0) = 1/2 + (1 + 1) = 2.5$. $D_2'(\pi/2) = \sum_{k=1}^{2} -k \cdot \sin k\pi/2 = -1 \cdot 1 - 2 \cdot 0 = -1$

2.4 Application: Fourier Series

2.4.4 Referencing the graph in Figure 2.9, it converges to the 2π-periodic extension of $f(x) = x$ except at points of the form $x = (2k + 1)\pi$ for integer k. At $x = (2k + 1)\pi$, it converges to the limiting average $\lim_{t \to 0^+} (f(x + t) + f(x - t))/2 = [(2k + 1)\pi + (-(2k + 1)\pi)]/2 = 0$.

Reading Questions for Section 2.4

1. What is the Fourier series coefficient a_0 (the constant coefficient that corresponds to $n = 0$)?

2. What are the Fourier series coefficients a_n and b_n when $n = 1, 2, 3, \ldots$?

3. What is the corresponding Fourier series?

4. What are the Fourier series coefficients (in complex number formulation) c_n, where $n \in \mathbb{Z}$?

5. What is the corresponding Fourier series?

6. What is the definition of the Dirichlet kernel?

7. To what type of function f does the Dirichlet-Jordan theorem apply?

8. For a function f under appropriate conditions, what does the Dirichlet-Jordan theorem identify as the limit of a f's Fourier series?

Exercises for Section 2.4

Exercises 1 and 2 derive the complex-valued expressions used for the derivation of the alternative Fourier coefficient format c_n. For each, use Euler's formula $e^{it} = \cos t + i \sin t$, where $i = \sqrt{-1}$ and $t \in \mathbb{R}$.

1. Prove $\frac{1}{2}(e^{it} + e^{-it}) = \cos t$.

2. Prove $\frac{1}{2i}(e^{it} - e^{-it}) = \sin t$.

Exercises 3–5 derive the complex-valued formulation of the Fourier series.

3. Show $c_0 = \frac{1}{2\pi} \int_{-\pi}^{\pi} f(t) e^{-i0t} \, dt$.

4. Prove $\frac{1}{2}(a_n - ib_n) = c_n$, $n = 1, 2, 3, \ldots$ where $c_n = \frac{1}{2\pi} \int_{-\pi}^{\pi} f(t) e^{-int} \, dt$ for $n \in \mathbb{Z}$.

5. Use the results from Exercises 3 and 4 to prove the Fourier series for f may be expressed as $\sum_{n=-\infty}^{\infty} c_n e^{inx}$.

Exercises 6–8 work with the function $f(x) = x^3$, $x \in (-\pi, \pi]$.

6. Since $f(x)$ is an odd function, in its Fourier series expansion $a_n = 0$ for all n. What are the values for the other Fourier coefficients $b_n = \frac{1}{\pi} \int_{-\pi}^{\pi} f(t) \sin(nt) \, dt$, where $n = 1, 2, 3, \ldots$?

7. What is the Fourier trigonometric series for $f(x)$?

8. Draw the graph of $f(x)$ over the interval $(-\pi, \pi)$ along with (from, say, your graphing calculator) the graph of the trigonometric series truncated after three terms: $b_1 \cdot \sin x + b_2 \cdot \sin 2x + b_3 \cdot \sin 3x$.

In Exercises 9–17, find the Fourier series for the function (as it is defined on the interval $(-\pi, \pi]$ and extended periodically with period 2π off of this interval).

9. $f(x) = |x|$

10. $g(x) = \begin{cases} 1 & \text{if } -\pi/2 < x \leq \pi/2 \\ -1 & \text{if } -\pi < x \leq -\pi/2 \text{ or } \pi/2 < x \leq \pi \end{cases}$

11. $h(x) = \begin{cases} 1-x & \text{if } 0 \leq x \leq \pi \\ 1+x & \text{if } -\pi < x < 0 \end{cases}$

12. $r(x) = 1$ 14. $t(x) = \sin x$ 16. $v(x) = \cos x$

13. $s(x) = x^2$ 15. $u(x) = |\sin x|$ 17. $w(x) = |\cos x|$

For the functions in Exercises 18–22, find the complex Fourier series coefficients $c_n = \frac{1}{2\pi} \int_{-\pi}^{\pi} f(x) e^{-inx} dx$, for $n \in \mathbb{Z}$. Use the fact that the general form for the antiderivative of e^{inx} is $\frac{1}{in} e^{inx} = \frac{-i}{n} e^{inx}$.

18. $f(x) = 1$ 20. $h(x) = x + c$, where $c \in \mathbb{R}$ 22. $v(x) = x^2$

19. $g(x) = x$ 21. $s(x) = |x|$

For each function in Exercises 23–30, defined on $(-\pi, \pi]$ and extended periodically with period 2π off of this interval, use the Dirichlet-Jordan theorem to find the values of the Fourier series at $-\pi$ and at π. Make sure you say why the function is of bounded variation, so that the Dirichlet-Jordan theorem applies.

23. $f(x) = 1$ 26. $r(x) = x^2$ 28. $t(x) = |\sin x|$

24. $g(x) = |x|$ 27. $s(x) = \sin x$ 29. $u(x) = \sec(x/4)$

25. $h(x) = x + c$, where $c \in \mathbb{R}$ 30. $w(x) = |\tan(x/4)|$

Advanced Exercises. Definition 2.4.2 states a function f is of bounded variation when it can be expressed as the difference of two nondecreasing functions. Exercises 31–34 show this definition for a function f on $[-\pi, \pi]$ is equivalent to the following: for any finite partition of x values formed as $-\pi = x_0 < x_1 < \cdots < x_m = \pi$, the sums $\sum_{j=1}^{m} |f(x_j) - f(x_{j-1})|$ are bounded.

31. Show a function f of bounded variation has, for a partition $-\pi = x_0 < x_1 < \ldots < x_m = \pi$, sums $\sum_{j=1}^{m} |f(x_j) - f(x_{j-1})|$ that are bounded.

32. Now assume a function f has the property: for any partition $-\pi = x_0 < x_1 < \ldots < x_m = \pi$, the sums $\sum_{j=1}^{m} |f(x_j) - f(x_{j-1})|$ are bounded. For any such sum, define f_p as the sum of the positive terms in the summation $\sum_{j=1}^{m} f(x_j) - f(x_{j-1})$, and define f_n as the negation of the sum of the negative terms. Then define $g(\pi)$ to be the supremum of f_p, taken over all possible sums from any given partition, and define $h(\pi)$ to be the supremum of f_n. Prove $g(\pi) - h(\pi) = f(\pi) - f(-\pi)$.

33. In the setting of the last problem, for $x \in (-\pi, \pi]$, define $g(x)$ as the supremum of f_p, where a term f_p is defined as the sum of the positive terms in a summation $\sum_{j=1}^{m} f(x_j) - f(x_{j-1})$ whose x_j terms come from a partition $-\pi = x_0 < x_1 < \cdots < x_m = x$. Similarly, define the function $h(x)$ as the supremum of f_n, where f_n is defined as the negation of the sum of the negative terms in the summation. Prove $g(x) - h(x) = f(x) - f(-x)$.

34. Show the functions $g(x)$ and $h(x)$ as defined in Exercise 33 are nondecreasing.

Exercises 35 and 36 prove one-sided limit facts that pertain to the Dirichlet-Jordan theorem.

35. Assume f is nondecreasing, and set $L = \inf\{f(t) : t > x\}$. Use Section 2.1's alternate characterization of the infimum to prove $\lim_{t \to 0^+} f(x + t) = L$. In the same way, prove $\lim_{t \to 0^+} f(x - t) = \sup\{f(t) : t < x\}$.

36. Use Exercise 35 to prove $\lim_{t \to 0^+} f(x + t)$ and $\lim_{t \to 0^+} f(x - t)$ always exist for a function f of bounded variation.

Notes for Chapter 2

Readers may also wish to examine the many excellent descriptions of the Riemann integral in a standard undergraduate format. These books often serve the course Real Analysis; authors include James Kirkwood [75] and Stephen Abbott [1]. Though not presented using step functions with the standard construction as formulated in this chapter and by Alan J. Weir [121], these texts' approach to the Riemann integral are equivalent to Section 2.1's.

Several texts provide strong overviews of Fourier series, including those by authors David Powers [92], Dunham Jackson [60], Robert Seeley [110], and John Polking, Albert Boggess, and David Arnold [90]. Harry Dym and Henry McKean [38] give a modern treatment, written at the graduate level and almost in outline form, but likely accessible to an engaged student completing this text. It provides a complete theory of Fourier series and associated integrals.

Further history for mathematical advances in Fourier series, beyond what was described in Section 2.4, is recommended for the interested reader. This chapter's discussion focused on physical processes (vibrating strings and heat) that expressed themselves in continuous time t. But discrete-time Fourier processes were equally important, and they have advanced in parallel to the continuous theory. Both were spurred by a desire to explain mathematically the physical world around us. For example, since the late 1700s, mathematicians (including Carl Friedrich Gauss) have published results on predicting the motion of a planet or some other heavenly object when given either a finite or infinite number of points in a discrete time sequence of the object's positions as it moves across the sky. In 1965, J. W. Cooley and John Tukey developed what is now commonly known as the fast Fourier transform (FFT), which numerically calculates the coefficients connected to discrete Fourier systems in an incredibly short time on a computer. Before Cooley and Tukey developed the FFT, many proposals to solve discrete real-life problems were theoretically appealing but impractical because of long calculating time. After Cooley and Tukey published the FFT, they were useful. Spurred by this realization, mathematicians produced a host of theoretical and applied results on discrete-time signal and system processing in the late 1900s and into today.

J. C. Burkill provided in his 1944 obituary of Lebesgue [19] a remarkably detailed and knowledgeable summary of the accomplishments and many significant moments in the life of Henri Lebesgue. He relates, for example, Lebesgue's earliest successes at applying his integral to Fourier series convergence issues, when in 1903 Lebesgue proved three substantial results, namely, "(1) that a trigonometric series representing a bounded function is a Fourier series..., (2) that the nth Fourier coefficient tends to zero (the Riemann-Lebesgue Lemma), and (3) that a Fourier series is integrable term by term" [19, p. 485]. Building on this remark, it would seem beneficial for this note to provide a more inclusive history of Fourier series convergence results. Dirichlet (who was a Ph.D. student of both Fourier and the famous complex function theorist Simeon Poisson, and who had, from as early as his teenage years, such well-known mathematicians and physicists for teachers and co-authors as Georg Ohm and Adrien-Marie Legendre) proved any function whose derivative is continuous has a Fourier series that converges to it everywhere. Dirichlet thought the result could be extended to any continuous function, and he struggled to prove it but failed. Many other mathematicians tried, including Weierstrass, Riemann, and Dedekind, who each publicly stated the conjecture that the Fourier series for a continuous function would converge at any point. The correct fact, surprising the mathematical world, came in 1876 when Paul David Gustav du Bois-Reymond constructed a continuous function having a Fourier series not convergent at one point. (du Bois-Reymond also proved any trigonometric series that converged to a continuous function f everywhere had to be the unique Fourier series for f.) Many other theorems have emerged. For example, the Hungarian mathematician Lipót Fejér examined a different type of summation, known as Cesàro summation and one that is linked with a probability mean (or weighted average). Using Cesàro summation makes the Fourier series of a continuous function converge everywhere (in fact, in a strong type of convergence that mathematicians call uniform). In 1966, Lennart Carleson proved the brilliant theorem named after him, which says (in a version to which Richard Hunt extended the result in 1968) a periodic function f in a function space L^p (for $p > 1$ and as we will define in Section 3.1) has a Fourier series that converges to f almost everywhere. Long before, in 1922 (when he was only 19 years old), Andrei Kolmogorov had shown the equivalent statement for L^1 false (terribly so!) by creating a function whose Fourier series diverges almost everywhere. He later improved that from almost everywhere to everywhere [73]. The L^1 case was examined for affirmative results—conditions on a function f that would guarantee its Fourier series converges pointwise to f—by the famous British mathematicians G. H. Hardy and J. H. Littlewood, who managed [54] to prove such convergence happened if the partial sums s_n of the function's Fourier series converged fast enough (in exact terms, when $\lim_{n\to\infty} s_n/\log n = 0$). In 1946 (at the Chekiang University in China), Fu Traing Wang generalized and improved Hardy and Littlewood's result to less stringent convergence speeds [119]. In 1966 (at Stanford University), Yitzhak Katznelson proved [72] that for any Lebesgue measure zero real set S, there exists a continuous function f whose Fourier series diverges across S (and perhaps at other points).

In the terrible times of genocide in World War II, there are many horrendous stories of mathematicians who suffered or were killed. (Conservative counts estimate that the Nazi regime killed more than eleven million civilians, including six million Jews, between 1935 and 1945.) Similar to Beppo Levi, Tullio Levi-Cevità and Vito Volterra were two brilliant Jewish mathematicians expelled from the academic community in Italy. Their fate was worse than Levi's, however, because they refused to sign the Mussolini government's 1931 oath to fascism. Volterra was one of only twelve professors who refused to sign (out of more than 1200 who were

Figure 2.11. Five mathematicians who have contributed significant results on the Fourier series convergence problem. From left to right: Carleson, Dirichlet, Jordan, du Bois-Reymond, and Fejér.

so ordered), and Levi-Cevità signed only after adding a note indicating significant reasons why he was personally in opposition. The government almost immediately then forced Volterra to resign from his academic position and enacted so-called race laws in 1938, especially targeted toward repressing Jews, which ousted Levi-Cevità (and Beppo Levi). Levi-Cevità died in Rome in 1941 at the age of 68 under forced isolation from academic discourse or other communication. In 1940, a more elderly Volterra (who was then 80), also died in Rome. Of course, the brutality toward Jews and the punishing treatment of nearly all academic mathematicians and other intellectuals, which Nazi cruelty and class hatred inflicted in certain other parts of Europe, was much worse than these Italian government acts (though they should not be diminished in their level of hatred and violence). For example, the effect on the Polish mathematical community was especially murderous. A further short comment on that sad activity—one of the worst in human history—will appear in Chapter 3's notes. Only a relatively few great Jewish mathematicians in Poland managed to escape persecution during this time. Hugo Steinhaus, who gave the first example of an everywhere-divergent Fourier series whose coefficients tend to zero [68] is, mercifully, one example. Steinhaus, the Ph.D. advisor to Stefan Banach (whose theory we will discuss in Chapter 3), also created a trigonometric series that converges everywhere but does not converge uniformly (a stronger type of convergence) on an interval [68].

In terms of other mathematical applications of the Lebesgue integral, Volterra's work on integral equations will be of particular interest in this text's study. We will investigate in Chapter 5 a few features of integral equations. For example, one type of Volterra equation is of the form $g(x) = f(x) + \int_c^x K(x,t)g(t)\,dt$, where $K(x,t)$ is the kernel function. The equation relates the functions f and g, where f is typically given (we may think of it as an input function) and the problem is to find the corresponding output function g. When the kernel function is of the form $K(x-t)$, then the integral equation is often called a convolution operator. Such equations have many applications. For example, when studying the physical properties of certain materials such as rubber, the Volterra equation relates an amount f of strain placed on the material with the amount g of stress within the material's structure.

Volterra began his studies of integral equations, including the Volterra integral equation, around 1900. In 1908, the Romanian mathematician Traian Lalescu wrote his dissertation on Volterra equations. Three years later, Lalescu wrote the first text on integral equations. The book [77] was widely acclaimed in the mathematical community and is still in print (in French). Other well-regarded texts on integral equations include Francesco Tricomi's [117], Peter Collins' [28], and David Porter and David Stirling's [91]. Section 2.1 mentioned the fact that Gaston Darboux's

Ph.D. students are universally highly regarded and well-known in mathematical circles. One of them, Émile Picard, is best known for his theorems in complex function theory. (One of Picard's theorems states, for example, that an everywhere-differentiable nonconstant complex function outputs every complex number, with possibly one exception.) Picard was also one of Lalescu's professors and his Ph.D. advisor.

We can't end this note without singling out the book that is widely recognized as the classic text on the subject of Fourier series. It is Antoni Zygmund's *Trigonometric Series* [124], written in two volumes and first published in 1935 at a height of excitement over Fejér's new discoveries and his way of thinking about Fourier series. Zygmund was only 35 when he wrote the text, and caught up in the exodus from his native Poland during World War II. He emigrated to the United States and taught at Mount Holyoke College until 1945. He eventually settled as a professor of mathematics at the University of Chicago, where he stayed until his death in 1992. Paul Cohen (mentioned in Section 1.1) was one of his Ph.D. students. Zygmund's most famous work was affectionately called "the Bible" by J. E. Littlewood [69]. The interested reader will want to pick up a copy in the library or buy a new edition (the book is still in print), at least to leaf through the pages to explore with Zygmund many curious and marvelous facts about functions and their trigonometric representations in terms of the Lebesgue integral and function spaces. An excellent review of Zygmund's *Trigonometric Series*, which was published in the 2004 *Bulletin of the AMS*, is also available online [69].

3
Function Spaces

This chapter describes several spaces of functions and their properties. Function spaces provide a setting in which to understand the way functions work, how functions in the space relate to one another, and how they have useful representations (for example, as infinite series in terms of various basis elements of the space). Function spaces also help us think about sophisticated mathematical objects such as *operators* (which map one function in a function space to another, similar to the way a function maps points to other points). In fact, a function space provides an underlying structure to the functions in its collection, in a similar way that we think of a collection of points (or vectors) in a space. The Euclidean space $\mathbb{R}^2$ of points is an example. Just as points in $\mathbb{R}^2$ are a well-defined distance apart, so are functions in certain function spaces a well-defined distance apart. Just as points (or vectors) in $\mathbb{R}^2$ have a magnitude (defined as the distance from the point to the origin), so do functions in certain function spaces have a well-defined magnitude. Just as vectors in $\mathbb{R}^2$ are at well-defined angles with each other, so are functions in certain function spaces at well-defined angles with each other.

This chapter will study function spaces equipped with a norm. The norm measures the size of a function in much the same way the absolute value measures the size of a number. The norm defines the function's *overall* size as it exists as an entity in its own right, which is different from the size of a function at a particular value x. One often used norm is defined in terms of the Lebesgue integral. That's why the study of function spaces began in the early 1900's—Lebesgue's integral provided a tool to define essential function norms. With his integral, such norms are possible and describe many function spaces as complete, in the sense that a convergent sequence of functions in the space converges to a limit function that is also in the space. Those are the types of function spaces we will study. Banach spaces are complete normed linear function spaces. This chapter introduces Banach spaces of functions.

Some Banach spaces have an additional structure, the inner product, that defines the angle between two functions in the space. The inner product then equips the Banach space with a Euclidean geometry that includes well-defined length and angles. We will explore an example of that type of Banach space, labeled a Hilbert space, in Section 3.2. Knowing how to define an angle between two functions will enable us to say, for example, when functions are at right angles with each other, much the same way we know when two lines in a space of points (or vectors) are at right angles with each other.

Each function space in this chapter will be an example of a *metric space*, which is a space equipped with a norm $\| \cdot \|$ (also called a metric). The norm is best thought of as a length, defined for functions in the space, where it must satisfy the properties

1. All functions have positive length except for the zero function, which is the only one with zero length.
2. The length of a difference of functions does not change relative to which function is subtracted from the other.
3. The length of the sum of two functions is never larger than the sum of the lengths of the functions. (This property is the triangle inequality.)

Stating these properties in terms of the norm notation, the metric must satisfy, for any elements f, g, and h of the metric space:

1. $\|f\| \geq 0$, and $\|f\| = 0$ exactly when $f = 0$;
2. $\|f - g\| = \|g - f\|$; and
3. $\|f + g\| \leq \|f\| + \|g\|$.

Figure 3.1. Stefan Banach as a young mathematician in 1919.

Banach spaces are named after the Polish mathematician Stefan Banach, who proved many foundational results about them. For example, the Hahn-Banach theorem describes a bounded linear functional. When defined on a subspace, the theorem says a bounded linear functional always extends out to the full function space. We will study linear functionals in Chapter 5. Banach is considered to be one of the founders of modern functional analysis, not only for the fundamental theorems that he proved, but also because he wrote in 1931 the first treatise on operator theory [8], a detailed and clear account (especially as it dates to the beginnings of the subfield's development) of how function spaces and associated operators illuminate the study of functions defined on infinite-dimensional spaces. The text has long been considered one of Banach's most important publications. It is where he first defined Banach spaces (which he called "spaces of type B"). Published in French just a year after its original circulation, it was translated to English in 1987. It is still in print.

After introducing Banach and Hilbert spaces associated with the Lebesgue integral, this chapter looks at infinite-dimensional function spaces that have an orthonormal basis structure. This additional structure allows for a general theory of Fourier series to be constructed. The chapter's last section presents an application of a function space's orthonormal basis to solve a family of differential equations used in quantum physics.

3.1 The Spaces L^p

An important Banach space is the L^p space. Actually, each real value $p \geq 1$ produces a different space. (There is also a space L^∞.) The L^p spaces share many properties. For any p with $1 \leq p < \infty$, the L^p norm is defined as $\|f\| = (\int |f|^p)^{1/p}$. For example, when $p = 1$ the norm of L^1 is $\|f\| = \int |f|$, and when $p = 2$, the norm of L^2 is $\|f\| = \sqrt{\int |f|^2}$. Any function in an L^p space must have finite L^p norm.

3.1 The Spaces L^p

One more condition helps determine when a function gains membership in an L^p space. It turns out that some functions behave badly, in ways that lead to problems when trying to take their Lebesgue integral. They are nonmeasurable functions. Functions that do not behave this way are measurable. We'll soon rigorously define the two categories of functions in terms of underlying properties. Because the Lebesgue integral is required to define L^p spaces (it is the fundamental construction for each L^p norm), nonmeasurable functions will have to be excluded from L^p spaces. In summary, a space L^p contains all measurable functions with finite L^p norm. The case $p = 1$ produces a collection of functions L^1 that is precisely the Lebesgue integrable functions L^1 already defined in Section 1.5.

This section describes L^p spaces. It starts with a discussion of measurable (and nonmeasurable) functions and then produces many examples of functions the L^p spaces contain. Important norm inequalities and other useful theorems conclude the section. The most significant of these properties, the Riesz-Fischer theorem, determines each L^p space as complete—we will think of convergence in terms of the L^p norm, and then any converging sequence of L^p functions will converge to a function in L^p.

3.1.1 Measurable Functions

Some functions behave badly in terms of the Lebesgue integral. To understand why, we'll consider properties that the Lebesgue measure ought to have. Poorly behaved functions will be ones that do not allow the measure to satisfy them. To motivate the Lebesgue measure's first foundational property, we recall the first function for which we defined the Lebesgue integral: a characteristic function of an interval. Section 1.3 defined $\int \mathcal{X}_I = m(I)$ for an interval I. In the same way, and in an expansion of the concept of a set's measure, we will always require the integral of a characteristic function $\int \mathcal{X}_A$ to equal $m(A)$ for any set A, not just for an interval. In fact, we will use this equation as the definition of Lebesgue measure for sets in Chapter 4. The trouble is that not all sets A behave well under this definition. That poor behavior then translates into the corresponding function $\mathcal{X}_A$ not behaving well. In addition, other basic properties of measure have to be satisfied. For example, we always require $m(A) + m(B) = m(A \cup B)$ for two disjoint sets; in fact, we require such a property to hold for any countable number of disjoint sets. Measures with this property are "countably additive," a property axiomatically fundamental to modern integration theory. We also always require $m(A) = m(A + t)$ for a set A and real number t, where $A + t$ is the translation set of A, defined as $A + t = \{a + t : a \in A\}$. Measures with this property are "translation invariant." But some sets A don't behave well in the sense that they do not allow the measure to satisfy one of these properties. They are the nonmeasurable sets.

For a nonmeasurable set A, the corresponding characteristic function $\mathcal{X}_A$ doesn't behave well in terms of the integral (and the foundational properties) and is a nonmeasurable function. The best way to understand this is to look at an example. What follows is a variation on the most famous nonmeasurable function, which was also the first one found. Named after Guiseppe Vitali, the Italian mathematician who constructed it in 1905, it is based on a nonmeasurable set we label V.

Example 3.1.1 **The Vitali Set V:** To construct the Vitali nonmeasurable function, we first construct a nonmeasurable set—the Vitali set V. Start by partitioning the interval $[0, 1]$ into an uncountable number of subsets. Two numbers v and w in $[0, 1]$ get into the same subset when

$v - w$ is a rational number. Hence the rationals in $[0, 1]$ form one subset. And, for example, the two irrational numbers $0.14159\ldots$ (which is $= \pi - 3$) and $0.04159\ldots$ (which is $\pi - 3.1$) belong to another. (These two numbers' difference is 0.1, which is rational. You can see that $\pi - 3.12$ is another element of the subset.) Every value in $[0, 1]$ is in exactly one of the subsets.

Now choose exactly one element from each of these subsets, and collect the choices into a set V. (The ability to do this step follows from the axiom of choice.) It's important to realize V depends on which elements you choose from the subsets. But once you make that choice, V is a beautifully and appropriately defined set. It exists and is bounded since it is contained in the interval $[0, 1]$. The problem, as we will see in a minute, is trying to calculate the Lebesgue measure of V—it can't be done.

To see that fact, list the rationals in $[0, 1]$ as $s_1, s_2, s_3, \ldots$. Take any one of these numbers s_j and form the set $S_j = \{s \oplus s_j : s \in V\}$, where, for two numbers s and s_j, the (modular) sum used is

$$s \oplus s_j = \begin{cases} s + s_j & \text{if } s + s_j \leq 1 \\ s + s_j - 1 & \text{if } s + s_j > 1. \end{cases}$$

We get a different set S_j for each rational s_j. First, the modular sum makes $0 \leq s + s_j \leq 1$, and so each S_j is a subset of $[0, 1]$. Second, each $x \in [0, 1]$ is in some S_j. For if x is in the subset of $[0, 1]$ that corresponds to the choice of $s \in V$, then $|x - s|$ equals some rational s_k. There are two possible cases. First, when $x - s = s_k$, then $x \in S_k$. Second, when $s - x = s_k$, then $s + (1 - s_k) - 1 = x$, and $x \in S_i$ where $s_i = 1 - s_k$. In either case, x is in some S_j. In fact, each $x \in [0, 1]$ is in at most one S_j. For if x were in $S_j \cap S_k$, then there would be two values $a, b \in V$ where (from the definition of S_j and S_k) $x = a + r$ and $x = b + t$ for rationals r and t. Subtracting, $a - b = t - r$, which is rational. But that means a and b are in the same subset originally partitioning $[0, 1]$. Since V consists of only one element from each subset, we must have $a = b$, which then means $t = r$, and so $S_j = S_k$. In short, each x is in exactly one S_j.

Putting the previous three points together implies $[0, 1] = \bigcup_{j=1}^{\infty} S_j$ for disjoint sets S_j. Each set S_j is a translation set of V, and so by translation invariance we should have $m(S_j) = m(V)$, $j = 1, 2, 3, \ldots$. Then by countable additivity, we should have

$$1 = m([0, 1]) = m\left(\bigcup_{j=1}^{\infty} S_j\right) = \sum_{j=1}^{\infty} m(S_j) = \sum_{j=1}^{\infty} m(V).$$

No matter what measure we might assign to $m(V)$, we have problems. If $m(V) = 0$, then this equation would imply $1 = 0$. If $m(V)$ is a positive or negative value, then the summation does not converge to a finite number, and so there would be no way for it to equal 1. It's impossible for the set V to have a correctly assigned measure. V is a nonmeasurable set. And then the characteristic function $f(x) = \mathcal{X}_V(x)$ is nonmeasurable, because the requirement for its Lebesgue integral to produce the measure of V is impossible to achieve: $\int \mathcal{X}_V(x)dx = m(V)$ does not exist. Since $m(V)$ is badly behaved, so is $\mathcal{X}_V$. ■

The reason behind the problems for a nonmeasurable function f is that the preimage V of a measurable set is a nonmeasurable set. In the last example, the Vitali set V was the preimage of $T = \{1\}$. (T is a measurable set because $m(T) = 0$, which is nicely defined as a measure.) A function f that has this problem (whenever there is a measurable set T where f's preimage of T is a nonmeasurable set) will not integrate correctly. These badly behaved functions are the nonmeasurable functions. Hence, just as $f(x) = \mathcal{X}_V(x)$ is nonmeasurable, so is the function $v(x) = \begin{cases} 1 & \text{if } x \in V \\ -1 & \text{if } x \notin V \end{cases}$. We don't want to include nonmeasurable functions in our function space—they wreak havoc.

We call $v(x)$ the *Vitali function*. This section will use it as an example when discussing membership in L^p spaces. Going forward, we will want to characterize measurable functions in a way that will be easy to work with. We've motivated the discussion by showing how nonmeasurable functions simply cannot be eligible for membership in L^p function space. Measurable functions will also be discussed in Chapter 4, when we define the Lebesgue measure of an arbitrary set. (From the last example, you can see why we will frame that definition in terms of measurable characteristic functions.)

The next definition makes the concept of a measurable function precise, though we have already described how measurable functions behave (in that they don't behave the way nonmeasurable functions do). The definition originated in the work of Riesz (see [98, p. 43]) and was further clarified by Weir (see [121, p. 119]). At first glance it will seem to be unrelated to some of the issues just described about measurable functions, but that's the beauty of the definition: it takes a difficult concept and produces a simple way of thinking about it. We'll prove its connections to the measure-theoretic issues already discussed as we proceed.

Definition 3.1.1 *For real-valued functions f and g with $g > 0$, the function $\text{mid}\{-g, f, g\}$ is defined as having the unique range value chosen from the three output values $-g$, f, and g (not necessarily distinct) that is between the other two.*

A real-valued function f is a measurable function exactly when $\text{mid}\{-g, f, g\}$ is in L^1 (is integrable in the sense of Definition 1.5.2) for every nonnegative function $g \in L^1$.

For example, if $f(x) = x$ and $g(x) = \mathcal{X}_{(-2,2)}(x)$, then $\text{mid}\{-g, x, g\}$ is -1 when $-2 < x < -1$, shifts to x when $1 \leq x \leq 1$ (because the function $f(x) = x$ lies between -1 and 1 in this range), shifts to 1 when $2 > x > 1$ (because $f(x) = x > 1$ in this range), and equals 0 otherwise (because then g and $-g$ both equal zero). We illustrate the graph of $\text{mid}\{-1, x, 1\}$ in Figure 3.2. The function $\text{mid}\{-1, x, 1\}$ is in L^1 (which is a requirement for $f(x) = x$ to be a measurable function), since $\text{mid}\{-1, x, 1\}$ is continuous and equals 0 off $(-2, 2)$. We will soon prove (in Theorem 3.1.1 part (ii)) $f(x) = x$ is measurable but $f \notin L^1$ since its integral is unbounded over $\mathbb{R}$.

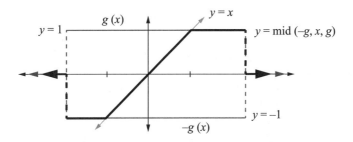

Figure 3.2. The graph of $\text{mid}\{-g, x, g\}$ in bold, superimposed atop $y = x$, $g = \mathcal{X}_{(-2,2)}$, and $-g$.

The hard part about Definition 3.1.1 is its formulation of a measurable function in terms of any nonnegative function $g \in L^1$, a generality that is so unspecific as to be intimidating at best and seemingly unmanageable at worst. Sometimes proofs of results (such as part (i) of the next theorem) smoothly manage this generality and can prove the result using any nonnegative integrable g, but other times (such as part (ii) of the next theorem) they do not. Fortunately, the definition has traction—it turns out we can narrow the consideration of all the functions g to only nonnegative step functions ϕ, or even to step functions of the form $c\mathcal{X}_I$, where c is nonnegative and I is a bounded interval. This ability follows from the LDCT: if $g \in L^1$ is nonnegative, then by Definition 1.5.1 we can find a sequence $\{\phi_n\}$ of nonnegative step functions converging to g almost everywhere, so that $\lim_{n\to\infty} \text{mid}\{-\phi_n, f, \phi_n\} = \text{mid}\{-g, f, g\}$. When the functions $f_n \equiv \text{mid}\{-\phi_n, f, \phi_n\}$ are integrable, then so are the functions $h_n \equiv \text{mid}\{-g, f_n, g\} = \max\{-g, \min\{f_n, g\}\}$. (See Exercise 35 of Section 2.2 for a proof of this.) Algebraically, we always have $|h_n| \leq g$. In addition, $\lim_{n\to\infty} \text{mid}\{-g, f_n, g\} = \text{mid}\{-g, f, g\}$, and so $\lim_{n\to\infty} h_n = \text{mid}\{-g, f, g\}$. The LDCT now applies and shows $\text{mid}\{-g, f, g\}$ is in L^1. We conclude $\text{mid}\{-g, f, g\}$ is integrable for nonnegative $g \in L^1$ whenever $\text{mid}\{-\phi_n, f, \phi_n\}$ is integrable for a nonnegative step function ϕ_n.

The further simplification is to the case where g is of the form $c\mathcal{X}_I$, where c is nonnegative and I is a bounded interval. To achieve it, we need show only the integrability of any $\text{mid}\{-c\mathcal{X}_I, f, c\mathcal{X}_I\}$ implies the integrability of $\text{mid}\{-\phi, f, \phi\}$ for any nonnegative step function ϕ. But that fact follows from Lemma 1.3.1's representation $\phi = \sum_{j=1}^{n} c_j \mathcal{X}_{I_j}$ where each interval I_j is disjoint, along with the algebraic fact that $\text{mid}\{-\phi, f, \phi\}$ then equals $\sum_{j=1}^{n} \mathcal{X}_{I_j} \cdot \text{mid}\{-c_j \mathcal{X}_{I_j}, f, c_j \mathcal{X}_{I_j}\}$. Whenever each term in the summation is integrable, so then must be $\text{mid}\{-\phi, f, \phi\}$.

So now proofs of measurability of f follow only from showing that $\text{mid}\{-c\mathcal{X}_I, f, c\mathcal{X}_I\}$ is integrable, for c a nonnegative constant and I a bounded interval. That's a much easier task than the more general one set before us in Definition 3.1.1. We'll take this easier route to determine a host of function types that are measurable. For example, we will show all continuous functions are measurable. So are all functions f in L^1. Several other results are important (and are presented in the exercises), but (motivated by a study of L^p) we note in particular that $|f|$ is measurable whenever f is, and $|f|^p$ is measurable for $p \in \mathbb{N}$ whenever f is. The following theorem formally lists and proves these facts.

Theorem 3.1.1 *For any real-valued function f:*

(i) If f is integrable, then it is measurable.
(ii) If f is continuous, then it is measurable.
(iii) If f is measurable, then so is $|f|$.
(iv) If f is measurable, then so is $|f|^p$, for any $p = 1, 2, 3, \ldots$.

Proof. To prove (i), assume f and g are integrable with g nonnegative and realize

$$\text{mid}\{-g, f, g\} = \max\{\min\{f, g\}, -g\}.$$

Since f, g, and $-g$ are integrable, so must each of the combinations formed in the minimum and maximum in this expression (see Exercise 35 of Section 2.2). Hence $\text{mid}\{-g, f, g\} \in L^1$.

For (ii), assume f is continuous. Form the function $h = f \cdot \mathcal{X}_I$, where I is a bounded interval. Then h is continuous almost everywhere (except possibly at the endpoints of I). Hence $h \in L^1$ by Theorem 2.1.7. But this fact along with part (i) then implies mid$\{-c\mathcal{X}_I, f, c\mathcal{X}_I\}$ is integrable, where c is a nonnegative constant, since mid$\{-c\mathcal{X}_I, f, c\mathcal{X}_I\}$ = mid$\{-c\mathcal{X}_I, f \cdot \mathcal{X}_I, c\mathcal{X}_I\}$.

For (iii), for any nonnegative constant c and interval I, we have

$$\text{mid}\{-c\mathcal{X}_I, |f|, c\mathcal{X}_I\} = |\text{mid}\{-c\mathcal{X}_I, f, c\mathcal{X}_I\}|.$$

But that function is integrable, since mid$\{-c\mathcal{X}_I, f, c\mathcal{X}_I\}$ is being assumed integrable, and by an application of Theorem 2.2.4.

For (iv), we note $f \cdot g$ is measurable whenever f and g are (see Exercise 18). An application p times of the result in part (iii) proves this more general fact. ∎

Further clarification of why we need every function in an L^p space to be measurable is provided by the next theorem, which provides a converse to Theorem 2.2.4.

Theorem 3.1.2 *Assume f is measurable. If $|f| \leq g$ where g is Lebesgue integrable, then f is Lebesgue integrable. Furthermore, if $|f|$ is Lebesgue integrable, then so is f.*

Proof. The first statement immediately follows from the fact that mid$\{-g, f, g\} = f$ whenever $|f| \leq g$. Therefore, whenever f is measurable, $f = $ mid$\{-g, f, g\}$ is integrable. The second statement then follows from the first, setting $g = |f|$. In short, if $|f| \in L^1$, then so is f when f is measurable. ∎

3.1.2 Definition of L^p

An L^p Banach space provides a measurement of a function's size. It is analogous to the way we think about sizes of points (how far away the points are from zero) or of vectors. For real numbers in $\mathbb{R}$, that structure is the absolute value $|x|$, the distance from a real number x to 0. The absolute value gives us lots of information. We say x is finite when $|x| < \infty$. The distance between real numbers x and y is $|x - y|$. The sequence $\{x_n\}$ converges to L when, given $\varepsilon > 0$, there exists $N > 0$ with $|x_n - L| < \varepsilon$ whenever $n > N$. All kinds of issues about real numbers are put in terms of the absolute value.

The same can be said about functions: we can use the Lebesgue integral to define a function's size, as long as it is measurable. The size has the properties of a metric as described in the introduction to this chapter. You can think of the metric, commonly written $\|f\|$ and called the *norm of f*, as measuring the distance from the function f to the zero function $g(x) \equiv 0$. It's not a pointwise distance $|f(x) - 0|$, which might differ at different x values, but an overall distance. Think of f as a single entity in the space of functions, whose single-valued distance from the function $g(x) \equiv 0$ is $\|f\|$.

How do we define the norm? Actually, there are many ways to do so, and each produces a different space of functions: a different collection to study and consider both overall and in terms of specific issues that a function theorist might want to investigate. The most widely considered of the spaces are the Lebesgue spaces L^p, where $p \geq 1$ is real. A function in an L^p space is assumed measurable, so the norm (constructed using the Lebesgue integral) is always well-defined. Part (i) of Theorem 3.1.1 provides a type of converse to this statement; namely, a function f in L^1 (for example) must be measurable. An L^p space has norm $\|f\| = (\int |f|^p)^{1/p}$. We will soon show this norm has the required properties outlined for norms in this section's introduction. A subscript notation on the norm symbol, writing it as $\|f\|_p$, indicates it is the L^p-norm. For example, L^2 has norm $\|f\|_2 = \sqrt{\int |f|^2}$.

The formula for the norm lets us define the collection of functions: we put a (measurable) function in L^p when its norm is finite. The following definition makes this notion more precise.

Definition 3.1.2 *For real $p \geq 1$, the function space L^p consists of all measurable functions f with finite L^p-norm. That is,*

$$L^p = \{f : f \text{ is measurable and } \|f\|_p < \infty\},$$

where $\|f\|_p = (\int |f|^p)^{1/p}$. An important note: an L^p space never distinguishes between two functions that are equal almost everywhere. They are understood to represent a single member of the function space.

For example, L^1 consists of measurable functions having finite norm, which translates to $\|f\|_1 = \int |f| < \infty$. Since a measurable function f has a finite Lebesgue integral exactly when $|f|$ does, Theorem 3.1.2 implies the space L^1 defined in this way (via the L^1-norm) is contained in the space of integrable functions L^1 in Definition 1.5.2. But because an integrable function is measurable, part (i) of Theorem 3.1.1 and Theorem 2.2.4 imply the space of integrable functions L^1 in Definition 1.5.2 is contained in the space L^1 defined via the L^1-norm. In conclusion, the two function spaces are the same. The use of the notation "L^1" is consistent.

As mentioned in Definition 3.1.2, if $f = g$ a.e. in L^p, then we think of them as the same function. The functions may differ, but we understand they represent the same element of an L^p space. The reason for this convention is that the norm, as defined by the Lebesgue integral, cannot distinguish between them: their Lebesgue integrals would be the same. When using the norm to determine distance, we'll see there is therefore no distance between two such functions.

The next example illustrates membership in L^1 for several functions.

Example 3.1.2 For a positive real number K, examine the truncated exponential function $f(x) = e^{-x}$, where $0 \leq x \leq K$ (and $f = 0$ otherwise). We know f is measurable, since it is continuous for x not 0 or K. Then the L^1 norm of f is

$$\|f\|_1 = \int_0^K |e^{-x}| \, dx = \int_0^K e^{-x} dx = -e^{-x} \Big|_0^K = 1 - e^{-K}.$$

Since this is finite, $f \in L^1$.

We can similarly define, for each positive integer n, $f_n(x) = e^{-x}$ when $0 \leq x \leq n$ (and $f_n = 0$ otherwise). The calculation just made shows these functions have integrals $\int f_n = 1 - e^{-n}$ bounded above by 1, and each f_n is in L^1 with $\|f_n\|_1 = 1 - e^{-n}$. Using the monotone convergence theorem to pass the limit across the integral sign, $1 = \lim_{n \to \infty} 1 - e^{-n} = \lim_{n \to \infty} \int f_n = \int \lim_{n \to \infty} f_n = \int_0^\infty e^{-x} dx = \int_0^\infty |e^{-x}| dx$, and so $f(x) = e^{-x}$, $x \geq 0$ is also an element of L^1, having $\|f\|_1 = 1$. ∎

How might we think about the distance between two functions f and g in L^p? Without a development of function spaces, we had to think very simply about this concept, defining distance in a straightforward but unsophisticated way. For example, distance could be thought

3.1 The Spaces L^p

of as the pointwise difference $|f(x) - g(x)|$, which generally changes from one point x to another. But now sizes of L^p functions f and g need not be in terms of range values at individual points, but as entities all to themselves in the space L^p. Correspondingly, the notion of distance between f and g can be much more sophisticated: defined by the norm, we can set the L^p-distance between function space elements f and g as the constant overall measurement $\|f - g\|_p = (\int |f - g|^p)^{1/p}$.

Example 3.1.3 Examine $f(x) = x$ and $g(x) = x^2$ over the interval $(0, 1)$. (Assume f and g vanish off $(0, 1)$.) What is the L^1-distance between f and g?

Use the norm to find this distance: $\|f - g\|_1 = \int |f - g| = \int_0^1 |x - x^2| dx$. (You can see how the L^1 distance between two functions is essentially just the area between them.) Since $x \geq x^2$ for $0 \leq x \leq 1$, the term inside the absolute value is nonnegative. Integrating, the distance is $(x^2/2 - x^3/3)\big|_0^1 = 1/2 - 1/3 = 1/6$. A similar calculation shows the L^1-distance between $f(x) = x$ and $h(x) = x^3$ (where the functions are also assumed to be supported only on $(0, 1)$) is $1/4$. Therefore, in L^1 and restricted to the interval $(0, 1)$, we consider x^2 to be closer to x than is x^3. ∎

The last example considered functions supported on the interval $(0, 1)$ (and vanishing elsewhere). Restricting attention to functions as they are defined only on an interval is common. An often-used approach considers the L^p space as functions only over a given interval. Such spaces are $L^p(a, b)$ (read aloud as "L p on a b"). The $L^p(a, b)$-norm is then $\|f\|_p = (\int_a^b |f(x)|^p dx)^{1/p}$. Similar notation works when the interval is closed or half-open. Example 3.1.2 showed $f(x) = e^{-x} \in L^1(0, \infty)$. The next question gives practice using L^p-norms to calculate sizes of functions and distances between them.

Question 3.1.1 Find the L^p-norm for the following functions in the given function space. Also, give a reason to justify each function's measurability.

1. $f(x) = \cos x$ in $L^1(-\pi, \pi)$
2. $h(x) = x^4$ in $L^2(0, 1)$
3. $g(x) = \cos x$ in $L^2(-\pi, \pi)$
4. $m(x) = 1/x^4$, $x \geq 1$ in $L^1(\mathbb{R})$

Now find the L^p-distance between the functions in the given function space:

5. $f(x) = \cos x$ and $g(x) = 1$ in $L^1(-\pi, \pi)$.
6. $f(x) = 2\sqrt{x}$ and $g(x) = 1 + x^2$ in $L^2(0, 4)$. ∎

To this point, the introduction to L^p spaces has focused on examples and questions that illuminate the definition in terms of the L^p-norm. We still need to show why $\|f\|_p$ satisfies the mathematical properties of a norm. Namely, for a constant c and functions f and g in an L^p space:

$$\|f\|_p \geq 0, \text{ and } \|f\|_p = 0 \text{ exactly when } f = 0;$$

$$\|c \cdot f\|_p = |c| \cdot \|f\|_p; \text{ and}$$

$$\|f + g\|_p \leq \|f\|_p + \|g\|_p \text{ (the triangle inequality).}$$

Theorem 3.1.3 will prove the properties. The proof will use the familiar triangle inequality: for $a, b \in \mathbb{R}$, $|a + b| \leq |a| + |b|$. It will also use Hölder's inequality.

Lemma 3.1.1 **Hölder's Inequality:** *Assume $p, q > 1$ satisfy $\frac{1}{p} + \frac{1}{q} = 1$. In this case, $f \in L^p$ and $g \in L^q$ implies that $f \cdot g \in L^1$ and $\int |f \cdot g| \leq \|f\|_p \cdot \|g\|_q$.*

Proof. The proof begins with an algebraic fact: for constants $a, b \geq 0$ and any λ, $0 < \lambda < 1$, $a^\lambda b^{1-\lambda} \leq \lambda a + (1-\lambda)b$.[1] Substituting $a = (\frac{|f|}{\|f\|_p})^p$, $b = (\frac{|g|}{\|g\|_q})^q$, $\lambda = 1/p$ (and hence $1 - \lambda = 1/q$), and integrating gives

$$\frac{1}{\|f\|_p \|g\|_q} \int |fg| \leq \frac{1}{p}\left(\left(\frac{1}{\|f\|_p}\right)^p \int |f|^p\right) + \frac{1}{q}\left(\left(\frac{1}{\|g\|_q}\right)^q \int |g|^q\right) = \frac{1}{p} \cdot 1 + \frac{1}{q} \cdot 1 = 1.$$

Multiplying both sides by $\|f\|_p \|g\|_q$, Hölder's inequality follows. ∎

Hölder's inequality is one of the most useful inequalities when working with function spaces linked to the integral. We apply it in the proof of the next theorem.

Theorem 3.1.3 *On an L^p space with $p \geq 1$, $\|f\|_p = (\int |f|^p)^{1/p}$ defines a norm.*

Proof. When $p = 1$, the result follows immediately from straightforward calculations and an application of the triangle inequality and other properties of absolute value. We leave it as an exercise. For any other value of p, $\|f\|_p \geq 0$ follows from the fact that $|f|^p \geq 0$, and so the pth root of its integral is nonnegative. When $\|f\|_p = 0$, then $|f|^p = 0$ a.e. by Corollary 2.2.1, and so $f = 0$ a.e. The last line of Definition 3.1.2 now applies: because $f = 0$ almost everywhere, we consider it as the zero function. Functions in an L^p space are indistinguishable if they are equal almost everywhere.

The second property of the norm follows in a straightforward way:

$$\|cf\|_p = \left(\int |cf|^p\right)^{1/p} = \left(|c|^p \int |f|^p\right)^{1/p} = |c| \cdot \|f\|_p \text{ for any constant } c \text{ and } f \in L^p.$$

To prove the third property[2] for $p > 1$, assume $f, g \in L^p$. Then so is $f + g$ because, algebraically, $|f+g|^p \leq 2^p(|f|^p + |g|^p)$. In addition, $|f+g|^p \leq |f||f+g|^{p-1} + |g||f+g|^{p-1}$. Hence

$$\int |f+g|^p \leq \int |f||f+g|^{p-1} + \int |g||f+g|^{p-1}.$$

Applying Hölder's inequality to each term on the right-hand side,

$$\int |f||f+g|^{p-1} \leq \|f\|_p \cdot \|(f+g)^{p-1}\|_q \text{ and } \int |g||f+g|^{p-1} \leq \|g\|_p \cdot \|(f+g)^{p-1}\|_q.$$

Here, $\|(f+g)^{p-1}\|_q = (\int |f+g|^{(p-1)q})^{1/q} = (\|(f+g)\|_p)^{p/q}$ because $q(p-1) = p$. Putting it together,

$$(\|(f+g)\|_p)^p = \int |f+g|^p \leq (\|f\|_p + \|g\|_p)(\|(f+g)\|_p)^{p/q}.$$

---

[1] This algebraic fact can be seen by examining the function $\sigma(t) = (1-\lambda) + \lambda t - t^\lambda$, $t \geq 0$, which has derivative $\sigma'(t) = \lambda(1 - t^{\lambda-1})$, and is hence has a minimum value at $t = 1$. Since $\sigma(t) \geq \sigma(1) = 0$, $(1-\lambda) + \lambda t \geq t^\lambda$. Replacing t by a/b establishes the fact for $b \neq 0$. (It is trivial when $b = 0$.)

[2] The corresponding third property, which we think of as analogous to the triangle inequality, is Minkowski's inequality when discussed for the L^p-norm. It is named after Hermann Minkowski, who also provided the first mathematical understanding of space-time, which is the geometry that Einstein used to describe special and general relativity.

3.1 The Spaces L^p

Dividing both sides by $(\|(f+g)\|_p)^{p/q}$ and noting $p - p/q = 1$, this says $\|f+g\|_p \le \|f\|_p + \|g\|_p$. ∎

Example 3.1.4 We illustrate Theorem 3.1.3 through direct calculation, showing the $L^2(0, 1)$-norm satisfies the three norm properties when $f(x) = x$, $g(x) = x^2$, and $c = -2$.

Solution. First, $\|f\|_2 = (\int_0^1 |x|^2 dx)^{1/2} = \sqrt{1/3} > 0$. Similarly, $\|g\|_2 = (\int_0^1 |x^2|^2 dx)^{1/2} = \sqrt{1/5} > 0$. Second, $\|c \cdot f\|_2 = (\int_0^1 |-2x|^2 dx)^{1/2} = (\int_0^1 4x^2 dx)^{1/2} = 2\sqrt{1/3} = c\|f\|_2$. A similar calculation holds for $\|c \cdot g\|_2$, which turns out to be $2\sqrt{1/5}$. Third, Minkowski's inequality is satisfied:

$$\|f+g\|_2 = \left(\int_0^1 |x + x^2|^2 dx\right)^{1/2} = \left(\int_0^1 x^2 + 2x^3 + x^4 dx\right)^{1/2} = \sqrt{1/3 + 1/2 + 1/5}$$
$$= \sqrt{31/30} = \sqrt{930}/30 < 30.5/30 < (17.3 + 13.4)/30 < (\sqrt{300} + \sqrt{180})/30$$
$$= (5\sqrt{3} + 3\sqrt{5})/15 = \sqrt{1/3} + \sqrt{1/5} = \|f\|_2 + \|g\|_2. \quad \blacksquare$$

Question 3.1.2 Show the $L^2(-\pi, \pi)$-norm satisfies the norm properties when $c = 1/\sqrt{2}$, $f(x) = \cos x$, and $g(x) = \sin x$. ∎

3.1.3 The L^p Spaces Are Complete

Just as convergence of a sequence of real numbers is defined in terms of the distance norm, so can we define convergence of function sequences in L^1 in terms of the L^1-norm. It differs from the pointwise convergence of Definition 1.4.4. This subsection shows it is complete: a sequence of L^p functions that are getting arbitrarily closer together (in the sense of the L^p-norm) converges to a function in the L^p space. The next definition formally presents this idea.

Definition 3.1.3 *We say a sequence of L^p functions $\{f_n\}_{n=1}^\infty$ converges to a limit function $f \in L^p$ when the following happens:*

Given $\varepsilon > 0$, there exists $N > 0$ such that $\|f_n - f\|_p < \varepsilon$ whenever $n > N$.

In this case, we write $\lim_{n \to \infty} f_n = f$ and we call f the L^p strong limit or the norm limit of the sequence f_n in L^p. (This convergence is also called, by some textbooks, convergence in mean.)

Example 3.1.5 We show the L^1 strong limit of $\{f_n(x)\} = \{x^n\}$, where $n = 1, 2, 3, \ldots$, is $f(x) = 0$.

Solution. $\|f_n - 0\|_1 = \int_0^1 |x^n - 0| dx = \int_0^1 x^n dx = x^{n+1}/(n+1)\big|_0^1 = 1/(n+1)$. Therefore, given $\varepsilon > 0$, we can choose $N = 1/\varepsilon - 1$, obtaining, whenever $n > N$

$$\|f_n - 0\|_1 = 1/(n+1) < 1/(N+1) = \varepsilon.$$

A similar calculation shows that the same is true for $p > 1$: For $p > 1$, $\{f_n(x)\}_{n=1}^\infty = \{x^n\}_{n=1}^\infty$ is a sequence of functions in $L^p(0, 1)$ and has L^p strong limit equal to $f(x) = 0$. We leave the proof as an exercise. ∎

Question 3.1.3 Graph $\{f_n(x)\} = \{n\mathcal{X}_{[0, \frac{1}{n^3}]}(x)\}$ for $n = 1, 2, 3, 4$. Show that $\{f_n\}_{n=1}^\infty$ converges to $f(x) = 0$ in the L^2-norm limit. Does the result hold generally in any L^p-norm limit, for $p \ge 1$? ∎

The next theorem is the main result of this section. It shows the well-behaved power of the L^p-norm limit for sequences of functions in an L^p space. The result was first proven in 1907 independently by Riesz and Ernst Fischer (see [98, p. 59] for details). The proof shows how the L^p-norm metric plays the analogous role of the absolute value (when thinking about pointwise limits) in the norm limit statement.

The proof of the Riesz-Fischer theorem depends upon the analytic fact in the following lemma. (See [121, p. 169].)

Lemma 3.1.2 *Let $f_k \in L^p$ for $k = 1, 2, 3, \ldots$. If $\sum \|f_k\|$ converges as a real-valued series, then $\sum f_k$ converges to a function g in L^p. Moreover, $\sum f_k(x)$ converges to $g(x)$ pointwise for almost all x.*

Proof. Let $\sum \|f_k\|_p = M$ and define the partial sums $s_n = \sum_{k=1}^{n} |f_k|$ for $n \geq 1$. Then by Theorem 3.1.3 and the triangle inequality (for any metric), $\{s_n\}$ is a nondecreasing sequence of nonnegative functions in L^p and $\|s_n\|_p \leq \sum_{k=1}^{n} \|f_k\|_p$. Thus $\|s_n\|_p \leq M$, which means $\int s_n^p \leq M^p$. By the monotone convergence theorem, the nondecreasing sequence $\{s_n^p\}$ must converge a.e. to a (nonnegative) function $h \in L^1$, and $\int h \leq M^p$. Thus $\{s_n\}$ converges a.e. to a nonnegative measurable function s, where $s^p = h$. This fact means $s \in L^p$, $\|s\|_p \leq M$, $\sum_{k=1}^{\infty} |f_k(x)|$ converges a.e. to s, and hence $\sum_{k=1}^{\infty} f_k(x)$ also converges (pointwise) a.e. to a function g that satisfies[3] $|g| \leq s$. By Theorem 3.1.1, g is measurable, and so $g \in L^p$ (with $\|g\|_p \leq M$). To show the convergence in L^p norm, define the partial sums $r_n = \sum_{k=1}^{n} f_k$ and note $\lim_{n \to \infty} r_n - g = 0$ a.e. By the triangle inequality, $|r_n - g| \leq |r_n| + |g| \leq 2s$, and so the Lebesgue dominated convergence theorem applies: $\lim_{n \to \infty} \int |r_n - g|^p = \int \lim_{n \to \infty} |r_n - g|^p = 0$. Hence $\|r_n - g\| \to 0$, which means $\sum f_k$ converges to g in norm. ∎

We now turn to the famous Riesz-Fischer theorem. Its proof discusses convergence for a sequence of L^p functions getting arbitrarily closer together in the sense of any L^p-norm. The closeness is expressed by saying the sequence is a *Cauchy sequence*. A Cauchy sequence $\{f_n\}$ satisfies: given $\varepsilon > 0$, there exists $N > 0$ so that $\|f_n - f_m\|_p < \varepsilon$ whenever $m, n \geq N$. Descriptively, the functions are within an arbitrarily small value ε of each other (where distance is measured by the L^p-norm) when they are out in the tail end of the sequence (when both m and n are at least N).

Theorem 3.1.4 The Riesz-Fischer theorem: *If $\{f_n\}$ is a Cauchy sequence of functions in an L^p space with $p \geq 1$ (so that, given $\varepsilon > 0$, there exists N so that $\|f_n - f_m\|_p < \varepsilon$ whenever $m, n \geq N$), then f_n converges to a function $f \in L^p$ in the L^p-norm limit.*

---

[3] A standard fact in real analysis is that if $\sum |a_n|$ converges, then so does $\sum a_n$.

3.1 The Spaces L^p

Proof. Let $\{f_n\}$ be a Cauchy sequence. Choosing $\varepsilon = 2^{-k}$ for $k = 1, 2, 3, \ldots$, we can find integers N_k (where $N_k < N_{k+1}$) so that

$$\|f_n - f_{N_k}\|_p < 2^{-k}, \text{ for } n \geq N_k.$$

Then the series

$$\|f_{N_1}\|_p + \|f_{N_2} - f_{N_1}\|_p + \|f_{N_3} - f_{N_2}\|_p + \cdots$$

is convergent by the comparison test. (It is dominated by $\|f_{N_1}\|_p + \sum_{k=1}^{\infty} 2^{-k} = \|f_{N_1}\|_p + 1$.)

We now apply Lemma 3.1.2, which says $f_{N_1} + f_{N_2} - f_{N_1} + f_{N_3} - f_{N_2} + \cdots$ converges in L^p-norm to a function f. But that means $\|f_{N_k} - f\|_p \to 0$ as $k \to \infty$, since the partial sums telescope.

Hence, given $\varepsilon > 0$, we may find k so that $\|f_{N_k} - f\|_p < \varepsilon$ and also $2^{-k} < \varepsilon$. Then, using the triangle inequality, $\|f_n - f\|_p \leq \|f_n - f_{N_k}\|_p + \|f_{N_k} - f\|_p < 2 \cdot \varepsilon$ for $n \geq N_k$. This shows f_n converges to f in L^p-norm. ∎

Example 3.1.6 Show the sequence of $L^2[0, 1]$ functions $\{f_n(x)\} = \{x/n\}$ is a Cauchy sequence that converges strongly (in L^1-norm) to a function $f \in L^1$. Identify the strong limit function f.

Solution. Assume $n > m$ and let $\varepsilon > 0$ be given. $\|f_n - f_m\| = \int_0^1 |f_n - f_m| = \int_0^1 (x/m - x/n) dx = 1/(2m) - 1/(2n) \leq 1/(2m) + 1/(2n)$ by the triangle inequality. Then $1/(2m) + 1/(2n) < \varepsilon$ when n and m are both larger than $1/\varepsilon$. Hence the sequence is Cauchy in the L^1-norm. By the Riesz-Fischer theorem, the sequence converges in L^1-norm to a function f in L^1. We identify it as $f(x) = 0$, since $\|f_n - 0\| = \int_0^1 x/n \, dx = 1/(2n)$, and so $\lim_{n \to \infty} \|f_n - 0\| = 0$ as required. ∎

Question 3.1.4 Show the sequence

$$\chi_{[0,1]}, \chi_{[0,1/2]}, \chi_{[1/2,1]}, \chi_{[0,1/2^2]}, \chi_{[1/2^2,2/2^2]}, \chi_{[2/2^2,3/2^2]}, \chi_{[3/2^2,1]}, \chi_{[0,1/2^3]}, \chi_{[1/2^3,2/2^3]}, \ldots$$

is Cauchy in L^2. Determine its L^2-norm limit and show, as guaranteed by the Riesz-Fischer theorem, the limit function is also in L^2. Does the sequence converge pointwise at any value x? Why or why not? ∎

This section has introduced the Banach spaces L^p, which are useful for studying properties of functions, especially as they relate to one another. The next section will demonstrate why L^2, equipped with norm $\|f\|_2 = \sqrt{\int |f|^2}$, is an especially useful function space. It has not only a norm, but also an inner product that acts the same way the dot product acts in a vector space $\mathbb{R}^n$. The inner product will give us a way to measure the angle between two functions in L^2, and will therefore allow us to think about when functions are orthogonal (or at right angles) to one another. That will lead to an understanding of orthogonal basis elements, which will provide a useful way to generalize the Fourier series that we studied in Section 2.4.

Solutions to Questions

3.1.1 (1) $\|f\| = \int_{-\pi}^{\pi} |\cos x| \, dx = \int_{-\pi}^{-\pi/2} (-\cos x) \, dx + \int_{-\pi/2}^{\pi/2} \cos x \, dx + \int_{\pi/2}^{\pi} (-\cos x) \, dx = 1 + 2 + 1 = 4$. (2) $\|g\|^2 = \int_{-\pi}^{\pi} |\cos x|^2 \, dx = \int_{-\pi}^{\pi} \frac{1}{2}(1 + \cos 2x) \, dx = \frac{1}{4}(2x + \sin 2x) \big|_{-\pi}^{\pi} = \pi$, so $\|g\| = \sqrt{\pi}$. (3) $\|h\|^2 = \int_0^1 |x^4|^2 \, dx = x^9/9 \big|_0^1 = 1/9$, so $\|h\| = 1/3$. (4) Applying either the monotone convergence theorem or the LDCT, $\|m\|^2 = \int_1^\infty |x^{-4}|^2 \, dx = 1/7$, so $\|h\| = 1/\sqrt{7}$. (5) $\|f - g\| = \int_{-\pi}^{\pi} |\cos x - 1| \, dx = \int_{-\pi}^{\pi} (1 - \cos x) \, dx = 2\pi$. (6) $\|f - g\|^2 = \int_0^4 |2\sqrt{x} - (1 + x^2)|^2 \, dx = \int_0^4 (4x - 4\sqrt{x} - 4x^{5/2} + 1 + 2x^2 + x^4) \, dx = 12164/105$. Hence $\|f - g\| = \sqrt{12154/105} \approx 10.76$.

3.1.2 First, $\|f\|^2 = \int_{-\pi}^{\pi} |\cos x|^2 \, dx = \pi$, so $\|f\| = \sqrt{\pi} > 0$, and $\|g\|^2 = \int_{-\pi}^{\pi} |\sin x|^2 \, dx = \pi$, so $\|g\| = \sqrt{\pi} > 0$. Second, $\|c \cdot f\|^2 = \int_{-\pi}^{\pi} \frac{1}{2} |\cos x|^2 \, dx = \pi/2$, so $\|c \cdot f\| = |c| \cdot \|f\|$. Similarly, $\|c \cdot g\|^2 = \int_{-\pi}^{\pi} \frac{1}{2} |\sin x|^2 \, dx = \pi/2$, so $\|c \cdot g\| = |c| \cdot \|g\|$. Finally, $\|f + g\|_2 = \sqrt{\int_{-\pi}^{\pi} |\cos x + \sin x|^2 \, dx} = \sqrt{\int_{-\pi}^{\pi} (1 + 2 \cos x \sin x) \, dx} = \sqrt{2\pi} < 3 < \sqrt{\pi} + \sqrt{\pi} = \|f\| + \|g\|$.

3.1.3 The graphs are step-function boxes with increasing height n and narrowing width $1/n^3$. $\|f_n - 0\|^2 = \int_0^{1/n^3} |n - 0|^2 \, dx = \int_0^{1/n^3} n^2 \, dx = 1/n$, which approaches 0 in the limit. (Given $\varepsilon > 0$, choose $N = 1/\varepsilon$.) The result does not hold in general. For example, $p = 3$ has $\|f_n - 0\|^3 = \int_0^{1/n^3} |n - 0|^3 \, dx = \int_0^{1/n^3} n^3 \, dx = 1$, which cannot be made arbitrarily small, even though the almost-everywhere pointwise limit of each f_n is $f(x) = 0$.

3.1.4 Given $\varepsilon > 0$, choose $N = \log_2(2/\varepsilon)$. Then, for $n, m > N$ and any k and j, we have $\|\mathcal{X}_{[k/2^n, (k+1)/2^n]} - \mathcal{X}_{[j/2^m, (j+1)/2^m]}\|^2 \le 1/2^n + 1/2^m < 2/2^N = \varepsilon$. The L^2-norm limit is $f(x) = 0$ (which is clearly in L^2), since $\|\mathcal{X}_{[k/2^n, (k+1)/2^n]} - 0\|^2 = \int_{k/2^n}^{(k+1)/2^n} dx = 1/2^n$, which approaches 0. But (surprisingly) the sequence does not converge pointwise to any value x, since, for $n \in \mathbb{N}$, there is always a value k for which x is between $k/2^n$ and $(k+1)/2^n$.

Reading Questions for Section 3.1

1. For functions f and g, with $g \ge 0$, how do we define the function mid$\{-g, f, g\}$?

2. When is a function f measurable?

3. What condition can guarantee $|f|$ is measurable?

4. Define L^p for any $p > 1$. What is its norm?

5. What does Hölder's inequality say? Name at least one place it is used to help prove a theoretical result.

6. What does it mean for a sequence $\{f_n\}$ of functions in L^p to be Cauchy?

7. State the Riesz-Fischer theorem. Why is it important?

Exercises for Section 3.1

For Exercises 1–4, find the L^p norm for the given function, considered as a member of the given L^p function space.

1. $f(x) = \cos x$ in $L^2(0, \pi)$

2. $g(x) = x^2$ in $L^2(0, 1)$

3. For $n \in \mathbb{Z}$, $h_n(x) = e^{inx}$ in $L^2(-\pi, \pi)$. (Hint: for $t \in \mathbb{R}$, $e^{it} = \cos t + i \sin t$. Also, $|x + iy|^2 = x^2 + y^2$ for $x, y \in \mathbb{R}$.)

4. $f(x) = x \cdot \cos x$ in $L^1(0, \pi)$

3.1 The Spaces L^p

For Exercises 5–8, find the L^p-norm distance between each pair of functions, considered as members of the L^p function space.

5. $f(x) = x$ and $g(x) = x^2$ in $L^2(0, 1)$

6. $f(x) = \cos x$ and $g(x) = \sin x$ in $L^1(0, \pi)$

7. $f(x) = e^{ix}$ and $g(x) = e^{i2x}$ in $L^2[-\pi, \pi]$. (*Hint:* see the comments included with Exercise 3.)

8. $f(x) = e^x$ and $g(x) = x$ in $L^1(0, 1)$

For Exercises 9–12, show the given function sequence converges in norm to the limit function in $L^2(0, 1)$.

9. $f_n(x) = x^n$ converges to $f(x) = 0$

10. $f_n(x) = 1 - x^n$ converges to $f(x) = 1$

11. $f_n(x) = x/n$ converges to $f(x) = 0$

12. $f_n(x) = \mathcal{X}_{I_n}(x)$ converges to $f(x) = 0$, where $I_n = [n \cdot 2^{-k} - 1, (n+1) \cdot 2^{-k} - 1]$ for k the unique integer satisfying $2^k \leq n < 2^{k+1}$. (The first three functions in the sequence are $\mathcal{X}_{[0,1]}$, $\mathcal{X}_{[0,1/2]}$, and $\mathcal{X}_{[1/2,1]}$.)

In exercises 13–17, calculate norms on an L^p space to show each result.

13. Show the Dirichlet function $D(x)$ as defined in Section 2.1.2 has L^1 norm equal to zero. Why does that allow you to conclude $D = 0$ almost everywhere?

14. On $L^1(-1, 1)$, $\| - 5 \cdot x^3 \|_1 = |-5| \cdot \|x^3\|_1$.

15. On $L^2(-1, 1)$, $\|\sqrt{2} \cdot x^2\|_2 = \sqrt{2}\|x^2\|_2$.

16. On $L^1(0, 1)$, $\|x + x^2\|_1 \leq \|x\|_1 + \|x^2\|_1$.

17. On $L^2(0, 1)$, $\|1 + x\|_2 \leq \|1\|_2 + \|x\|_2$.

Prove the statements in Exercises 18–20.

18. If f and g are measurable, then so is $f \cdot g$. Use the following outline:
    (a) Enumerate the rationals as $\{r_k\}_{k=1}^\infty$. Show, for $t \in \mathbb{R}$, that $\max\{2r_k t - r_k^2 : 1 \leq k \leq n\} \to t^2$ as $n \to \infty$.
    (b) If f is a measurable function, then use the last result to show f^2 is also measurable.
    (c) Assume f and g are measurable. Use the last result along with the polarization identity
    $$f \cdot g = \frac{1}{4}[(f+g)^2 - (f-g)^2]$$
    to prove $f \cdot g$ is a measurable function. (See, e.g., [121, p. 124].)

19. For a function f, positive constant c, and bounded interval I,
    $$\mathrm{mid}\{-c\mathcal{X}_I, |f|, c\mathcal{X}_I\} = |\mathrm{mid}\{-c\mathcal{X}_I, f, c\mathcal{X}_I\}|.$$

20. For $p \in \mathbb{N}$, prove $\{f_n(x) = x^n\}_{n=1}^{\infty}$ is a sequence of functions in $L^p(0, 1)$ and has L^p strong limit equal to $f(x) = 0$.

3.2 The Hilbert Space Properties of L^2 and ℓ^2

This section looks at two function spaces, L^2 and ℓ^2, that have an extra structure beyond those for other L^p spaces (such as the norm, linearity, and strong limits). The additional structure is an inner product. It has many marvelous properties. The inner product will be well-defined on L^2 and ℓ^2 but is not, for example, on other L^p spaces. The inner product is analogous to the dot product (the scalar product) that a multivariate calculus course describes for finite-dimensional vector spaces. The inner product will be in many ways much more powerful. In fact, Section 3.3 will use it as a foundation to the structure of L^2 (or ℓ^2), in a way that will allow us to think of L^2 as infinite-dimensional, with basis elements (functions) that can be constructed through repeated applications of the inner product. The following definition allows us to begin this journey in regards to L^2.

Definition 3.2.1 *The inner product of two functions f and g in $L^2(\mathbb{R})$ is $\langle f, g \rangle = \int_{-\infty}^{\infty} f(x)g(x)\,dx$.*

In $L^2(\mathbb{R})$, the inner product $\langle f, g \rangle$ is a real number. It allows us to think of L^2 as a generalization of two-dimensional Euclidean space, which is the familiar Cartesian plane. (In the same way, L^2 is a generalization of three-dimensional Euclidean space, where points are determined by three Cartesian coordinates. Any finite number n of dimensions works the same way.) We can see the generalization beginning to appear through properties of the inner product and resulting geometric properties of the space. For example, a version of the Pythagorean theorem holds for functions in L^2. We will prove that. Simpler, fundamental properties of the inner product on L^2 need to be listed first, as they are used to prove more advanced results.

Fundamental Properties of the Inner Product on L^2

For real functions $f, g, h \in L^2$ and real constants a and b:

1. $\langle f, f \rangle = \|f\|^2$,
2. $\langle f, f \rangle \geq 0$ and equals 0 only when $f = 0$ a.e.,
3. $\langle f, g \rangle = \langle g, f \rangle$, and
4. $\langle af + bg, h \rangle = a\langle f, h \rangle + b\langle g, h \rangle$.

The properties follow immediately from the inner-product definition, except for the second one, which is a restatement of Theorem 3.1.3 (in regards to the first property of a norm). Exercises 1–3 deal with the proofs of Properties 1, 3, and 4. Another property, the Schwarz inequality, is a special case of Hölder's inequality.

Theorem 3.2.1 The Schwarz Inequality: *For $f, g \in L^2$, $|\langle f, g \rangle| \leq \|f\| \cdot \|g\|$.*

3.2 The Hilbert Space Properties of L^2 and ℓ^2

Proof. By Theorem 2.2.4, $|\langle f, g \rangle| = |\int f(x)g(x)\,dx| \leq \int |f(x)g(x)|\,dx$. Now apply Hölder's inequality with $p = 2$. ∎

The Schwarz inequality shows the inner product can be used to define an angle between two nonzero elements f and g of L^2. Dividing by $\|f\| \cdot \|g\|$, we see $\dfrac{|\langle f, g \rangle|}{\|f\|\|g\|} \leq 1$, and so $-1 \leq \dfrac{\langle f, g \rangle}{\|f\|\|g\|} \leq 1$. We can therefore define a unique angle θ, where $0 \leq \theta \leq \pi$, that satisfies

$$\cos \theta = \frac{\langle f, g \rangle}{\|f\|\|g\|}.$$

The angle θ is interpreted as the angle between the nonzero functions f and g as elements of the L^2 space.

The concept of an angle between nonzero functions establishes a geometry for the space. For example, we can determine when two functions f and g are at right angles, in a way that is analogous to when two lines in Euclidean space are at right angles. In this case, θ would equal $\pi/2$ and $\cos \theta$ would be $\cos \pi/2 = 0$, resulting in $0 = \dfrac{\langle f, g \rangle}{\|f\|\|g\|}$. In short, $\langle f, g \rangle = 0$ when f and g are at right angles. The next example and question begin to explore such issues, and they illustrate some straightforward inner product calculations for functions in L^2.

Example 3.2.1 Find the inner product of $f(x) = x^2$ when $x \in [0, 2]$ (and $f = 0$ otherwise) and $g(x) = 2x + 1$ when $x \in [1, 3]$ (and $g = 0$ otherwise). Then find the angle between f and g (when these functions are thought of as elements of L^2). Also, find a function h of the form $h(x) = ax + b$, $x \in [0, 2]$, that is at right angles with f.

Solution. The multiplicative product of f and g is $f(x)g(x) = x^2(2x+1) = 2x^3 + x^2$ when $x \in [1, 2]$ (and $fg = 0$ otherwise). Hence $\langle f, g \rangle = \int_1^2 f(x)g(x)\,dx = \int_1^2 (2x^3 + x^2)\,dx = (x^4/2 + x^3/3)\big|_1^2 = 8 + 8/3 - 1/2 - 1/3 = 59/6$. Since the norms of f and g are $\|f\| = (\int_0^2 |x^2|^2\,dx)^{1/2} = (\int_0^2 x^4\,dx)^{1/2} = \sqrt{32/5}$ and $\|g\| = (\int_1^3 |2x+1|^2\,dx)^{1/2} = (\int_1^3 4x^2 + 4x + 1\,dx)^{1/2} = (4x^3/3 + 2x^2 + x\big|_1^3)^{1/2} = \sqrt{36 + 18 + 3 - 4/3 - 2 - 1} = \sqrt{158/3}$, the angle θ between f and g satisfies $\cos \theta = \dfrac{\langle f, g \rangle}{\|f\|\|g\|} = \dfrac{59/6}{\sqrt{32/5}\sqrt{158/3}} \approx .5356$.

Therefore (as an angle between 0 and π radians formed by f and g), $\theta \approx \arccos(.5356) \approx 58$ degrees, or about $\theta = .32\pi$ radians. Finally, if the linear function $h(x) = ax + b$ is at right angles with f, then $\langle f, h \rangle = 0$. That implies $0 = \int fh = \int_0^2 x^2 \cdot (ax + b)\,dx = (ax^4/4 + bx^3/3)\big|_0^2 = 4a + 8b/3$. Solving for b gives $b = -3a/2$. In conclusion, any nonzero h of the form $h(x) = a(x - 1.5)$, such as $h(x) = x - 1.5$, is perpendicular as an L^2 element to $f(x) = x^2$ for $x \in [0, 2]$. ∎

Question 3.2.1 Find the inner product of $f(x) = 3x^2 + 1$ when $x \in [1, 3]$ (and $f = 0$ otherwise) and $g(x) = 2x$ when $x \in [1, 3]$ (and $g = 0$ otherwise). What is the angle between f and g when they are thought of as elements of L^2? Finally, what function h of the form $h(x) = ax + b$ when $x \in [1, 3]$ is at right angles (when thought of as elements of L^2) with f? ∎

We have seen $\langle f, g \rangle = 0$ when f and g are at right angles as elements of L^2. That fact has many wonderful ramifications. Several geometric properties about L^2 follow, and a host of them describe the geometric structure of L^2 as similar to the familiar geometry of Euclidean space.

The next two theorems are powerful examples. Each of their titles indicate the well-known Euclidean property they imitate.

Theorem 3.2.2 The Pythagorean Theorem: *If two L^2 functions f and g are at right angles, then*
$$\|f\|^2 + \|g\|^2 = \|f + g\|^2.$$

Proof. Use the fundamental properties of the inner product to expand the right-hand side:
$$\|f + g\|^2 = \langle f+g, f+g \rangle = \langle f, f+g \rangle + \langle g, f+g \rangle$$
$$= \langle f, f \rangle + \langle f, g \rangle + \langle g, f \rangle + \langle g, g \rangle = \langle f, f \rangle + 2\langle f, g \rangle + \langle g, g \rangle.$$

Since f and g are at right angles, $\langle f, g \rangle = 0$, and the result follows. ∎

Theorem 3.2.3 The Parallelogram Law: *For functions $f, g \in L^2$,*
$$\|f + g\|^2 + \|f - g\|^2 = 2\|f\|^2 + 2\|g\|^2.$$

Proof. Use the fundamental properties of the inner product to expand each term on the left-hand side:
$$\|f \pm g\|^2 = \langle f \pm g, f \pm g \rangle = \langle f, f \rangle \pm 2\langle f, g \rangle + \langle g, g \rangle.$$

Now add the expressions together (the middle terms have opposite signs and will cancel out) to obtain the result. ∎

Some explanation may be needed as to why these last two theorems are suggestively named after familiar geometric laws. The Euclidean geometry's Pythagorean theorem is a rule about right triangles. How can functions be thought of in relationship to each other according to right triangles? The answer lies in thinking of functions as vectors, objects that have both direction and length. Vectors may be represented as arrows in diagrams. The simple diagrams can shed great light on geometric relationships between vector elements of the space. In this way, the directions of functions f and g in L^2, thought of as vectors, may be determined in relationship to one another: the angle between f and g is θ, where $\cos\theta = \langle f, g \rangle/(\|f\| \cdot \|g\|)$. Also, the vector that originates at the base of f and runs to the tip of g may be interpreted as the function $f + g$. Hence f, g, and $f + g$ form a triangle (in the plane of functions formed by the vectors f and g) as in Figure 3.3 (a).

As we've already realized, the angle between f and g is a right angle when $\langle f, g \rangle = 0$. In this case, we interpret the Pythagorean Theorem (the square of the hypotenuse equaling the sum of the squares of the legs) as $\|f + g\|^2 = \|f\|^2 + \|g\|^2$, which is precisely the equation guaranteed in Theorem 3.2.2. Similarly, the parallelogram law from Euclidean geometry says twice the sum of the squares of the sides of a parallelogram equals the sum of the squares of the parallelogram's diagonals (the line segments that connect opposite corners). In Figure 3.3(b), we interpret the function $-f$ as a vector of the same length but in opposite direction to f. Since $\|f\| = \|-f\|$ (and the same for g), and since the diagonals outlined by the functions

3.2 The Hilbert Space Properties of L^2 and ℓ^2

Figure 3.3. In (a), the functions f, g, and $f+g$ form a triangle of vectors in L^2. In (b), they (along with $-f$, $-g$, and $f-g$) form a parallelogram and its diagonals.

$f, g, -f$, and $-g$ may be interpreted as the vectors $f+g$ and $f-g$, the law manifests itself as the equation in Theorem 3.2.3: $\|f+g\|^2 + \|f-g\|^2 = 2\|f\|^2 + 2\|g\|^2$.

Any normed, linear, inner product space that is complete is a *Hilbert space*. Named after the German mathematician David Hilbert, a Hilbert space possesses the key features of Euclidean space that allow techniques of calculus to apply in general ways. For example, the completeness requirement stipulates that limits of Cauchy sequences exist in the space—thereby guaranteeing a richness of limits that allows calculus techniques to apply, and a geometry of Euclidean space to provide structure.

Just as L^2 is defined with an inner product (and hence is a Hilbert space), so also is $L^2(a, b)$ for an interval $(a, b) \in \mathbb{R}$. That inner product is $\langle f, g \rangle = \int_a^b f(x) \cdot g(x)\, dx$, for any $f, g \in L^2(a, b)$. This formulation satisfies the fundamental properties of an inner product; for example, $\langle f, f \rangle = \|f\|^2$ for f in the function space $L^2(a, b)$. The following definition describes $L^2(a, b)$ as a Hilbert space, just as Definition 3.2.1 did for L^2.

Definition 3.2.2 *The inner product of functions f and g in $L^2(a, b)$ is*

$$\langle f, g \rangle = \int_a^b f(x) \cdot g(x)\, dx.$$

As for any Hilbert space inner product, in $L^2(a, b)$ the angle θ between f and g satisfies

$$\|f\| \cdot \|g\| \cdot \cos\theta = \langle f, g \rangle.$$

Though f and g are real-valued, L^2 spaces of (measurable) complex-valued functions also exists. Without going into the details of the development, we note that the inner product is defined as $\langle f, g \rangle = \int f \cdot \overline{g}$, where $\overline{g}$ is the complex conjugate of g. Using only real-valued functions, the next example illustrates the inner product construction on a space $L^2(a, b)$.

Example 3.2.2 Let $f(x) = 1$ and $g(x) = \sin x$ in $L^2(0, \pi)$. Then $\langle f, g \rangle = \int f \cdot g = \int_0^\pi 1 \cdot \sin x\, dx = 2$. Furthermore, since $\|f\| = (\int_0^\pi 1^2\, dx)^{1/2} = \sqrt{\pi}$ and, similarly, $\|g\| = \sqrt{\pi/2}$, the angle θ formed by f and g satisfies $\sqrt{\pi} \cdot \sqrt{\pi/2} \cdot \cos\theta = 2$. Solving, $\cos\theta \approx .9003$, or $\theta \approx .143\pi$ radians.

Similarly, let $f(x) = x$ and $g(x) = x^2 + 1$ in $L^2(0, 1)$. Then $\langle f, g \rangle = \int_0^1 x \cdot (x^2 + 1)\, dx = x^4/4 + x^2/2 \,\big|_0^1 = 3/4$. Since $\|f\| = \sqrt{3}/3$ and $\|g\| = \sqrt{28/15}$, the angle θ satisfies $\sqrt{3}/3 \cdot \sqrt{28/15} \cdot \cos\theta = 3/4$. Solving, $\theta \approx \pi/10$ radians. ∎

Question 3.2.2 In 1–4, determine the values of the inner products on $L^2(0, 1)$. Then find an approximate radian measurement for the angle between the two functions when thought of as elements of $L^2(0, 1)$.

1. $\langle x^2, x+1 \rangle$ 2. $\langle x+1, x+1 \rangle$ 3. $\langle \sin x, x \rangle$ 4. $\langle e^x, x+1 \rangle$ ∎

There are many other Hilbert spaces besides L^2 and $L^2(a, b)$.

Example 3.2.3 The space ℓ^2 is the collection of infinitely long vectors with complex number entries of the form $\vec{v} = \begin{bmatrix} a_0 \\ a_1 \\ a_2 \\ \vdots \end{bmatrix}$. To get into ℓ^2, $\vec{v}$ must have finite norm $\|\vec{v}\| = (\sum_{j=0}^{\infty} |a_j|^2)^{1/2} < \infty$. Elements of ℓ^2 are called "square-summable" in recognition of $\sum_{j=0}^{\infty} |a_j|^2 < \infty$. The inner product of any real-valued vectors $\vec{v}$ and $\vec{w} = \begin{bmatrix} b_0 \\ b_1 \\ b_2 \\ \vdots \end{bmatrix}$ is $\langle \vec{v}, \vec{w} \rangle = \langle \begin{bmatrix} a_0 \\ a_1 \\ a_2 \\ \vdots \end{bmatrix}, \begin{bmatrix} b_0 \\ b_1 \\ b_2 \\ \vdots \end{bmatrix} \rangle = \sum_{j=0}^{\infty} a_j \cdot b_j$. When the vectors have complex entries, the inner product is $\langle \vec{v}, \vec{w} \rangle = \langle \begin{bmatrix} a_0 \\ a_1 \\ a_2 \\ \vdots \end{bmatrix}, \begin{bmatrix} b_0 \\ b_1 \\ b_2 \\ \vdots \end{bmatrix} \rangle = \sum_{j=0}^{\infty} a_j \cdot \overline{b}_j$, where $\overline{b}_j$ is the complex conjugate of b_j.

If $\vec{v} = \begin{bmatrix} 1 \\ 1/2 \\ 1/6 \\ \vdots \\ 1/j! \\ \vdots \end{bmatrix}$ and $\vec{w} = \begin{bmatrix} 2 \\ -1 \\ 2/3 \\ \vdots \\ 2(-1)^{j+1}/j \\ \vdots \end{bmatrix}$, then both vectors are elements of ℓ^2 because they have finite norm. $\|\vec{v}\| = (\sum_{j=0}^{\infty} 1/(j!)^2)^{1/2}$, which is finite by the ratio test. $\|\vec{w}\| = (\sum_{j=0}^{\infty} |(2(-1)^{j+1}/j)|^2)^{1/2} = 2(\sum_{j=0}^{\infty} 1/j^2)^{1/2}$, which has a convergent p-series with $p = 2$. The inner product of $\vec{v}$ and $\vec{w}$ is

$$\langle \vec{v}, \vec{w} \rangle = \sum_{j=0}^{\infty} \frac{1}{j!} \cdot \frac{2(-1)^{j+1}}{j} = 2 \sum_{j=0}^{\infty} \frac{(-1)^{j+1}}{j \cdot j!}.$$ ∎

Question 3.2.3 Show, for complex a with $|a| < 1$, the vector $\vec{k}_a = \begin{bmatrix} 1 \\ a \\ a^2 \\ a^3 \\ \vdots \end{bmatrix}$ is an element of ℓ^2. (Hint: a geometric series shows $\|\vec{k}_a\|$ is finite.) ∎

It turns out the space ℓ^2 is easily expressed as equivalent to the Hardy space H^2, named after the British mathematician G. H. Hardy. We will discuss equivalent Hilbert spaces in later sections. For example, two spaces that are equivalent in a way that preserves inner products for corresponding elements will be called isometrically isomorphic. In this case, ℓ^2 turns out to be isometrically isomorphic to the space of complex functions analytic in a region containing 0. (The functions are the ones that have a Maclaurin power series in the region.) The equivalence will be realized by expressing an element $\vec{v} = \begin{bmatrix} a_0 \\ a_1 \\ \vdots \end{bmatrix}$ as the function $f(z) = \sum_{j=0}^{n} a_j z^n$. For example, the element $\vec{k}_a$ in the last example (where $|a| < 1$) corresponds to the analytic function $f(z) = \sum_{j=0}^{n} a^n z^n = 1/(1-az)$, where $|z| < 1/|a|$. Since $|a| < 1$, $f(z)$ is always well-defined in terms of this infinite geometric series for $|z| < 1$. We will study H^2 further in Chapter 5.

This section developed many examples of a Hilbert space: $L^2(a, b)$, L^2, and the isometrically isomorphic spaces ℓ^2 and H^2. As Hilbert spaces have an inner product, they possess a geometric

3.2 The Hilbert Space Properties of L^2 and ℓ^2

structure similar to that found in any n-dimensional Euclidean vector space. We will later develop Hilbert space generalizations of the L^2 spaces, after we develop a more general version of measure. These more general Hilbert spaces will be useful; they will provide a framework for applications in both Chapters 4 and 5.

Solutions to Questions

3.2.1 $\langle f, g \rangle = \int_1^3 (3x^2 + 1) \cdot 2x \, dx = \int_1^3 (6x^3 + 2x) \, dx = 128$. Also, $\|f\| = (\int_1^3 |3x^2 + 1|^2 \, dx)^{1/2} = (\int_1^3 9x^4 + 6x^2 + 1 \, dx)^{1/2} = 12\sqrt{17/5}$ and $\|g\| = (\int_1^3 |2x|^2 \, dx)^{1/2} = (\int_1^3 4x^2 \, dx)^{1/2} = 2\sqrt{26/3}$. The angle θ between f and g satisfies $\cos \theta = \frac{\langle f, g \rangle}{\|f\| \|g\|} = \frac{128}{12\sqrt{17/5} \cdot 2\sqrt{26/3}} \approx .9825$. So $\theta \approx \arccos(.9825) \approx 10.73$ degrees (around $.06\pi$ radians). Finally, $0 = \langle f, h \rangle = \int_1^3 (3x^2 + 1)(ax + b) \, dx = 64a + 28b$, so $b = -16a/7$. Thus any nonzero $h(x) = a(x - 16/7)$ works.

3.2.2 (1) $\langle x^2, x + 1 \rangle = \int_0^1 (x^3 + x^2) \, dx = 7/12$. Hence $\frac{\langle x^2, x+1 \rangle}{\|x^2\| \|x+1\|} = \frac{7/12}{\sqrt{1/5}\sqrt{7/3}} \approx .8539$. So $\theta \approx \arccos(.8539) \approx 0.5473 \approx .174\pi$. (2) $\langle x + 1, x + 1 \rangle = \|x+1\|^2$, so $\frac{\langle x+1, x+1 \rangle}{\|x+1\| \|x+1\|} = 1$, and $\theta = \arccos(1) = 0$. Of course there is a zero-degree angle between identical functions! (3) $\langle \sin x, x \rangle = \int_0^1 x \sin x \, dx = \sin 1 - \cos 1 \approx .3012$. Hence $\frac{\langle \sin x, x \rangle}{\|\sin x\| \|x\|} \approx \frac{.3012}{.7226\sqrt{1/2}} \approx .5895$, and $\theta \approx \arccos(.5895) \approx 0.94034 \approx .299\pi$. (4) $\langle e^x, x + 1 \rangle = \int_0^1 (x + 1)e^x \, dx = e$. Hence $\frac{\langle e^x, x+1 \rangle}{\|e^x\| \|x+1\|} \approx \frac{e}{\sqrt{(e^2-1)/2}\sqrt{7/3}} \approx .9956$, and $\theta \approx 0.9341 \approx .297\pi$.

3.2.3 $\|\vec{k}_a\|^2 = \sum_{j=0}^{\infty} |a^j|^2 = \sum_{j=0}^{\infty} (|a|^2)^j = \frac{1}{1-|a|^2} < \infty$.

Reading Questions for Section 3.2

1. What is the inner product on L^2? On $L^2(a, b)$ for a real interval (a, b)?

2. What does the Schwarz inequality say about functions f and g in L^2?

3. How does the inner product on a Hilbert space define an angle θ between elements f and g?

4. How do you know $\frac{|\langle f, g \rangle|}{\|f\|_2 \|g\|_2} \leq 1$ for functions f and g in L^2?

5. Name and state two theorems describing geometric properties in L^2 that mimic geometric properties in Euclidean space.

6. Describe the set of elements in the Hilbert space ℓ^2. Give at least two examples.

7. What is the inner product of vectors $\vec{u}$ and $\vec{v}$ in ℓ^2?

Exercises for Section 3.2

In Exercises 1–3, prove the property of the inner product on L^2 as described in Definition 3.2.1.

1. $\langle f, f \rangle = \|f\|^2$

2. $\langle f, g \rangle = \langle g, f \rangle$

3. $\langle af + bg, h \rangle = a \langle f, h \rangle + b \langle g, h \rangle$

In Exercises 4–13, find the values of the inner product $\langle f, g \rangle$ and the norms $\|f\|$ and $\|g\|$ in the given L^2 space. Then determine an approximate radian measurement for the angle between f and g, as the functions are thought of as elements of the Hilbert space.

4. $f(x) = \sqrt{2x^2 + 1}$ and $g(x) = x$ in $L^2(0, 4)$

5. $f(x) = 1$ and $g(x) = \sqrt{x - 1}$ in $L^2(1, 5)$

6. $f(x) = \sin x$ and $g(x) = \cos x$ in $L^2(-\pi, \pi)$

7. $f(x) = \sin(2x)$ and $g(x) = \sin x$ in $L^2(-\pi, \pi)$

8. $f(x) = e^{ix}$ and $g(x) = e^{2ix}$ in $L^2(-\pi, \pi)$, where $i = \sqrt{-1}$. (The exponential function involving the constant i in the power integrates just as does the exponential function with a real constant in the power. For example, $\int e^{ix} \, dx = (1/i)e^{ix} + C = -ie^{ix} + C$.)

9. $f(x) = e^{inx}$ and $g(x) = e^{imx}$ in $L^2(-\pi, \pi)$, where m and n are integers with $m \neq n$. (*Hint:* see the note at the end of the last exercise.)

10. $f(x) = x$ and $g(x) = \sin x$ in $L^2(-\pi, \pi)$

11. $f(x) = x^n$ and $g(x) = x^m$, where $m, n \in \mathbb{N}$ and $n \neq m$ in $L^2(1, 5)$

12. $f(x) = \mathcal{X}_{[1,2]}(x)$ and $g(x) = \sqrt{x-1} \cdot \mathcal{X}_{[1,2]}(x)$ in L^2

13. $f(x) = \mathcal{X}_{[a,b]}(x)$ and $g(x) = \mathcal{X}_{[c,d]}(x)$ in L^2, where $a < b < c < d$

Exercises 14–17 discuss functions on $L^2(-\pi, \pi)$.

14. Find the $L^2(-\pi, \pi)$-norm of 1, $\sin x$, and $\cos x$. (*Hint:* apply the double angle formulas $\cos 2x = 1 - 2\sin^2 x = 2\cos^2 x - 1$.)

15. Show $\langle 1, \sin x \rangle = 0$, $\langle 1, \cos x \rangle = 0$, and $\langle \sin x, \cos x \rangle = 0$.

16. What is the distance between $\sin x$ and $\cos x$ in $L^2(-\pi, \pi)$?

17. As n approaches infinity, show the strong limit of $\frac{\sin(nx)}{n}$ is 0 in $L^2(-\pi, \pi)$.

Exercises 18–25 discuss vectors in ℓ^2.

18. Show $\langle \vec{k}_a, \vec{k}_b \rangle = 1/(1 - ab)$ for $a, b \in \mathbb{R}$ with $|a| < 1$ and $|b| < 1$, where $\vec{k}_a$ and $\vec{k}_b$ are defined as elements of ℓ^2 as in Question 3.2.3.

19. Find the norm of $\begin{bmatrix} 1 \\ 1/2 \\ 1/4 \\ \vdots \\ 1/2^n \\ \vdots \end{bmatrix}$ in ℓ^2.

20. Determine, for $a \in \mathbb{R}$ with $|a| < 1$, $\|\vec{k}_a\|$, where $\vec{k}_a \in \ell^2$ is defined as in Question 3.2.3.

21. Show $\langle \vec{e}_j, \vec{e}_k \rangle = 0$ for $j \neq k$ and $j, k \in \mathbb{N}$ when $\vec{e}_n = \begin{bmatrix} 0 \\ \vdots \\ 0 \\ 1 \\ 0 \\ \vdots \end{bmatrix}$, where the 1 appears in the nth entry of the column vector. (For example, $\vec{e}_1 = \begin{bmatrix} 1 \\ 0 \\ 0 \\ \vdots \end{bmatrix}$.)

22. Using $\cos\theta = \langle f, g\rangle/(\|f\| \cdot \|g\|)$, find the angle θ between $\vec{e}_j$ and $\vec{e}_k$ for $j \neq k$ and $j, k \in \mathbb{N}$, where $\vec{e}_n$ is defined as in the last exercise. Use an approximate radian measurement.

23. Using $\cos\theta = \langle f, g\rangle/(\|f\| \cdot \|g\|)$, find the angle θ between $\vec{k}_{1/2}$ and $\vec{k}_{1/3}$ as elements of ℓ^2, where $\vec{k}_a$ is defined as in Question 3.2.3. Use an approximate radian measurement.

24. Show the column vector $\vec{v} = \begin{bmatrix} 1 \\ 1/3 \\ 1/5 \\ \vdots \end{bmatrix}$ is an element of ℓ^2 by proving it has a finite norm. You do not need to evaluate it.

25. Show the column vector $\vec{w} = \begin{bmatrix} 1/2 \\ 1/4 \\ 1/6 \\ \vdots \end{bmatrix}$ is an element of ℓ^2 by proving it has a finite norm. You do not need to evaluate it. Then determine an infinite series representation for the ℓ^2 inner product $\langle \vec{v}, \vec{w}\rangle$, where $\vec{v}$ is as in the last exercise.

Exercises 26–30 illustrate the Pythagorean theorem and the parallelogram law in Hilbert spaces.

26. Prove the Pythagorean theorem of Theorem 3.2.2 holds in $L^2(0, 1)$ for the functions $f(x) = x^2$ and $g(x) = 3x^2 - 9/5$. Exhibit the values for $\|f\|$, $\|g\|$, and $\|f + g\|$.

27. If $f(x) = \sin x$ and $g(x) = \cos x$ in $L^2(0, 2\pi)$, determine the values for $\|f\|$, $\|g\|$, and $\|f + g\|$, and show the Pythagorean theorem holds.

28. Show the parallelogram law of Theorem 3.2.3 holds in $L^2(0, 1)$ for $f(x) = x^2$ and $g(x) = 3x^3 + 5$.

29. Let $f(x) = e^x$ and $g(x) = x$. What does the parallelogram law say for these functions as elements of $L^2(0, 1)$?

30. For given functions f and g in a Hilbert space $L^2(a, b)$, prove if the Pythagorean theorem is true for f and g, then it is also true for cf and cg, where c is a constant. What would $\langle cf, cg\rangle$ be?

3.3 Orthonormal Basis for a Hilbert Space

A vector space is a collection of objects called vectors that can be multiplied by constants (for example, real numbers) and added together to produce another vector. The constants are called *scalars* since multiplication by any constant scales the vectors, giving them different lengths. A collection of vectors is defined as a full-fledged *vector space* when associated addition and multiplication operations have certain properties. In particular, the addition of vectors has to be commutative and associative. There must also exist a unique zero vector, where $\vec{0} + \vec{v} = \vec{v}$ for any vector $\vec{v}$. The zero vector also allows you to define the negative of a vector, where $\vec{v} + (-\vec{v}) = \vec{0}$. Finally, the operation of multiplication of a vector by a scalar has to distribute properly across addition. If these properties hold, the set of vectors with the operations forms a vector space.

Vector spaces are sometimes equipped with an inner product, which is an operation we defined in Section 3.2 for L^2. An inner product on a vector space with real-valued scalars follows the same fundamental properties listed after Definition 3.2.1. Here's a wonderful property of such a vector space: it has a basis, where you can write any vector in the space as a linear combination of the basis elements. That also holds for complete vector spaces that require an infinite number of vectors to exist in their basis. Symbolically, $\vec{v} = c_1\vec{e}_1 + c_2\vec{e}_2 + \cdots$ for constants c_n and basis elements $\vec{e}_n$, where $n = 1, 2, \ldots$; we will define carefully what the sum's equality means. This section describes bases formally. It focuses on the concept of a basis for a Hilbert space (thought of as a vector space). It will also show how bases for L^2 can be constructed. Elements of an L^2 space (in fact, of any Hilbert space) will turn out to have a useful (infinite series) representation in terms of the space's basis. The representation will form a straightforward generalization of the Fourier series representation for functions that was described in Section 2.4.

3.3.1 A Vector Space Basis

What is meant by a basis for a vector space? You may have studied (in multivariate calculus or linear algebra) the structure that a basis puts on a finite-dimensional vector space. Our discussion here will include those situations, but we will want to consider infinite-dimensional structures, which we can apply to L^2 and other Hilbert spaces that help us study functions. The concept of a basis therefore needs to work in that higher-dimensional situation, too, and that leads to the next definition. It perhaps will look different from a definition of a basis you may have seen for finite-dimensional vector spaces, but it very effectively works there, too.

Definition 3.3.1 *A set of elements $\{b_n\}$ in a vector space equipped with an inner product is orthogonal when $\langle b_m, b_n \rangle = 0$ for two distinct elements b_m and b_n. An orthogonal basis for a Hilbert space $\mathcal{H}$ is a set of orthogonal elements $\{b_n\}$, where each $b_n \in \mathcal{H}$ has the property:*

For $f \in \mathcal{H}$, if $\langle f, b_n \rangle = 0$ for every vector b_n in the set, then $f = 0$.

The basis is orthonormal if, in addition, each basis element has $\langle b_n, b_n \rangle = 1$. The (well-defined) dimension of the Hilbert space is the cardinality of any of its orthogonal basis sets.

The definition can be very easy to use, as the next example illustrates.

Example 3.3.1 A finite-dimensional vector space has an easily identifiable basis. For example, the three-dimensional Euclidean space $\mathbb{R}^3$ is the set of vectors of the form $[a_1, a_2, a_3]$, where $a_n \in \mathbb{R}$. The space is equipped with an inner product (the dot product) defined as $\langle [a_1, a_2, a_3], [b_1, b_2, b_3] \rangle = a_1 b_1 + a_2 b_2 + a_3 b_3$. With this inner product and the norm $\|[a_1, a_2, a_3]\| = \sqrt{a_1^2 + a_2^2 + a_3^2}$, $\mathbb{R}^3$ is a (complete) Hilbert space. A frequently used basis (not unique) is the set $\{[1, 0, 0], [0, 1, 0], [0, 0, 1]\}$. A quick check verifies it is orthogonal under the inner product. Also, for $f = [a_1, a_2, a_3] \in \mathbb{R}^3$, if the inner product of f with the first basis element is zero, then $a_1 = a_1 + 0 + 0 = \langle [a_1, a_2, a_3], [1, 0, 0] \rangle = 0$. Taking the inner product of f similarly with the second and third basis elements, we see a_2 and a_3 would equal zero if these inner products were also zero. Therefore, by Definition 3.3.1, the set forms an orthogonal basis for $\mathbb{R}^3$. ∎

Question 3.3.1 Show $\{\vec{e}_n\}_{n=1}^{\infty}$, where $\vec{e}_n = \begin{bmatrix} 0 \\ \vdots \\ 0 \\ 1 \\ 0 \\ \vdots \end{bmatrix}$ is the column vector with all zero entries except for a single entry of 1 that appears in its nth component, forms a basis for ℓ^2. Say why this implies ℓ^2 has dimension equal to (countable) infinity. Is the basis orthonormal? Why or why not? ∎

3.3.2 The Gram-Schmidt Process

The function space L^2 is a normed vector space, where the functions are the vectors and the L^2-norm is defined as in Definition 3.2.1. As L^2 has an inner product, we can consider if an orthogonal set forms a basis in the sense of Definition 3.3.1. This concept is geometric (the orthogonality describes a property of right angles between the space's vectors), and so it is another aspect of the Euclidean geometry of a Hilbert space. The norm and inner product calculate the distance between vectors and angles between vectors. The inner product geometry then provides an explicit, constructive way to generate a complete set of orthogonal basis elements for the vector space. How does it work? The standard method used is the *Gram-Schmidt process*. The next definition outlines it. This section then comments on its usefulness. Several points first need to be outlined.

Expanding Definition 3.3.1, functions $f_0, \ldots, f_n$ in L^2 are said to be (mutually) orthogonal when the inner product of any two distinct functions equals zero: $\int f_j \cdot f_k = \langle f_j, f_k \rangle = 0$ when $j \neq k$. The Gram-Schmidt process produces mutually orthogonal functions.

Given a finite orthogonal collection of L^2 functions $f_0, \ldots, f_n$, the space of functions

$$M = \{a_0 f_0 + a_1 f_1 + \cdots + a_n f_n : a_1, \ldots, a_n \in \mathbb{R}\}$$

is a (finite) linear subspace of L^2. It forms a linear subspace because any linear combination $af + bg$ of elements f and g and scalars a and b is also an element of the space, and it is closed[4] under the strong limit of Definition 3.1.3. (The norm limit of any convergent sequence of functions in M will also be in M.) The set M is the *span* of $f_0, \ldots, f_n$. The Gram-Schmidt process creates a set of orthogonal functions $f_0, \ldots, f_n$ out of a set of (nonzero) functions in L^2, say $g_0, \ldots, g_n$, where the span of $f_0, \ldots, f_n$ is the same as the span of $g_0, \ldots, g_n$. The following definition describes the process.

Definition 3.3.2 *The **Gram-Schmidt Process** consists of the following algorithm performed on a given finite set of vectors* $\{g_0, \ldots, g_n\}$:

$$\text{Start by letting } f_0 = g_0.$$

$$\text{Then set } f_1 = g_1 - \frac{\langle g_1, f_0 \rangle}{\|f_0\|^2} f_0.$$

$$\text{Next, } f_2 = g_2 - \frac{\langle g_2, f_1 \rangle}{\|f_1\|^2} f_1 - \frac{\langle g_2, f_0 \rangle}{\|f_0\|^2} f_0.$$

Continue this process until $f_0, \ldots, f_n$ *have been defined (the same number of functions as the number of independent functions g in the original set). At each step, use the rule*

$$f_j = g_j - \frac{\langle g_j, f_{j-1} \rangle}{\|f_{j-1}\|^2} f_{j-1} - \frac{\langle g_j, f_{j-2} \rangle}{\|f_{j-2}\|^2} f_{j-2} - \cdots - \frac{\langle g_j, f_0 \rangle}{\|f_0\|^2} f_0.$$

It is not difficult to check that the collection of functions $\{f_j\}$ forms an orthogonal set, so f_j is orthogonal to f_k for $j \neq k$. For example, $f_1 = g_1 - \frac{\langle g_1, f_0 \rangle}{\|f_0\|^2} f_0$, hence $\langle f_1, f_0 \rangle = \langle g_1, f_0 \rangle - \frac{\langle g_1, f_0 \rangle}{\|f_0\|^2} \langle f_0, f_0 \rangle = \langle g_1, f_0 \rangle - \langle g_1, f_0 \rangle = 0$. That's an important property: the Gram-Schmidt process produces a finite collection of functions all at right angles with one another. It

---

[4] See, for example, [86, p. 267].

thereby establishes a geometry for the space—a right-angled orthogonal framework, the same way the Euclidean plane has right-angled orthogonal Cartesian axes.

Furthermore, we can normalize these orthogonal functions, making them all have length one. That manipulation is easy: simply divide each one by its norm, setting $e_j = f_j/\|f_j\|$ for $j = 0, 1, 2, \ldots, n$. We call the resulting collection orthonormal, indicating they are orthogonal to each other (the first part of the word) and normalized to have length one (the second part of the word). So we get a set $\{e_0, e_1, \ldots, e_n\}$ of orthonormal functions in L^2. Because the process can be applied indefinitely to a sequence of functions $g_0, g_1, g_2, \ldots, g_n$, there is no upper limit on n.

Example 3.3.2 We apply the Gram-Schmidt process to the collection of functions $\{g_0, g_1, g_2, g_3\}$ in $L^2(0, \infty)$, where $g_n(x) = x^n e^{-x/2}$ (for $n = 0, 1, 2, 3$) to form an othonormal set of functions $\{e_0, e_1, e_2, e_3\}$.

First, we apply Definition 3.3.2 to find a set of functions $f_0(x)$, $f_1(x)$, $f_2(x)$, and $f_3(x)$ orthogonal in $L^2(0, \infty)$.

We start by letting $f_0(x) = g_0(x) = x^0 e^{-x/2} = e^{-x/2}$. Then we set $f_1 = g_1 - \frac{\langle g_1, f_0 \rangle}{\|f_0\|^2} f_0$. Since $\|f_0(x)\|^2 = \int_0^\infty |e^{-x/2}|^2 \, dx = \int_0^\infty e^{-x} \, dx = 1$, and $\langle g_1, f_0 \rangle = \int_0^\infty x e^{-x/2} \cdot e^{-x/2} \, dx = \int_0^\infty x e^{-x} \, dx = 1$ (see Question 2.2.5), we see $f_1(x) = x^1 e^{-x/2} - (1/1) e^{-x/2} = (x - 1) e^{-x/2}$.

Next, $f_2 = g_2 - \frac{\langle g_2, f_1 \rangle}{\|f_1\|^2} f_1 - \frac{\langle g_2, f_0 \rangle}{\|f_0\|^2} f_0$. Since we have $\|f_1\|^2 = 1$, $\langle g_2, f_1 \rangle = 4$, and $\langle g_2, f_0 \rangle = 2$ (see the next question on the calculation of these values), the next orthogonal function turns out to be $f_2(x) = x^2 e^{-x/2} - (4/1)(x - 1) e^{-x/2} - (2/1) e^{-x/2} = (x^2 - 4x + 2) e^{-x/2}$.

Finally, $f_3 = g_3 - \frac{\langle g_3, f_2 \rangle}{\|f_2\|^2} f_2 - \frac{\langle g_3, f_1 \rangle}{\|f_1\|^2} f_1 - \frac{\langle g_3, f_0 \rangle}{\|f_0\|^2} f_0$. Working out the values for the inner products and norms, $f_3(x) = (x^3 - 9x^2 + 18x - 6) e^{-x/2}$.

The only step that remains is normalization. We have seen $\|f_0(x)\| = 1$, and so f_0 is already normalized. We set $e_0(x) = f_0(x) = e^{-x/2}$. Similarly, $e_1(x) = f_1(x)/\|f_1(x)\| = (x - 1) e^{-x/2}$. Since $\|f_2(x)\| = 2$, we set $e_2(x) = f_2(x)/\|f_2(x)\| = (1/2)(x^2 - 4x + 2) e^{-x/2}$. Finally, it turns out $\|f_3\| = 6$, and so we set $e_3(x) = (1/6) f_3(x) = (1/6)(x^3 - 9x^2 + 18x - 6) e^{-x/2}$.

This example's normalized functions have numerous applications and are the first few terms in a collection of (weighted) Laguerre functions. Each function's polynomial factor is a Laguerre polynomial. (Actually, for technical reasons, the odd Laguerre polynomials are the negatives of the polynomial factors.) For example, we've just shown the degree-two Laguerre polynomial is $L_2(x) = (1/2)(x^2 - 4x + 2)$. Similarly, $L_1(x) = 1 - x$ and $L_3(x) = (1/6)(-x^3 + 9x^2 - 18x + 6)$. We examine Laguerre polynomials further in this section's exercise set and in the next section. ∎

Question 3.3.2 Show the Laguerre functions e_0, e_1, e_2, and e_3 obtained in the last example are orthogonal in $L^2(0, \infty)$. Also show why $\|f_2\|^2 = 4$, $\|f_1\|^2 = 1$, $\langle g_2, f_1 \rangle = 4$, $\langle g_2, f_0 \rangle = 2$, $\langle g_3, f_1 \rangle = 18$, and $\langle g_3, f_2 \rangle = 36$ in that example. ∎

Question 3.3.3 Apply the Gram-Schmidt process to the functions g_0, g_1, and g_2 in $L^2(-1, 1)$ defined as $g_n(x) = x^n$ (for $n = 0, 1, 2$) to obtain functions e_0, e_1, and e_2 that are orthonormal in $L^2(-1, 1)$. These functions (up to a constant multiple) are the first three Legendre polynomials, and the full (infinite) collection of Legendre polynomials is generated as the process is applied for $n = 0, 1, 2, \ldots$. The Legendre polynomials arise in several applied problems. For example, they solve the Legendre ordinary differential equation, which appears when finding the solution

3.3 Orthonormal Basis for a Hilbert Space

to the Laplacian in polar coordinates. (The Laplacian is the partial differential equation whose solutions are the harmonic functions—a type of function that arises in physics.) ∎

The Gram-Schmidt process has many wonderful properties. Though we will prove a few of them, we will describe more. Our hope is to illuminate the powerful geometric framework the Gram-Schmidt process provides to Hilbert spaces and to spark additional interest in learning about the process. The Gram-Schmidt process can be used to develop and prove many Hilbert space theorems.

The Gram-Schmidt process can continue indefinitely: it is defined for a finite collection of functions, but there is no upper bound on their number. We can always add more functions to the collection to produce additional orthonormal functions g_n without any upper limit on n. In this way, we can consider the Gram-Schmidt process to be applied to a countably infinite collection of functions $\{g_0, g_1, g_2, \ldots\}$. For example, we might start with the monomials in an $L^2(a, b)$ space, applying the Gram-Schmidt process to the collection $\{1, x, x^2, x^3, \ldots\}$ to produce a countably infinite orthonormal family of polynomials. This type of construction is useful, as we will see a basis for a general L^2 function space is infinite-dimensional.

Example 3.3.3 Example 3.3.2 described how the Gram-Schmidt process produces the first four Laguerre polynomials when the starting collection of functions are the first four monomials multiplied by $e^{-x/2}$. We could have continued the process, applying it to the countably infinite collection $\{x^n e^{-x/2}\}_{n=0}^{\infty}$. The result is the infinite set of Laguerre functions (and the Laguerre polynomials that they contain as factors). Example 3.3.2 described the first four, which were labeled $e_0, e_1, e_2,$ and e_3. They continue with

$$e_4(x) = (1/24)(x^4 - 16x^3 + 72x^2 - 96x + 24)e^{-x/2},$$
$$e_5(x) = (-1/120)(x^5 - 25x^4 + 200x^3 - 600x^2 + 600x - 120)e^{-x/2},$$
$$e_6(x) = (1/720)(x^6 - 36x^5 + 450x^4 - 2400x^3 + 5400x^2 - 4320x + 720)e^{-x/2}, \text{ etc.}$$

It turns out the nth degree Laguerre polynomial may be defined directly (avoiding the Gram-Schmidt process) as

$$e^{-x} L_n(x) = \frac{1}{n!} \left(\frac{d}{dx}\right)^n (x^n e^{-x}), \ n = 0, 1, 2, \ldots,$$

where the term $(\frac{d}{dx})^n$ means "take the nth derivative." This type of formula is a *Rodrigues formula* and can be constructed for any orthonormal set of polynomials derived from the Gram-Schmidt process on monomials in various L^2 spaces.[5]

A basis is not always formed when applying the Gram-Schmidt process to a countably infinite collection of functions. A marvelous result (which is investigated in the advanced exercises) is that the infinite set of Laguerre functions forms an (infinite-dimensional) orthogonal basis for $L^2(0, \infty)$ in the sense of Definition 3.3.1. ∎

---

[5] Named after the French mathematician Olinde Rodrigues, who developed a general theory of the Rodrigues formula for certain types of differential equations in his 1815 dissertation, a Rodrigues formula always produces a sequence of functional expressions by repeated differentiation of some other function. A set of orthogonal polynomials (that serve, for example, as basis functions for a Hilbert space) can often be generated with a Rodrigues formula, just as for this case of the Laguerre polynomials.

Question 3.3.4 Show the first-degree (weighted) Laguerre polynomial e_1 obtained in Example 3.3.2 is the same, up to a constant, as $L_1(x)e^{-x/2}$ obtained from the definition of $L_n(x)$ given in Example 3.3.3. In other words, show $e_1 = c_1 L_1(x)e^{-x/2}$, where $c_1 \in \mathbb{R}$ and

$$e^{-x} L_1(x) = \frac{1}{1!} \left(\frac{d}{dx}\right)^1 (x^1 e^{-x}).$$

Then show the analogous result for $n = 2$: $e_2 = c_2 L_2(x)e^{-x/2}$. ∎

A Hilbert space with a countable number of basis elements (as opposed to an uncountable number) is *separable*. Our discussions are restricted to separable Hilbert spaces. That is not overly restrictive; in fact, for a long time the definition of a Hilbert space typically included the requirement that it be separable. Real-world phenomena that lend themselves to being modeled with nonseparable Hilbert spaces exist, including the study of specific topics (such as a bosonic subatomic field) in quantum mechanics (see, for example, [114]). These applications have appeared only fairly recently as highly specialized concerns of academic specialties. A broad spectrum of applications lend themselves well to a study of separable Hilbert spaces. And finite-dimensional Hilbert spaces (which are obviously separable) are in and of themselves of great interest in many fields.

3.3.3 The Projection Theorem

Since the Gram-Schmidt process's coefficients are of the form $\dfrac{\langle \_, f_j \rangle}{\|f_j\|^2}$, an approximation theorem emerges: *Given a function f in L^2, there exists a best approximation to f that is in the (closed) span of a collection of functions $g_0, g_1, \ldots, g_n$.* The span is the set of linear combinations: $M = \{a_0 g_0 + \cdots + a_n g_n : a_1, \ldots, a_n \in \mathbb{R}\}$. The best approximation to f by an element g in M minimizes the value of $\|f - g\|$ and is (cf. [86, p. 297])

$$g = c_0 f_0 + c_1 f_1 + \cdots + c_n f_n, \text{ where } c_j = \frac{\langle f, f_j \rangle}{\|f_j\|^2}, \text{ and where the functions } f_0, \ldots, f_n$$

are produced when the Gram-Schmidt process is applied to $\{g_0, g_1, \ldots, g_n\}$.

In this case we say "$f - g \in M^\perp$" (we read this statement as "f minus g is in M-perp"), which means $\langle f - g, h \rangle = 0$ for every $h \in M$. From earlier commentary on the inner product, you know to interpret this inner product is zero by saying, "$f - g$ is orthogonal to every element h in M." In addition, $\|f\|^2 = \|g\|^2 + \|f - g\|^2$. In this way, the Gram-Schmidt process establishes a powerful geometry in a Hilbert space. For any element f of the space and collection of functions $g_0, g_1, \ldots, g_n$, we can use the Gram-Schmidt process to divide the Hilbert space into two subspaces. We label one subspace M, which is the span of the collection of the g_i functions and out of which we can always find a best approximation g to f, and another subspace $M^\perp$, where $f - g$ is in $M^\perp$ and is orthogonal to every element in M. The best approximation g is labeled the "orthogonal projection of f onto M." The construction of the orthogonal subspaces M and $M^\perp$ will have many applications. The next example illustrates the construction.

Example 3.3.4 The function $f(x) = x^2$, $0 < x < 1$ is in $L^2(0, \infty)$. Define a subspace $M = \{a_0 g_0 + a_1 g_1 + a_2 g_2 : a_0, a_1, a_2 \in \mathbb{R}\}$ with $g_n(x) = x^n e^{-x/2}$, $n = 0, 1, 2$. We determine the best approximation to $f(x)$ in M and write $f = g + h$, where $g \in M$ and $h \in M^\perp$.

3.3 Orthonormal Basis for a Hilbert Space

Solution. We first use the Gram-Schmidt process to construct the functions f_0, f_1, and f_2. From Example 3.3.2, we have $f_0(x) = g_0(x) = e^{-x/2}$, $f_1(x) = (x-1)e^{-x/2}$, and $f_2(x) = (x^2 - 4x + 2)e^{-x/2}$. Furthermore, Example 3.3.2 calculated their norms: $\|f_0\| = \|f_1\| = 1$, and $\|f_2\| = 2$.

According to the formula for g, the best approximation of $f(x) = x^2 \mathcal{X}_{(0,1)}$ by a function g in subspace M of $L^2(0, \infty)$ is

$$g = \sum_{n=0}^{2} \frac{\langle x^2 \mathcal{X}_{(0,1)}, f_n(x) \rangle}{\|f_n(x)\|^2} \cdot f_n(x),$$

where the inner product and norm are taken in $L^2(0, \infty)$. The resulting L^2 inner products are

$$\langle x^2 \mathcal{X}_{(0,1)}, e^{-x/2} \rangle = \int_0^1 x^2 e^{-x/2} \, dx = 16 - 26/\sqrt{e},$$

$$\langle x^2 \mathcal{X}_{(0,1)}, (x-1)e^{-x/2} \rangle = \int_0^1 (x^3 - x^2)e^{-x/2} \, dx = 80 - 132/\sqrt{e}, \text{ and}$$

$$\langle x^2 \mathcal{X}_{(0,1)}, (x^2 - 4x + 2)e^{-x/2} \rangle = \int_0^1 (x^4 - 4x^3 + 2x^2)e^{-x/2} \, dx = 416 - 686/\sqrt{e}.$$

The best approximation of $x^2 \mathcal{X}_{(0,1)}$ in M is therefore provided by

$$g(x) = [(16 - 26/\sqrt{e})/1]e^{-x/2} + [(80 - 132/\sqrt{e})/1](x-1)e^{-x/2}$$
$$+ [(416 - 686/\sqrt{e})/4](x^2 - 4x + 2)e^{-x/2}$$
$$= [8(13x^2 - 42x + 18) - (171.5x^2 - 554x + 237)/\sqrt{e}\,]e^{-x/2}.$$

Then we form the function

$$h(x) = f(x) - g(x) = x^2 \mathcal{X}_{(0,1)} - [8(13x^2 - 42x + 18) - (171.5x^2 - 554x + 237)/\sqrt{e}\,]e^{-x/2}$$

as an element of $M^\perp$. We can show $h \in M^\perp$, for example, by showing it is orthogonal in $L^2(0, \infty)$ to the functions $g_n(x) = x^n e^{-x/2}$, $n = 0, 1, 2$. To this end, we note h is orthogonal to g_2 because

$$\langle h, g_2 \rangle = \int_0^1 x^2 \cdot x^2 e^{-x/2} \, dx - \int_0^\infty 8(13x^2 - 42x + 18)e^{-x/2} \cdot x^2 e^{-x/2} \, dx$$
$$+ \int_0^\infty (171.5x^2 - 554x + 237)/\sqrt{e} \cdot e^{-x/2} \cdot x^2 e^{-x/2} \, dx$$
$$= 768 - 1266/\sqrt{e} - 8(13 \cdot 24 - 42 \cdot 6 + 18 \cdot 2)$$
$$+ (171.5 \cdot 24 - 554 \cdot 6 + 237 \cdot 2)/\sqrt{e} = 0.$$

Wow! That calculation looks (and it is) incredible, but it's not just luck—the Gram-Schmidt process sets it up so that it works out. The other two calculations, showing h is orthogonal to g_0 and g_1, follow similarly.

The theory then also implies the minimum value for $\|f - g\|$ for any g in M is the value we get for the best approximation g, where

$$\|f - g\| = \left\{ \int_0^\infty (x^2 \mathcal{X}_{(0,1)} - [8(13x^2 - 42x + 18) \right.$$
$$\left. - (171.5x^2 - 554x + 237)/\sqrt{e}\,] e^{-x/2})^2 \, dx \right\}^{1/2}.$$

An algebraic mess reveals that value as $\|f - g\| = \sqrt{164640/\sqrt{e} - 49919.8 - 135749/e} \approx \sqrt{.1416} \approx .3762$. The details are left for the interested (and hard-working) reader. ∎

Question 3.3.5 In $L^2(0, \infty)$, define the two-dimensional subspace $M = \{a_0 e^{-x/2} + a_1 x e^{-x/2} : a_0, a_1 \in \mathbb{R}\}$.

1. Determine the best $L^2(0, \infty)$ approximation in M to $f(x) = x^2 \mathcal{X}_{(0,1)}$. Then write $f = g + h$ with $g \in M$ and $h \in M^\perp$, and find $\|f - g\|$.
2. Repeat part 1 with $f(x) = x \mathcal{X}_{(0,1)}$. ∎

The previous example and question describe the best approximation construction, through the Gram-Schmidt process, as described in

> **Theorem 3.3.1 The Projection Theorem:** *If M is a closed[6] linear subspace of L^2, then $f \in L^2$ can be uniquely expressed as $f = g + h$, where $g \in M$ and $h \in M^\perp$.*

Proof. Choose the element g of M so that $\|f - g\|$ is smallest. (This choice is possible, even when M is infinite-dimensional, because M is closed.) Then define $h = f - g$ to get the desired representation of f as $f = g + h$.

Now if a function s is in both M and $M^\perp$, then (by the definition of $M^\perp$), $\langle s, s \rangle = 0$, and so by the inner product properties, $s = 0$. Hence $M \cap M^\perp = \{0\}$. This implies the representation for f is unique, for if another such representation is, say $f = \hat{g} + \hat{h}$, then $g - \hat{g} = h - \hat{h} \in M \cap M^\perp$. Because 0 is the only function in both M and $M^\perp$, we see $g = \hat{g}$ and $h = \hat{h}$. ∎

The notation $M^\perp$ and its reading as "M–perp" suggests a geometry that is indeed present. We have essentially broken the Hilbert space L^2 into a union of disjoint subsets: $L^2 = M \cup M^\perp$. The orthogonality of the subspaces provides even more structure: M and $M^\perp$ are essentially at right angles to one another as subspaces, a concept that is understood through the inner product as $\langle g, h \rangle = 0$ for $g \in M$ and $h \in M^\perp$. A crude picture, but one that is helpful and provides accurate intuition, is given in Figure 3.4. Here, M is drawn as a two-dimensional plane (though the picture represents a full generality where M might have any dimension as a subset of L^2) and $M^\perp$ is a one-dimensional linear subspace intersecting M at right angles. The functions f, g, and h are represented as vectors in their appropriate spaces, and, as the projection theorem says, $f = g + h$ with $g \in M$ of minimal length.

---

[6] Here, closed means a Cauchy sequence $\{f_1, f_2, \ldots\}$ of elements in M, where $\|f_n - f_m\|$ gets close to 0 as n and m get large, will converge in the strong limit to a function in M.

3.3 Orthonormal Basis for a Hilbert Space

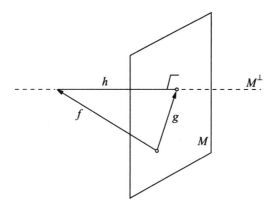

Figure 3.4. For a subspace M of L^2, a function $f \in L^2$ (represented as a vector) decomposes as $f = g + h$, with $g \in M$ minimal and $h \in M^\perp$.

Our examples thus far have looked only at finite-dimensional subspaces M (generated by a finite number of functions). The projection theorem also holds when M is infinite-dimensional. A technical concern when working with infinite-dimensional subsets is that closure is no longer automatic. (It was when M was finite-dimensional.) When using an infinite number of basis elements, the projection theorem requires us to work with a closed subspace. We can guarantee a closed set by adding functions to M that might form strong limits of Cauchy sequences. Adding these limit points to M forms the closure of M, labeled $\overline{M}$, which is closed. Then the projection theorem expresses $f \in L^2$ as $f = g + h$, where $g \in \overline{M}$ and $h \in \overline{M}^\perp$. It is not hard to show $\overline{M}^\perp = M^\perp$ (we leave the proof as an exercise), and so we can always write $f \in L^2$ as $f = g + h$, where $g \in \overline{M}$ and $h \in M^\perp$.

Example 3.3.5 In $L^2(0, 1)$, define $M = \{g(x) : g(x) = c_0 + c_1 x\}$, where c_0 and c_1 are real constants. Use the projection theorem and the Gram-Schmidt process to describe $M^\perp$.

Solution. Functions $g(x)$ in M are linear combinations $g = c_0 g_0 + c_1 g_1$ of $g_0(x) = 1$ and $g_1(x) = x$. Applying the Gram-Schmidt process to $\{g_0, g_1\}$ produces $f_0(x) = 1$ and $f_1(x) = x - \frac{\langle x, 1 \rangle}{\|1\|^2} 1$. Here, $\langle x, 1 \rangle = \int_0^1 x \cdot 1 \, dx = 1/2$, and $\|1\|^2 = \int_0^1 1^2 \, dx = 1$. Hence $f_1(x) = x - 1/2$. Normalizing f_0 and f_1 into functions of $L^2(0, 1)$-length equal to 1, an orthonormal basis for M is formed by $e_1(x) = 1$ and $e_2(x) = \sqrt{3}(2x - 1)$.

We now show, as expected, $M^\perp = \{h(x) : h(x) = f(x) - \langle f, e_1 \rangle e_1 - \langle f, e_2 \rangle e_2,$ where $f \in L^2(0, 1)\}$. The projection theorem implies $h \in M^\perp$ is of the form $f - g$, where $f \in L^2(0, 1)$ and $g \in M$. For such f, the projection theorem's formula for g is $g = \langle f, e_1 \rangle e_1 + \langle f, e_2 \rangle e_2$. Then, for $r = a e_1 + b e_2 \in M$,

$$\langle f - g, r \rangle = \langle f, a e_1 + b e_2 \rangle - \langle g, a e_1 + b e_2 \rangle = a \langle f, e_1 \rangle + b \langle f, e_2 \rangle - a \langle g, e_1 \rangle - b \langle g, e_2 \rangle$$
$$= a \langle f, e_1 \rangle + b \langle f, e_2 \rangle - a \langle \langle f, e_1 \rangle e_1 + \langle f, e_2 \rangle e_2, e_1 \rangle - b \langle \langle f, e_1 \rangle e_1 + \langle f, e_2 \rangle e_2, e_2 \rangle$$
$$= a \langle f, e_1 \rangle + b \langle f, e_2 \rangle - a \langle f, e_1 \rangle \langle e_1, e_1 \rangle - b \langle f, e_2 \rangle \langle e_2, e_2 \rangle = 0. \qquad \blacksquare$$

Question 3.3.6 What is $M^\perp$ for the subspace M of $L^2(0, 1)$ defined as $M = \{f(x) : f(x) = c_0 + c_1 x + c_2 x^2\}$ for real constants c_0, c_1, and c_2? Use the projection theorem and the Gram-Schmidt process to express your result. $\blacksquare$

Infinite-dimensional subspaces M are often important to study, and we have already discussed the projection theorem in terms of such subspaces. In Section 2.4 we saw one example (the standard Fourier series) of how functions in an infinite-dimensional L^2 space can be viewed as an infinite sum of coefficients times basis elements (taken from an infinite collection). Fourier series use basis elements composed of sines and cosines. Now we have a way to think about Fourier series more generally, using whatever orthogonal basis elements we might have that generate the L^2 space. The description of functions in terms of their Fourier series is exactly the type of expression that the projection theorem, applied to possibly infinite-dimensional subspaces M, helps us consider.

For an L^2 space and an orthogonal set of functions $\{f_0, \ldots, f_n\}$ resulting from the Gram-Schmidt process, the values $c_j = \frac{\langle f, f_j \rangle}{\|f_j\|^2}$ are the (generalized) Fourier coefficients for f, taken with respect to the orthogonal set. For $L^2(-\pi, \pi)$, when $f_{2j} = \sin jt$, $j = 1, 2, \ldots$ and $f_{2j+1} = \cos jt$, $j = 0, 1, 2, \ldots$, you get the Fourier coefficients of Section 2.4 for f. That's because $\langle f, f_{2j} \rangle = \int_{-\pi}^{\pi} f(x) \cdot \sin jt \, dt$ and $\|f_{2j}\|^2 = \int_{-\pi}^{\pi} \sin^2 jt \, dt = \pi$. Hence the value for c_j exactly matches the coefficients $(1/\pi) \int_{-\pi}^{\pi} f(t) \cdot \sin jt \, dt$ from the Fourier sine series' basic formulation in Section 2.4.1. The cosine basis elements work similarly to produce the cosine Fourier coefficients from Section 2.4.1. In the same manner and for $L^2(-\pi, \pi)$, when $f_j = e^{ijx}$ with $j = 0, \pm 1, \pm 2, \ldots$, you get the (complex) Fourier coefficients for f that were developed in Section 2.4.

From Definition 3.3.1, we know a (finite) orthonormal basis for M is a set of n distinct, orthogonal functions $e_1, \ldots, e_n$ in M that have $\|e_j\| = 1$, and where the only $f \in M$ orthogonal to all of them is $f = 0$. The Gram-Schmidt process produces such a collection of functions $e_1, \ldots, e_n$, and so they automatically form a basis for the subspace M of their linear combinations. In short, $\{e_1, \ldots, e_n\}$ is a basis set for $M = \{\sum_{j=1}^{n} c_j e_j\}$, for constants c_j. In a similar manner, a complete orthonormal basis for L^2 is a set of orthonormal functions $e_1, e_2, \ldots$ in L^2 such that the closure of their span $\overline{M}$ equals all of L^2. The span of $\{e_j\}_{j=1}^{\infty}$ is the set of linear combinations of the e_j's. It is the subspace $M = \bigvee \{e_j\} = \{f : f = \sum a_j e_j\}$, where the sum is taken over only a finite number of the e_j's (but any finite number of them). The closure of the span, $\overline{M} = \overline{\bigvee \{e_j\}}$, is the smallest closed subset of L^2 that contains M.

These constructs allow us to define the Fourier series of $f \in L^2$ with respect to an orthonormal basis $\{e_n\}_{n=1}^{\infty}$. It is $\sum_{j=1}^{\infty} \langle f, e_j \rangle e_j$. As we have mentioned, the orthonormal basis generates (via its linear combinations) an infinite-dimensional set that is not automatically closed. In that sense, for an orthonormal set $\{e_j\}_{j=1}^{\infty}$, the Fourier series $\sum_{j=1}^{\infty} \langle f, e_j \rangle e_j$ might not converge (in the norm limit) to f. But the projection theorem implies it converges to the nearest available point, g in $\overline{M}$, where M is the subspace of linear combinations of the e_j's. There is good news: for many L^2 spaces, the norm limit convergence of the Fourier series to its function f is automatic. In particular, if we are studying an L^2 space defined on an interval (a, b) of $\mathbb{R}$, such as $L^2(0, \infty)$, then the Fourier series $F_n = \sum_{j=1}^{n} \langle f, e_j \rangle e_j$ formed from an orthonormal basis $\{e_j\}_{j=1}^{\infty}$ does converge in the norm limit to f: we have $\lim_{n \to \infty} \| \sum_{j=1}^{n} \langle f, e_j \rangle e_j - f \| = 0$. The interval (a, b) can be of any type: open or closed, finite, half-infinite, or all of $\mathbb{R}$. And (as is typically done)

3.3 Orthonormal Basis for a Hilbert Space

when the Gram-Schmidt process is used to generate the orthonormal basis, the basis elements are often connected with special function theory, a branch of mathematics. Special functions are often thought of as named functions. The next example illustrates this.

Example 3.3.6 As we have just commented, there exists a complete orthonormal basis for $L^2(a, b)$ depending on what interval (a, b) is studied. For example (when $(a, b) = (0, \infty)$ is the set of positive reals), Example 3.3.3 suggested the Gram-Schmidt process generates a complete orthonormal basis[7] consisting of Laguerre functions (with the Laguerre polynomials a classic type of special function forming the main components of these basis elements). The basis is of the form $\{c_n L_n(x) e^{-x/2}\}_{n=0}^{\infty}$ where c_n is a normalizing constant and $L_n(x)$ is the Laguerre polynomial (of order 0), defined by

$$e^{-x} L_n(x) = \frac{1}{n!} \left(\frac{d}{dx}\right)^n (x^n e^{-x}), \; n = 0, 1, 2, \ldots .$$

Here, $L_0(x) = \frac{1}{0!}(x^0 e^{-x}) \cdot e^x = 1/0! = 1$, $L_1(x) = \frac{1}{1!} \frac{d}{dx}(x^1 e^{-x}) \cdot e^x = (e^{-x} - xe^{-x}) \cdot e^x = 1 - x$, and so on. These functions are identified as particular special functions, arising in the 1800's (the age of "classical mathematics") as satisfying a system of Laguerre differential equations. The differential equations are $xy'' + (1-x)y' + cy = 0$ where c is a real constant (see [111, p. 207]).

The moral of the story: many classical special functions and polynomials arise naturally as the building blocks for orthonormal bases for different $L^2(a, b)$ spaces. The Laguerre polynomials form one such case. This section's exercises and Section 3.4 will investigate other examples. ∎

Be careful: convergence of the Fourier series $\sum_{n=1}^{\infty} \langle f, e_j \rangle e_j$ to the function f happens in the sense of the norm limit within the $L^2(a, b)$ space. There is never a guarantee that the Fourier series will converge to f pointwise. In other words, for $x \in (a, b)$, it is not guaranteed that $\lim_{n \to \infty} \sum_{j=1}^{n} \langle f, e_j \rangle e_j(x) = f(x)$. The notes at the end of Chapter 2 discussed the history of some of the discoveries in this area. For example, the most generally heralded positive result is due to Lennart Carleson in 1966, who proved every periodic function $f \in L^2$ has a Fourier series that converges to f pointwise for almost all x. But even continuous functions are not guaranteed to have a Fourier series that converges pointwise everywhere.

3.3.4 Equivalence of Hilbert Spaces

This last subsection describes when two differently defined Hilbert spaces (they might contain completely different types of elements) can be mapped to one another in a way that shows they are actually equivalent. The map must set up an equivalence of the algebraic structure of the vector space, and must preserve the way the inner product and norm act.

We might name the two Hilbert spaces $\mathcal{H}$ and $\mathcal{K}$. The map that provides the equivalence between them is a Hilbert space isomorphism (also called an isometric isomorphism). The

---

[7] The infinite set of Laguerre functions, as generated by the Gram-Schmidt process, fully generates $L^2(0, \infty)$, but this is not at all obvious. Another way to say this is that $M^{\perp} = \{0\}$ when M is the subspace of linear combinations of the Laguerre functions.

isomorphisms pair each element of the first space with an element of the second so that every element is paired, so a Hilbert space isomorphism T is a one-to-one onto map from $\mathcal{H}$ to $\mathcal{K}$. Also, for h in the first Hilbert space $\mathcal{H}$, $\|T(h)\|_{\mathcal{K}} = \|h\|_{\mathcal{H}}$. Finally, the map is linear: for scalars a and b and $h, g \in \mathcal{H}$, $T(ah + bg) = aT(h) + bT(g)$. In short, the Hilbert space isomorphism preserves the norm structure and the linearity structure of the space $\mathcal{H}$ as its elements are mapped to $\mathcal{K}$.

Definition 3.3.3 *A Hilbert space isomorphism is a one-to-one linear map T from one Hilbert space $\mathcal{H}$ onto another $\mathcal{K}$ that satisfies $\|T(h)\|_{\mathcal{K}} = \|h\|_{\mathcal{H}}$. When such an isomorphism T exists, we say $\mathcal{H}$ and $\mathcal{K}$ are isometrically isomorphic to one another.*

An automatic consequence of a Hilbert space isomorphism preserving the norm structure is its preserving the inner product structure.

Theorem 3.3.2 *A Hilbert space isomorphism T from a Hilbert space $\mathcal{H}$ to another $\mathcal{K}$ satisfies*
$$\langle T(g), T(h)\rangle_{\mathcal{K}} = \langle g, h\rangle_{\mathcal{H}}, \text{ for } g, h \in \mathcal{H}.$$

Proof. The proof follows from the relationship between the norm and the inner product that exists for any element f in any Hilbert space: $\|f\|^2 = \langle f, f\rangle$. Using this and expanding the inner product via its fourth fundamental property,

$$\|g+h\|^2 - \|g-h\|^2 = \langle g+h, g+h\rangle - \langle g-h, g-h\rangle$$
$$= (\langle g,g\rangle + \langle g,h\rangle + \langle h,g\rangle + \langle h,h\rangle) - (\langle g,g\rangle - \langle g,h\rangle - \langle h,g\rangle + \langle h,h\rangle).$$

Applying the inner product's third fundamental property (which says $\langle g, h\rangle = \langle h, g\rangle$) and solving for $\langle g, h\rangle$, we get

$$\langle g, h\rangle = \frac{1}{4}(\|g+h\|^2 - \|g-h\|^2).$$

In the same way, $\langle T(g), T(h)\rangle = \frac{1}{4}(\|T(g) + T(h)\|^2 - \|T(g) - T(h)\|^2)$. But we can now apply the fact that the isomorphism T is linear, as well as the fact that it preserves norms, and so

$$\langle T(g), T(h)\rangle = \frac{1}{4}(\|T(g+h)\|^2 - \|T(g-h)\|^2) = \frac{1}{4}(\|g+h\|^2 - \|g-h\|^2) = \langle g, h\rangle. \blacksquare$$

Not surprisingly, when there exists a Hilbert space isomorphism between two Hilbert spaces, the spaces have the same dimension. Conversely, when two Hilbert spaces have the same finite or countably infinite dimension (in the sense of Definition 3.3.1), then there is a straightforward way to construct a Hilbert space isomorphism between them. Map orthonormal basis elements to orthonormal basis elements. Then every linear combination of basis elements is determined according to the linearity property of the Hilbert space isomorphism. The orthonormality of the basis elements on both the domain and range sides of the map guarantees that the isomorphism preserves the norm. In this way, the mapping of basis elements determines how the isomorphism works in its entirety.

3.3 Orthonormal Basis for a Hilbert Space

Example 3.3.7 Show $L^2(0, \infty)$ is isometrically isomorphic to ℓ^2.

Solution. We form a Hilbert space isomorphism T mapping $L^2(0, \infty)$ onto ℓ^2 by mapping one set of orthonormal basis elements onto the other. Example 3.3.6 identified the orthonormal basis elements of $L^2(0, \infty)$ as $\{e_n = L_n(x)e^{-x/2}\}_{n=0}^{\infty}$, where $L_n(x)$ is the nth normalized Laguerre polynomial. Example 3.2.3 introduced the Hilbert space ℓ^2. Its basis elements are the vectors $\{\vec{w}_n = \begin{bmatrix} 0 \\ \vdots \\ 0 \\ 1 \\ 0 \\ \vdots \end{bmatrix}\}_{n=0}^{\infty}$, where the 1 appears in the nth entry spot, $n = 0, 1, 2, \ldots$. The vectors $\vec{w}_n$ can express any ℓ^2 vector $\vec{v} = \begin{bmatrix} a_0 \\ a_1 \\ a_2 \\ \vdots \end{bmatrix}$ in the corresponding Fourier series: $\vec{v} = \sum_{n=0}^{\infty} a_n \cdot \vec{w}_n$.

Now that the two spaces' corresponding orthonormal bases are identified, the isomorphism T pairs the basis elements: define $T(e_n) = \vec{w}_n$ for $n = 0, 1, 2, \ldots$. And then T's linearity means it extends to any element $f \in L^2(0, \infty)$: if f has Fourier series $f(x) = \sum_{n=0}^{\infty} a_n \cdot L_n(x)e^{-x/2}$, then $T(f) = \sum_{n=0}^{\infty} a_n \cdot T(L_n(x)e^{-x/2}) = \sum_{n=0}^{\infty} a_n \cdot \vec{w}_n = \begin{bmatrix} a_0 \\ a_1 \\ a_2 \\ \vdots \end{bmatrix}$. ∎

More can be said about the Hilbert space isomorphism constructed in the last example. We know, for example, it preserves norms (and hence the square of the norms):

$$\|f\|_{L^2}^2 = \|T(f)\|_{\ell^2}^2, \quad \text{for } f \in L^2(0, \infty).$$

Expanding, if $f \in L^2(0, \infty)$ has Fourier series $f = \sum_{n=0}^{\infty} a_n \cdot L_n(x)e^{-x/2}$, then

$$\int_0^{\infty} |f(x)|^2 \, dx = \sum_{n=0}^{\infty} |a_n|^2.$$

This wonderful fact, which has many theoretical and real-life applications, is *Parseval's identity*. It holds not just for this case of $L^2(0, \infty)$, but in general for any separable L^2 space whose orthonormal basis elements can be identified and mapped via a Hilbert space isomorphism to the basis element $\vec{w}_n$ of ℓ^2. Whenever $f \in L^2(a, b)$ has Fourier series $f = \sum_{n=0}^{\infty} a_n \cdot e_n$, where $\{e_n\}_{n=0}^{\infty}$ forms a basis for $L^2(a, b)$, then (using a Hilbert space isomorphism T mapping $L^2(a, b)$ onto ℓ^2) Parseval's identity results: $\int_a^b |f(x)|^2 \, dx = \sum_{n=0}^{\infty} |a_n|^2$.

Why should we be concerned about when Hilbert space isomorphisms can be constructed between different Hilbert spaces? Parseval's identity is an example of how such isomorphisms will give us different ways to represent and think about functions (as elements of Hilbert spaces). In Chapter 5 we will also see that Hilbert space isomorphisms will be a key tool in the study of bounded linear operators, which are types of maps of functions to functions.

Solutions to Questions

3.3.1 Because the ℓ^2 inner product is the sum of the products of corresponding entries, $\langle e_n, e_m \rangle = 0$ when $n \neq m$, and $\langle e_n, e_m \rangle = 0$ when $n = m$. For any vector $\vec{v} \in \ell^2$ with nth element v_n, if $v_n = \langle \vec{v}, e_n \rangle = 0$ for all $n = 1, 2, \ldots$, then all the entries in $\vec{v}$ would be 0, and $\vec{v} = \vec{0}$. By Definition 3.3.1, $\{e_n\}_{n=1}^{\infty}$ forms a basis for ℓ^2. Since $|\mathbb{N}| = \aleph_0$, the dimension of ℓ^2 is $\aleph_0$.

3.3.2 There are six pairs of inner products to consider, for example, $\langle e_1, e_2 \rangle = \frac{1}{2} \int_0^\infty (x^3 - 5x^2 + 6x - 2)e^{-x}\, dx$. By the LDCT, the integral is $-\frac{1}{2} \lim_{n\to\infty} (x^3 - 2x^2 + 2x)e^{-x}\big|_0^n = 0$. We leave the other five for the reader. Similarly, $\|f_2\|^2 = \int_0^\infty (x^2 - 4x + 2)^2 e^{-x}\, dx = \lim_{n\to\infty} -(x^4 - 4x^3 + 8x^2 + 4)e^{-x}\big|_0^n = 4$. Also, $\langle g_2, f_1 \rangle = \int_0^\infty (x^3 - x^2)e^{-x}\, dx = \lim_{n\to\infty} -(x^3 + 2x^2 + 4x + 4)e^{-x}\big|_0^n = 4$. The others follow similarly.

3.3.3 As in Definition 3.3.2, $f_0(x) = g_0(x) = x^0 = 1$. Then $f_1 = g_1 - \frac{\langle g_1, f_0 \rangle}{\|f_0\|^2} f_0$. Since $\|f_0(x)\|^2 = \int_{-1}^1 |1|^2\, dx = 2$, and $\langle g_1, f_0 \rangle = \int_{-1}^1 x\, dx = 0$, $f_1(x) = x^1 - \frac{0}{2} \cdot 1 = x$. Next, $f_2 = x^2 - \frac{\langle x^2, x \rangle}{\|x\|^2} x - \frac{\langle x^2, 1 \rangle}{\|1\|^2} 1 = x^2 - 0 \cdot x - (\int_{-1}^1 x^2\, dx)/2 = (1/3)(3x^2 - 1)$. Finally, $\|x\|^2 = 2/3$ and $\|x^2 - 1/3\|^2 = 8/45$, so $e_0 = 1/\sqrt{2}$, $e_1 = \sqrt{3/2} \cdot x$, and $e_2 = \sqrt{5/8}(3x^2 - 1)$.

3.3.4 Using the product rule, $(\frac{d}{dx})^1(x^1 e^{-x}) = (1-x)e^{-x}$. Hence the definition in Example 3.3.3 has $L_1(x) = 1 - x$, as desired. Similarly, $(\frac{d}{dx})^2(x^2 e^{-x}) = (\frac{d}{dx})^1[(2x - x^2)e^{-x}] = (x^2 - 4x + 2)e^{-x}$, so the definition produces $L_2(x) = \frac{1}{2}(x^2 - 4x + 2)$.

3.3.5 (1) Apply $g = c_0 f_0 + c_1 f_1$, where $c_j = \frac{\langle x^2 \chi_{(0,1)}, f_j \rangle}{\|f_j\|^2}$ for $j = 0, 1$. From Example 3.3.2, $f_0 = g_0 = e^{-x/2}$ and $f_1 = (x-1)e^{-x/2}$. Note $\|f_0\| = \|f_1\| = 1$, and so $c_0 = \int_0^1 x^2 e^{-x/2}\, dx = 16 - 26/\sqrt{e}$ and $c_1 = \int_0^1 x^2(x-1)e^{-x/2}\, dx = 80 - 132/\sqrt{e}$. The best approximation is $g(x) = [16 - 26/\sqrt{e} + (80 - 132/\sqrt{e})(x-1)]e^{-x/2}$. Define $h \in M^\perp$ as $h(x) = x^2 \chi_{(0,1)} - [16 - 26/\sqrt{e} + (80 - 132/\sqrt{e})(x-1)]e^{-x/2}$. We get $\|h(x)\|^2 \approx 1/5 - .1136 + .0568 = 0.1432$. Part (2) is similar.

3.3.6 A function $h = f - g$ is in $M^\perp$, for any $f \in L^2(0,1)$. Applying the Gram-Schmidt process to $1, x$, and x^2 in $L^2(0,1)$, we get $f_0 = 1$, $f_1 = x - 1/2$, and $f_2(x) = x^2 - x + 1/2 - 1/3 = x^2 - x + 1/6$. Also, $\|1\|^2 = 1$, $\|x - 1/2\|^2 = 1/12$, and $\|x^2 - x + 1/6\|^2 = 1/180$. By the projection theorem, $g(x) = [\int_0^1 f(x)\, dx] + [12 \int_0^1 f(x) \cdot (x - 1/2)\, dx] \cdot (x - 1/2) + [180 \int_0^1 f(x) \cdot (x^2 - x + 1/6)\, dx] \cdot (x^2 - x + 1/6)$.

Reading Questions for Section 3.3

1. When is a set of Hilbert space elements $\{b_n\}$ orthogonal?

2. What is a basis for a Hilbert space $\mathcal{H}$? What is the dimension of $\mathcal{H}$?

3. Describe the Gram-Schmidt process as it would be applied to a set of vectors $\{g_0, g_1, g_2\}$.

4. How do you normalize a vector f in a Hilbert space?

5. List the first three Laguerre polynomials (of order 0).

6. What does it mean for $\{e_n\}_{n=0}^\infty$ to be a complete orthonormal basis for L^2?

7. On which $L^2(a, b)$ Hilbert spaces do the Legendre polynomials form a complete orthogonal basis?

8. What is the general form of a Fourier series for a function f in a Hilbert space having an orthonormal basis $\{e_n\}_{n=0}^\infty$?

9. Explicitly identify a complete orthonormal basis for $L^2(0, \infty)$.

10. What does Parseval's identity say about a function $f \in L^2(a, b)$?

3.3 Orthonormal Basis for a Hilbert Space

Exercises for Section 3.3

Exercises 1–3 work with monomials in $L^2(-1, 1)$.

1. Show $f_n = x^n$ and $f_m = x^m$ are orthogonal in $L^2(-1, 1)$ exactly when $n + m$ is odd.

2. For a nonnegative integer n, prove $f_n = x^n$ is an element of $L^2(-1, 1)$ but does not have norm equal to 1.

3. For a nonnegative integer n, normalize $f_n = x^n$ in $L^2(-1, 1)$ so it has norm equal to 1.

The monomials $g_n(x) = x^n$ with $n = 0, 1, 2, \ldots$ can generate a complete orthogonal set (an orthogonal basis) for $L^2(a, b)$ whenever the interval (a, b) is bounded (see [38, p. 49]). In other words, the Gram-Schmidt process will generate an orthonormal basis. Exercises 4–6 work within this setting.

4. Examine $L^2(-1, 1)$, showing $g_0 = 1$, $g_1 = x$, $g_2 = x^2$, $g_3 = x^3$, and $g_4 = x^4$ produce, as they are run through the Gram-Schmidt process, orthonormal basis elements $e_0 = \sqrt{2}/2$, $e_1 = \sqrt{3/2}x$, $e_2 = \sqrt{5/8}(3x^2 - 1)$, $e_3 = \sqrt{7/8}(5x^3 - 3x)$, and $e_4 = \sqrt{9/128}(35x^4 - 30x^2 + 3)$. (This problem is an expansion of Question 3.3.3.) The functions are the first five (normalized) Legendre polynomials.

5. In $L^2(-\pi, \pi)$, find three orthonormal basis elements that are polynomials of degree at most two. Compare your results to the basis elements for $L^2(-1, 1)$ listed in the last exercise.

6. For $L^2(a, b)$ where the interval (a, b) is bounded, use $g_0 = 1$, $g_1 = x$, and $g_2 = x^2$ to produce, as they are run through the Gram-Schmidt process, orthonormal basis elements e_0, e_1, and e_2. Explicitly determine e_0, e_1, and e_2 in terms of a and b.

Exercises 7–10 discuss the dimension of separable Hilbert spaces.

7. Prove the space $\mathcal{H}$ of column vectors $\vec{v} = \begin{bmatrix} a_1 \\ a_2 \\ a_3 \\ a_4 \end{bmatrix}$, having real valued entries, forms a Hilbert space with inner product $\langle \begin{bmatrix} a_1 \\ a_2 \\ a_3 \\ a_4 \end{bmatrix}, \begin{bmatrix} b_1 \\ b_2 \\ b_3 \\ b_4 \end{bmatrix} \rangle_\mathcal{H} = a_1 b_1 + a_2 b_2 + a_3 b_3 + a_4 b_4$.

8. Explicitly determine an orthonormal basis for the Hilbert space $\mathcal{H}$ of column vectors in the last exercise, using the inner product on each pairwise combination to prove your result is orthonormal. Then use Definition 3.3.1 to determine the dimension of the space.

9. Using the Gram-Schmidt process on elements $1, x, x^2$, and x^3, show the Hilbert space $\mathcal{K} = \{a_0 + a_1 x + a_2 x^2 + a_3 x^3 : a_0, a_1, a_2, a_3 \in \mathbb{R}\}$ is four-dimensional in the sense of Definition 3.3.1. (For this space, use the inner product $\langle a_0 + a_1 x + a_2 x^2 + a_3 x^3, b_0 + b_1 x + b_2 x^2 + b_3 x^3 \rangle = a_0 b_0 + a_1 b_1 + a_2 b_2 + a_3 b_3$.)

10. Two finite-dimensional Hilbert spaces $\mathcal{H}$ and $\mathcal{K}$ are isometrically isomorphic precisely when they have the same dimension. Use this to say why the Hilbert spaces $\mathcal{H}$ and $\mathcal{K}$ in the last two exercises are isometrically isomorphic, and define a Hilbert space isomorphism T (describing its action on a general element of $\mathcal{H}$ and showing T satisfies Definition 3.3.3) that maps $\mathcal{H}$ onto $\mathcal{K}$.

For every nonnegative integer $n = 0, 1, 2, \ldots$, the nth degree Hermite polynomial is defined by the Rodrigues formula $H_n(x) = (-1)^n e^{x^2} \frac{d^n}{dx^n} [e^{-x^2}]$. Exercises 11–16 are about the Hermite polynomials.

11. Compute the Hermite polynomial H_n for $n = 0, 1, 2, 3,$ and 4.

12. For every nonnegative integer $n = 0, 1, 2, \ldots$, the nth Hermite polynomial $H_n(x)$ is a solution of the nth Hermite differential equation $y'' - 2xy' + 2ny = 0$. Prove this result for $n = 0, 1, 2, 3,$ and 4, using your results from the previous exercise.

13. Find the $L^2(\mathbb{R})$ norms for $f_n = e^{-x^2/2} \cdot H_n(x)$ for $n = 0, 1, 2, 3, 4$. Multiply the functions f_n by a constant to form e_n, where $n = 0, 1, 2, 3, 4$, so their $L^2(\mathbb{R})$ norms satisfy $\|e_n\| = 1$.

14. For the following choices of n and m, show $\langle e_n, e_m \rangle = 0$ when $n \neq m$, and where e_j is defined as in the last exercise. (a) $n = 1$ and $m = 2$. (b) $n = 0$ and $m = 1$. (c) $n = 0$ and $m = 2$.

15. Use the result of the last exercise to show $M \equiv \vee\{e_0, e_1, e_2\}$ forms a closed linear subspace of $L^2(\mathbb{R})$ having $e_0, e_1,$ and e_2 as a complete orthonormal basis for M. Also, identify what the elements of M are.

16. Use the Rodrigues formula to prove $\frac{d}{dx}[H_n(x)e^{-x^2}] = -H_{n+1}(x)e^{-x^2}$.

The Hermite polynomials form in $L^2(\mathbb{R})$ another collection of polynomials that arise out of the Gram-Schmidt process applied to the monomial functions $1, x, x^2, x^3, \ldots$, just as Exercise 4 described the case for the Legendre polynomials in $L^2(-1, 1)$. However, none of the monomials are elements of $L^2(\mathbb{R})$. Therefore, to input them directly into the Gram-Schmidt process, we use a weighted inner product for $L^2(\mathbb{R})$, namely $\langle f, g \rangle = \int_0^\infty f(x)g(x) e^{-x^2} dx$. Just as this section constructed basis elements for $L^2(0, \infty)$, we may instead weight the input functions, applying the Gram-Schmidt process to $1 \cdot e^{-x^2/2}, xe^{-x^2/2}, x^2 e^{-x^2/2}, \ldots$, which allows us to use the standard inner product for $L^2(\mathbb{R})$, $\langle f, g \rangle = \int_\mathbb{R} f \cdot g$. We use this latter option in Exercises 17–20. In them, you may wish to use the fact $\int_{-\infty}^\infty e^{-x^2} dx = \sqrt{\pi}$.

17. Prove the Gram-Schmidt process applied to $g_0 = e^{-x^2/2}$ and $g_1 = xe^{-x^2/2}$ in $L^2(\mathbb{R})$ produces the first two normalized Hermite functions $e_n(x) = c_n H_n(x)e^{-x^2/2}$ with $n = 0, 1$, where $H_n(x)$ is the nth Hermite polynomial defined above and c_n is a constant that makes $\|e_n\| = 1$.

18. Find the third normalized Hermite function e_2 in $L^2(\mathbb{R})$ produced by the Gram-Schmidt process applied to g_0 and g_1 from the previous exercise, along with $g_2(x) = x^2 e^{-x^2/2}$.

19. In $L^2(\mathbb{R})$, define the subspace $M = \{g(x) : g(x) = c_0 e^{-x^2/2} + c_1 x e^{-x^2/2}\}$, where c_0 and c_1 are arbitrary (real) constants. Use the projection theorem and the Gram-Schmidt process to describe $M^\perp$.

20. What is $M^\perp$ for the subspace M in $L^2(\mathbb{R})$ defined as $M = \{g(x) : g(x) = (c_0 + c_1 x + c_2 x^2)e^{-x^2/2}\}$ for real constants $c_0, c_1,$ and c_2? Use the projection theorem and the Gram-Schmidt process to determine your answer.

Exercises 21–24 are linked to the Legendre polynomials discussed in Question 3.3.3.

21. Using the Gram-Schmidt process on the monomials $1, x, x^2, \ldots, x^5$ in $L^2(-1, 1)$, determine the fifth-degree Legendre polynomial $e_5(x)$. (This is a continuation of Exercise 4.)

22. The Rodrigues formula $P_n(x) = \frac{1}{2^n n!} \frac{d^n}{dx^n}(x^2 - 1)^n$ generates the Legendre polynomials for $n = 0, 1, 2, \ldots$. Confirm the elements $e_0, e_1, \ldots, e_5$ produced in Exercises 4 and 20 match

3.3 Orthonormal Basis for a Hilbert Space

the normalizations of this formula's output values P_n. In other words, show $e_n(x) = c_n P_n(x)$, $n = 0, 1, \ldots, 5$, where c_n is a (normalizing) constant. It turns out that $c_n = \sqrt{(2n+1)/2}$.)

23. In $L^2(-1, 1)$, define the subspace $M = \{g(x) : g(x) = c_0 e_0 + c_1 e_1\}$, where e_0 and e_1 are as in Exercise 4 and c_0 and c_1 are real constants. Use the projection theorem and the Gram-Schmidt process to describe $M^\perp$.

24. In $L^2(-1, 1)$, let M be the subspace of polynomials of degree 5 or less. Use the projection theorem and the Gram-Schmidt process to describe $M^\perp$.

Exercises 25–26 deal with ℓ^2.

25. Show $\{\vec{e}_n = \begin{bmatrix} 0 \\ \vdots \\ 0 \\ 1 \\ 0 \\ \vdots \end{bmatrix}\}_{n=1}^\infty$, where the 1 is in the nth entry, is an orthonormal basis for ℓ^2.

26. Find the norm of $\vec{v} = \sum_{n=1}^\infty (1/2^{n-1}) \vec{e}_n$, where $\vec{e}_n$ is as in the previous exercise.

Exercises 27–29 are results for standard constructions of Fourier sine and cosine series.

27. Show $\{(\cos nx)/\sqrt{\pi}\}_{n=1}^\infty$ and $\{(\sin nx)/\sqrt{\pi}\}_{n=1}^\infty$ form orthonormal sets in $L^2(-\pi, \pi)$.

28. Show $\{1/\sqrt{2\pi}\} \cup \{(\cos nx)/\sqrt{\pi}, (\sin nx)/\sqrt{\pi} : n = 1, 2, 3, \ldots\}$ forms a single set of orthonormal functions in $L^2(-\pi, \pi)$.

29. For a real a, $a > 0$, show $\{1/\sqrt{2a}\} \cup \{(\cos n\pi x/a)/\sqrt{a}, (\sin n\pi x/a)/\sqrt{a} : n = 1, 2, 3, \ldots\}$ forms a single set of orthonormal functions in $L^2(-a, a)$.

Exercises 30 and 31 describe how orthogonal polynomials are connected to L^2 spaces.

30. For $n = 1, 2, 3, \ldots$, the set of $e_n(x) = \sqrt{(2n+1)/2} P_n(x)$, where $P_n(x)$ is the Legendre polynomial defined by the Rodrigues formula $P_n(x) = \frac{1}{2^n n!} \frac{d^n}{dx^n}(x^2 - 1)^n$, forms an orthonormal basis for $L^2(-1, 1)$. For the Hilbert spaces listed, describe (in the same way we have just expressed for $L^2(-1, 1)$) the corresponding set $\{e_n\}_{n=0}^\infty$ of orthonormal basis elements defined in terms of a classical set of polynomials.
    (a) $L^2(0, \infty)$ (b) $L^2(\mathbb{R})$

31. For the following Hilbert spaces $\mathcal{H}$, construct a Hilbert space isomorphism T that maps $\mathcal{H}$ onto $\mathcal{K} = \ell^2$. Define T in terms of its action on basis elements.
    (a) $L^2(0, \infty)$ (b) $L^2(\mathbb{R})$ (c) $L^2(-1, 1)$

Exercises 32–36 combine to prove $\overline{M}^\perp = M^\perp$.

32. Use the fact $M \subseteq \overline{M}$ to prove $\overline{M}^\perp \subseteq M^\perp$.

33. Let f be a function in $\overline{M}$. For $\varepsilon > 0$, why must there exist an element $\hat{f} \in M$ such that $\|f - \hat{f}\| < \varepsilon$?

34. For a function $g \in M^\perp$, use the Schwarz inequality of Theorem 3.2.1 to prove, for $\varepsilon > 0$, that $|\langle g, f - \hat{f} \rangle| \leq \|g\| \cdot \varepsilon$, where f and $\hat{f}$ are as in the previous exercise.

35. Why is $\langle g, \hat{f} \rangle = 0$, where g and $\hat{f}$ are as in the last exercise? Use this fact and the result from the previous exercise to prove, for $\varepsilon > 0$, that $|\langle g, f \rangle| \leq \|g\| \cdot \varepsilon$ for any $g \in M^\perp$ and $f \in \overline{M}$.

36. Use the previous exercise to argue $\langle g, f \rangle = 0$ for $g \in M^\perp$ and $f \in \overline{M}$. Then say why this implies $M^\perp \subseteq \overline{M}^\perp$.

Advanced Exercises. Exercises 37–46 study the orthonormal trigonometric functions e_n, where $n = 0, 1, 2, \ldots$, $e_{2n+1}(x) = (\cos nx)/\sqrt{\pi}$, and $e_{2n} = (\sin nx)/\sqrt{\pi}$ in $L^2(-\pi, \pi)$. (See Exercise 27, which shows these functions are orthonormal in $L^2(-\pi, \pi)$.) The following exercises show they form an orthonormal basis for $L^2(-\pi, \pi)$. The proof is presented in [123, p. 218]). It proceeds by way of contradiction, so it begins by assuming there exists a nonzero function $f \in L^2(-\pi, \pi)$ such that $f \perp e_n$ for $n = 0, 1, 2, \ldots$ and searches for a contradiction. Exercises 37 to 46 use consistent notation: terms that appear in one exercise may have been defined in previous exercises.

37. As a special case, assume $f \not\equiv 0$ (so without loss of generality there must be some element x_0 such that $f(x_0) > 0$) is a continuous function in $L^2(-\pi, \pi)$. (If $x \notin [-\pi, \pi]$, then we assume $f(x) = 0$.) A fact from calculus says f attains its maximum M on $[-\pi, \pi]$ at some value x_0, so $f(x_0) = M > 0$. Use Definition 1.4.3 to say why there exists $\delta > 0$ so $f(x) \geq M/2$ whenever $x \in [-\pi, \pi]$ with $|x - x_0| < \delta$.

38. Set $g(x) = 1 + \cos(x - x_0) - \cos \delta$. Explain why $g(x) > 1$ for $x \in [-\pi, \pi]$ with $|x - x_0| < \delta$. (Such x are in the interval $(x_0 - \delta, x_0 + \delta)$.)

39. Explain why $|g(x)| \leq 1$ for $x \in [-\pi, \pi]$ with $x \notin (x_0 - \delta, x_0 + \delta)$.

40. Show there is a constant $C > 1$ such that $g(x) \geq C > 1$ for $x \in (x_0 - \delta/2, x_0 + \delta/2)$.

41. To prove the system of functions $\{e_n\}_{n=0}^{\infty}$ is a basis, we show any f that satisfies $\langle f, e_n \rangle = 0$ for all n must have $f = 0$ almost everywhere. Hence we assume $\langle f, e_n \rangle = 0$. The trigonometric form of the elements e_n would then say $\langle f, h \rangle = 0$ for any trigonometric polynomial h (a polynomial formation of the elements e_n). Use the fact that g^N is a trigonometric polynomial for $N = 1, 2, \ldots$ to show $\int_{-\pi}^{\pi} f(x) \cdot g(x)^N \, dx = 0$.

42. Let S be the set of elements of $[-\pi, \pi]$ not in $I \equiv (x_0 - \delta, x_0 + \delta)$. Use Exercise 39 to show $|\int_S f \cdot g^N| \leq 2\pi \cdot M \cdot 1^N = 2\pi M$.

43. Use Exercise 40 to show $\int_I f \cdot g^N \geq \int_{x_0 - \delta/2}^{x_0 - \delta/2} f(x) \cdot g(x)^N \, dx \geq \frac{1}{2} M \cdot \delta \cdot C^N$.

44. Use the results of the last two exercises to prove $\lim_{N \to \infty} \int_{-\pi}^{\pi} f(x) \cdot g(x)^N \, dx = \infty$. Say why this is a contradiction to the result of Exercise 41, and conclude there can be no nonzero continuous function perpendicular to every e_n.

45. Now assume $f \in L^2(-\pi, \pi)$ is not necessarily continuous, and assume $\langle f, e_n \rangle = 0$ for $n = 1, 2, \ldots$. From Section 2.5 this fact is equivalent to $\langle f(x), e^{inx} \rangle = 0$ for $n = 0, \pm 1, \pm 2, \ldots$. It turns out we may also assume, without loss of generality, that f is periodic, so $f(-\pi) = f(\pi)$. Define $F(x) = \int_0^x f(t) \, dt$. (Hölder's inequality guarantees F is well defined.) Using

integration by parts, prove $\langle F, e^{inx} \rangle = \frac{1}{in} \int_{-\pi}^{\pi} f(x) e^{inx} \, dx = 0$ for $n = 1, 2, \ldots$. Conclude, for some constant C, that $F(x) = C$ a.e.

46. Use the last exercise to prove $f = F' = 0$ a.e. This completes the proof that the orthonormal trigonometric functions e_n form an orthonormal basis for $L^2(-\pi, \pi)$.

3.4 Application: Quantum Mechanics

A Hilbert space provides an excellent setting for the mathematics that models quantum mechanics. In 1926, the Austrian physicist Erwin Schrödinger constructed the now famous Schrödinger equation to explain how a physical quantum state changes over time. For example, a physicist might wonder about the state (values reflecting the physical existence) of an electron within a hydrogen atom. It follows the Schrödinger equation, at least according to Schrödinger's model. It can therefore be modeled mathematically by the solution to the equation. This section shows how Hilbert spaces, especially $L^2(\mathbb{R})$, serve well as the settings to describe solutions to the Schrödinger equation. For additional details on the mathematics of quantum mechanics, the interested reader may wish to consult texts by George Simmons [111] and David Griffiths [48, p. 112], which much of the material presented here follows.

3.4.1 The Hermite Differential Equation

One of the simplest quantum situations to model is the controlled system for the *classical harmonic oscillator*. Here, a particle of mass m has a constrained movement that forces it to flow back and forth along the x-axis (the position axis). A force of the form $-kx(t)$, where k is a constant and t is time, is a restoring force and binds the particle to an equilibrium position $x = 0$.[8] The differential equation that describes its motion through time t is $m \cdot x''(t) = -kx(t)$. It is easy to check (by substitution into it) that a solution to the equation is of the form $x(t) = x_0 \cos(\sqrt{k/m} \cdot t)$, where x_0 is the position of the particle at time $t = 0$. Because this solution oscillates according to the cosine oscillations (between the values $-x_0$ and x_0), we describe the particle's position as a harmonic oscillation. Furthermore, analyzing the period of the cosine function, we see the period of the oscillation is $T = 2\pi\sqrt{m/k}$, and the frequency $f = 1/T$ satisfies $k = 4\pi^2 m f^2$. Applying physical properties of total energy, kinetic energy v (which is $v = k[x(t)]^2/2$), and potential energy, an expression for total energy E is the constant $E = kx_0^2/2$.[9]

The identities for E, v, and k feed into the Schrödinger equation that describes this harmonic oscillator. That second-order Schrödinger differential equation, which we only present and do

---

[8] In the mathematical model of a harmonic oscillator, the equilibrium position is chosen to be a position about which the object oscillates in a symmetric manner. Physicists naturally choose $x = 0$ because of its simplicity.

[9] The detailed explanation of these results for energy follows from standard physics equations, the kinetic energy is $(1/2)mv^2 = (1/2)m(x'(t))^2$ (here, v is velocity) and the potential energy is $(1/2)kx^2$. Applying $x(t) = x_0 \cos(\sqrt{k/m} \cdot t)$, the sum of potential and kinetic energy is $E = (1/2)x_0^2[m \sin^2(\sqrt{k/m} \cdot t) \cdot (k/m) + k \cos^2(\sqrt{k/m} \cdot t)]$. Since $\sin^2\theta + \cos^2\theta = 1$, this gives $E = (1/2)x_0^2 k$. In conclusion, there is a tradeoff between kinetic and potential energy to produce the constant total energy E: as one rises the other must fall by an equivalent amount.

not attempt to derive, becomes

$$\psi''(x) + (8\pi^2 m/h^2)(E - k[x(t)]^2/2)\psi(x) = 0,$$

where h is Planck's constant. The equation's solutions $\psi(x)$ are typically assumed to follow appropriate physical properties (which will be imposed upon the system as we proceed), and they are sometimes described as the Schrödinger wave functions or Schrödinger eigenfunctions.[10].

How is the equation connected to L^2 spaces? The standard physical conditions imposed on the wave functions ψ are $\lim_{|x|\to\infty} \psi(x) = 0$ and $\int_{-\infty}^{\infty} |\psi(x)|^2 dx = 1$. (The latter condition follows from a natural linkage of the physics to probability theory.) You can see how the second condition creates a corresponding formulation of the mathematics in terms of L^2: any wave function must be in L^2.

Much more can be said when a change of variables is applied. Let $u = 2\pi\sqrt{fm/h} \cdot x$. Then the Schrödinger equation becomes $\psi''(u) + (2p + 1 - u^2)\psi(u) = 0$, where p is a constant (here $p = E/(hf) - 1/2$). Furthermore, we might define ψ in terms of not only u but also a new variable $g(u)$ that depends on u, according to $\psi(u) = g(u) \cdot e^{-u^2/2}$. Applying the product rule, we see the description of ψ (according to this product of g with the exponential) results in the following first and second derivatives: $\psi'(u) = (g'(u) - ug(u)) \cdot e^{-u^2/2}$ and $\psi''(u) = (g''(u) - 2ug'(u) + (u^2 - 1)g(u)) \cdot e^{-u^2/2}$. Substituting and simplifying, the equation becomes

$$g''(u) - 2u \cdot g'(u) + 2pg(u) = 0.$$

This general equation, the Hermite differential equation, was studied in 1864 by the French mathematician Charles Hermite, who managed to find its solutions using classical techniques (see [57]). Summarizing our discussion so far, we can determine the Schrödinger equation for the harmonic oscillator. Through a change of variables (and by applying a product format to the structure of the solution), the Schrödinger equation is transformed into the Hermite differential equation, which was solved more than 60 years before Schrödinger formulated any of these results—in fact, 23 years before Schrödinger was born.

How is the Hermite differential equation solved? The classical approach is to consider the solution as a Maclaurin series: $g(u) = \sum_{n=0}^{\infty} a_n u^n$. Substituting it into the Hermite equation $g''(u) - 2u \cdot g'(u) + 2pg(u) = 0$, along with the derivatives $g'(u) = \sum_{n=1}^{\infty} na_n u^{n-1}$ and $g''(u) = \sum_{n=2}^{\infty} n(n-1)a_n u^{n-2}$, and then simplifying by reindexing, produces a two-term recursion formula:

$$a_{n+2} = \frac{2(p-n)}{(n+1)(n+2)} a_n, \text{ for } n = 0, 1, 2, \ldots.$$

Letting $a_0 = 1$ produces the following first (even) infinite series solution, which we call g_1:

$$g_1(u) = 1 - \frac{2p}{2!}u^2 + \frac{2^2 p(p-2)}{4!}u^4 - \frac{2^3 p(p-2)(p-4)}{6!}u^6 + \cdots.$$

---

[10] The equation may also be viewed as deriving from the Hamiltonian set of differential equations for a harmonic oscillator, which takes the form $dx/dt = \partial F/\partial y$, $dy/dt = -\partial F/\partial x$. Here y is momentum and $F = y^2 + c^2 x^2$ for a constant c. An excellent reference that develops this viewpoint is [86, p. 543]

3.4 Application: Quantum Mechanics

Letting $a_1 = 1$ produces a second (odd) infinite series solution, which we call g_2:

$$g_2(u) = u - \frac{2(p-1)}{3!}u^3 + \frac{2^2(p-1)(p-3)}{5!}u^5 - \frac{2^3(p-1)(p-3)(p-5)}{7!}u^7 + \cdots .$$

The solutions are not constant multiples of each other (one is a series in even powers, while the other is in odd powers). A fundamental theorem of differential equations (see, for example, Simmons' text [111, p. 84]) says any solution to the Hermite differential equation is of the form $c_1 g_1(u) + c_2 g_2(u)$ for constants c_1 and c_2. The point is that an analysis of g_1 and g_2 will provide a full description of any solution to the Hermite equation.

This analysis begins by applying the constraints listed in our discussion of the Schrödinger equation; namely, that $\lim_{|x| \to \infty} \psi(x) = 0$ and $\int_{-\infty}^{\infty} |\psi(x)|^2 dx = 1$. Recalling the substitution and the product format that transformed the equation from one in $\psi(x)$ to one in $g(u)$, the first constraint is seen to imply $\lim_{|u| \to \infty} g(u)e^{-u^2/2} = 0$. A surprisingly cleancut fact results, which is the next theorem.

Theorem 3.4.1 *When $\lim_{|u| \to \infty} g_1(u)e^{-u^2/2} = 0$, the infinite series solution $g_1(u)$ to the Hermite equation is a polynomial. (The infinite series has only a finite number of nonzero terms.) The same is true for the series $g_2(u)$: when $\lim_{|u| \to \infty} g_2(u)e^{-u^2/2} = 0$, $g_2(u)$ is a polynomial.*

Proof. We prove the result for $g_1(u)$; the proof for $g_2(u)$ is essentially identical.[11] The Maclaurin series for $e^{u^2/2}$ is $\sum_{n=0}^{\infty} 1/(2^n n!) \cdot u^{2n}$. Denoting the coefficients as $b_{2n} = 1/(2^n n!)$, the ratio of adjacent terms' coefficients is then $b_{2(n+1)}/b_{2n} = 1/(2n+2)$. We have written $g_1(u)$ as $g_1(u) = \sum_{n=0}^{\infty} a_{2n} \cdot u^{2n}$. From its two-term recursion formula, its ratio of adjacent terms' coefficients is $a_{2n+2}/a_{2n} = 2(2n-p)/[(2n+1)(2n+2)]$. Therefore, assuming $g_1(u)$ is not a polynomial,

$$\frac{a_{2n+2}/a_{2n}}{b_{2n+2}/b_{2n}} = \frac{2(2n-p)(2n+2)}{(2n+1)(2n+2)},$$

which approaches 2 as $n \to \infty$. Hence $a_{2n+2}/b_{2n+2} > 1.5 \cdot [a_{2n}/b_{2n}]$ for all large n, say $n \geq N$ where N is some properly chosen positive integer. That means, through repeated applications, that $a_{2N+2k}/b_{2N+2k} > 1.5^k \cdot [a_{2N}/b_{2N}]$ for $k = 1, 2, \ldots$. Eventually (for $k \geq K$ for some K), the terms exceed 1, and so the series' corresponding coefficients of u^{2n} compare as $a_{2n} > b_{2n}$ for all large n.

Comparing ratios, that means

$$g_1(u)e^{-u^2/2} = \sum_{n=0}^{\infty} a_{2n} u^{2n} \bigg/ \sum_{n=0}^{\infty} b_{2n} u^{2n} = \frac{a_0 + a_2 u^2 + a_4 u^4 + \cdots + a_{2n} u^{2n} + \cdots}{b_0 + b_2 u^2 + b_4 u^4 + \cdots + b_{2n} u^{2n} + \cdots}$$

can't satisfy $\lim_{|u| \to \infty} g(u)e^{-u^2/2} = 0$. The only way out of the conundrum is to allow $g_1(u)$ to be a polynomial, which means, from the two-term recursion formula, that $2n - p = 0$ for some

---

[11] Simmons' text [111, p. 211] provides a more thorough reference to this result.

(smallest) n. In short, the highest power of the polynomial $g_1(u)$ is p, which must therefore be a nonnegative integer. ∎

Figure 3.5. Charles Hermite at age 65.

The result of the theorem is powerful: it says basic solutions $\psi(u) = g(u)e^{-u^2/2}$ to the classical harmonic oscillator's Schrödinger equation (when it is written as $g''(u) - 2u \cdot g'(u) + 2pg(u) = 0$) always have $g(u)$ equal to polynomial solutions to the Hermite equation. After Hermite described them as solutions, constant multiples (that produce a coefficient 2^n on the highest power x^n) of them were called, naturally, Hermite polynomials. The functions $\psi(u)$ they generate were termed Hermite functions. A basic way to express the polynomials comes from the two series that terminate and are of the form given above for either $g_1(u)$ or $g_2(u)$. The resulting polynomials of degree n are generally labeled $H_n(u)$. We can find them by examining g_1 and g_2 with (as the proof of the last theorem has stipulated) p equal to the polynomial's power. We get (applying the constant multiples so the leading coefficient is 2^n) the following: $H_0(u) = 1$, $H_1(u) = 2u$, $H_2(u) = 4u^2 - 2$, $H_3(u) = 8u^3 - 12u$, etc. Hence, for example, the first several basic solutions to the classical harmonic oscillator's Schrödinger equation are $\psi_0(u) = e^{-u^2/2}$, $\psi_1(u) = 2ue^{-u^2/2}$, $\psi_2(u) = (4u^2 - 2)e^{-u^2/2}$, $\psi_2(u) = (8u^3 - 12)e^{-u^2/2}$, etc.

But we've already seen them. They were the Hermite polynomials described in Exercises 11 through 16 of Section 3.3. The exercises indicated that the Hermite functions also result from the Gram-Schmidt process applied to $g_n(x) = x^n e^{-x^2/2}$ in $L^2(\mathbb{R})$. They additionally indicated an alternative way to produce the Hermite polynomials. Namely, we can use the Rodrigues formula,

$$H_n(u) = (-1)^n \cdot e^{u^2} \cdot \frac{d^n(e^{-u^2})}{du^n}.$$

Example 3.4.1 We calculate the first three Hermite polynomials, using the Rodrigues formula.

Solution. The zeroth derivative is the function itself, and therefore $H_0(u) = (-1)^0 \cdot e^{u^2} \cdot \frac{d^0(e^{-u^2})}{du^0} = e^{u^2} \cdot e^{-u^2} = 1$. Since $n = 1$ uses the first derivative $d(e^{-u^2})/du = (-2u)e^{-u^2}$, we get

$$H_1(u) = (-1)e^{u^2}(-2u)e^{-u^2} = 2u.$$

Similarly, $H_2(u) = (-1)^2 e^{u^2} d^2(e^{-u^2})/du^2 = e^{u^2}[(-2)e^{-u^2} + 4u^2 e^{-u^2}] = 4u^2 - 2$. As the proof of Theorem 3.4.1 implied must happen, the Rodrigues formula calculations resulted in the even-degree polynomials being even functions, and the odd-degree polynomials being odd functions. ∎

How do we know this expression from Rodrigues formula generates the Hermite polynomials needed to solve Schrödinger's equation for any degree n? A calculation shows that the Rodrigues formula's functions $H_n(u)$ satisfy the Hermite differential equation $H_n''(u) - 2u \cdot H_n'(u) + 2nH_n(u) = 0$. For if we differentiate the Rodrigues formula (using

3.4 Application: Quantum Mechanics

the product rule on the right-hand side), we get $H'_n(u) = 2uH_n(u) - H_{n+1}(u)$, and so (differentiating) $H''_n(u) = 2uH'_n(u) + 2H_n(u) - H'_{n+1}(u)$. Another straightforward calculation shows $\frac{d^{n+1}(e^{-u^2})}{du^{n+1}} + 2u\frac{d^n(e^{-u^2})}{du^n} + 2n\frac{d^{n-1}(e^{-u^2})}{du^{n-1}} = 0$. Multiplying by $(-1)^{n+1}e^{u^2}$ gives $H_{n+1}(u) - 2uH_n(u) + 2nH_{n-1}(u) = 0$. Therefore $H'_n(u) = 2uH_n(u) - H_{n+1}(u) = 2uH_n(u) - 2uH_n(u) + 2nH_{n-1}(u) = 2nH_{n-1}(u)$. Substitution yields $H''_n(u) = 2uH'_n(u) - 2nH_n(u)$. This equality shows $H_n(u)$ (as generated by the Rodrigues formula) satisfies the Hermite differential equation. The uniqueness of the polynomial solutions (up to a constant) shows the polynomials H_n must be (up to a constant) the desired Hermite polynomials.

Question 3.4.1 Find the next four Hermite polynomials $H_3(u)$, $H_4(u)$, $H_5(u)$, and $H_6(u)$ generated by Rodrigues' formula for H_n. ∎

3.4.2 The Hermite Functions Form an Orthogonal Basis for $L^2(\mathbb{R})$

There is more to this delightful story about Hermite polynomials, and our description now switches to use the standard variable x. As exercises in Section 3.3 have suggested, the collection of functions $\{H_n(x)e^{-x^2/2}\}$ is orthogonal in $L^2(\mathbb{R})$. In fact, they form an orthogonal basis. The easiest proof follows from the Rodrigues formula representation of the Hermite polynomials. The next theorem uses it to show the Hermite functions are orthogonal in $L^2(\mathbb{R})$.

Theorem 3.4.2 *The collection of Hermite functions* $\{\phi_n(x) = (\sqrt{\pi}2^n n!)^{-1/2} H_n(x) e^{-x^2/2}\}$ *are orthonormal in $L^2(\mathbb{R})$, in that they satisfy*

$$\int_{-\infty}^{\infty} H_n(x)e^{-x^2/2} \cdot H_m(x)e^{-x^2/2} \, dx = \begin{cases} 0 & \text{when } n \neq m \\ \sqrt{\pi}2^n n! & \text{when } n = m. \end{cases}$$

When used in this sense, we sometimes call the function $w(x) = e^{-x^2}$ the corresponding weight function for the Hermite polynomials.

Proof. Without loss of generality, $m \leq n$. By the Rodrigues formula,

$$\int_{-\infty}^{\infty} H_m(x) H_n(x) e^{-x^2} \, dx = \int_{-\infty}^{\infty} H_m(x) (-1)^n \frac{d^n(e^{-x^2})}{dx^n} \, dx.$$

Repeated integration by parts shows the right-hand side equals $\int_{-\infty}^{\infty} e^{-x^2} \frac{d^n(H_m(x))}{dx^n} \, dx$. The integral vanishes if $m < n$, since the nth derivative of the mth degree polynomial $H_m(x)$ is 0 in that case. If $m = n$, then the integral equals $n! c_n \int_{-\infty}^{\infty} e^{-x^2} \, dx$, where c_n is the leading coefficient of $H_n(x)$. A calculation shows $c_n = 2^n$. Since $\int_{-\infty}^{\infty} e^{-x^2} \, dx = \sqrt{\pi}$, the result follows. ∎

Question 3.4.2 By direct calculation, verify the last theorem when $n = 1$ and $m = 2$; i.e., show $\int_{-\infty}^{\infty} 2xe^{-x^2/2} \cdot (4x^2 - 2)e^{-x^2/2} \, dx = 0$, $\int_{-\infty}^{\infty} (2x)^2 e^{-x^2} \, dx = 2\sqrt{\pi}$, and $\int_{-\infty}^{\infty} (4x^2 - 2)^2 e^{-x^2} \, dx = 2! 2^2 \sqrt{\pi}$. You may wish to use the fact that $\int_{-\infty}^{\infty} e^{-x^2} \, dx = \sqrt{\pi}$, which this section's advanced exercises prove. ∎

As before, we use a standard notation $f(x) \sim \sum_{n=0}^{\infty} a_n \phi_n(x)$ for the generalized Fourier series expansion of f in terms of the orthonormal collection of weighted Hermite polynomials $\{\phi_n(x)\}_{n=0}^{\infty}$. Here the Fourier coefficient a_n is defined as $a_n = \int_{-\infty}^{\infty} f(x) \phi_n(x) \, dx =$

$(\sqrt{\pi}2^n n!)^{-1/2} \int_{-\infty}^{\infty} f(x) H_n(x) e^{-x^2/2} dx$. Because the formula may be thought of as an $L^2(\mathbb{R})$-inner product, we sometimes write $a_n = \langle f, \phi_n \rangle$.

> **Lemma 3.4.1** Suppose a (real-valued) function g has generalized Fourier series expansion $g(x) \sim \sum_{n=0}^{\infty} a_n \phi_n(x)$, and that $e^{-x^2/2} F(x)$ is a bounded function with limit 0 as $x \to \pm\infty$, where $F(x) \equiv \int_{-\infty}^{x} g(t) e^{t^2/2} dt$ and where the integral is assumed to converge for all x. Then
> $$e^{-x^2/2} F(x) \sim b_0 + \sum_{n=1}^{\infty} b_n \phi_n(x), \text{ where } b_n = (2n)^{-1/2} a_{n-1} \text{ for } n = 1, 2, 3, \ldots.$$

Proof. Using integration by parts (with $V(x) = F(x)$ and $U(x) = H_{n-1}(x) e^{-x^2} dx$) and applying the fact $\frac{d}{dx}[H_n(x) e^{-x^2}] = -H_{n+1}(x) e^{-x^2}$ (which follows from the Rodrigues formula), for $n = 1, 2, \ldots$ we have

$$b_n = \int_{-\infty}^{\infty} e^{-x^2/2} F(x) \phi_n(x) dx = (\sqrt{\pi} 2^n n!)^{-1/2} \int_{-\infty}^{\infty} F(x) H_n(x) e^{-x^2} dx$$

$$= (\sqrt{\pi} 2^n n!)^{-1/2} \int_{-\infty}^{\infty} g(x) e^{x^2/2} H_{n-1}(x) e^{-x^2} dx$$

$$= (2n)^{-1/2} \int_{-\infty}^{\infty} g(x) (\sqrt{\pi} 2^{n-1}(n-1)!)^{-1/2} H_{n-1}(x) e^{-x^2/2} dx = (2n)^{-1/2} a_{n-1}. \blacksquare$$

The last theorem can be expanded. The set of weighted Hermite polynomials forms a complete orthonormal system (a basis) for $L^2(\mathbb{R})$. This result is our main theorem in this subsection:

> **Theorem 3.4.3** The collection of weighted Hermite polynomials $\{(\sqrt{\pi} 2^n n!)^{-1/2} H_n(x) e^{-x^2/2}\}$, with $n = 0, 1, 2, \ldots$, forms a basis for $L^2(\mathbb{R})$. More explicitly, if $f \in L^2(\mathbb{R})$ has L^2-inner product $\langle f, \phi_n(x) \rangle = 0$ for $n = 0, 1, 2, \ldots$, then $f = 0$ almost everywhere.

We give a direct proof of the theorem that follows from two lemmas.[12] It expands a technique that Antoni Zygmund and Richard Wheeden used [123, pp. 217–219] to prove the trigonometric system complete for $L^2(-\pi, \pi)$. (Advanced Exercises of Section 3.3 outline their proof.)

> **Lemma 3.4.2** The weighted Hermite polynomials are a basis for $L^2(\mathbb{R})$ if and only if the system $\{x^n e^{-x^2/2}\}_{n=0}^{\infty}$ is complete in $L^2(\mathbb{R})$.

Proof. A monomial x^n with nonnegative integer power can be written as a finite linear combination $x^n = a_0 H_0(x) + a_1 H_1(x) + \cdots + a_n H_n(x)$ of Hermite polynomials. Therefore the linear

---

[12] The material presented first appeared in *The American Mathematical Monthly* [65]. The author thanks James Rovnyak and Scott Chapman for their assistance with the publication.

3.4 Application: Quantum Mechanics

span of Hermite polynomials $\{f : f = \sum_{n=0}^{\infty} c_n H_n(x)\}$, where the coefficients c_n are real-valued constants, is the same as the linear span of monomials $\{f : f = \sum_{n=0}^{\infty} c_n x^n\}$. ∎

Lemma 3.4.3 *Suppose $f(x)$ is a bounded continuous function with $f(x) \sim \sum_{n=0}^{\infty} a_n \phi_n(x)$ and has $a_n = 0$ for $n = 0, 1, 2, \ldots$. Then $f \equiv 0$.*

Proof. Proceed by way of contradiction using Lemma 3.4.2: assume there is a bounded continuous function f not identically zero with $\langle f(x), x^n e^{-x^2/2} \rangle = 0$ for all nonnegative integer powers n (again, this inner product is defined as for $L^2(\mathbb{R})$). Since $f \not\equiv 0$, there is a real value x_0 such that (without any loss of generality) $f(x_0) > 0$. Since f is continuous, there exist positive values δ and c such that $f(x) > c$ whenever $|x - x_0| < \delta$. The proof uses these values to construct an algebraic contradiction.

Define a constant C as $C = (1 + 3\delta^2/8)^2$. Because δ is greater than 0, $C > 1$. Define

$$s_n(x) = (1 + \delta^2/2 - (x - x_0)^2/2)^{2n} e^{-x^2/2}$$

for any nonnegative integer n. We will show $0 = \langle s_n, f \rangle$ provides the contradiction. Three facts result:

1. Whenever $|x - x_0| < \delta$, for a nonnegative integer n the function $s_n(x) > e^{-x^2/2}$, since $1 + \delta^2/2 - (x - x_0)^2/2 > 1 + \delta^2/2 - \delta^2/2 = 1$.
2. Whenever $|x - x_0| < \delta/2$, for a nonnegative integer n we have $s_n(x) > C^n e^{-x^2/2}$, since $(1 + \delta^2/2 - (x - x_0)^2/2)^{2n} > (1 + \delta^2/2 - \delta^2/8)^{2n} = (1 + 3\delta^2/8)^{2n} = C^n$.
3. The function $s_n(x)$ is always nonnegative because it is a product of a squared term and an exponential. Hence, if $|x - x_0| > \delta$, then $-(x - x_0)^2/2 < -\delta^2/2$, which means $0 \leq s_n(x) < (1 + \delta^2/2 - \delta^2/2)^{2n} e^{-x^2/2} = e^{-x^2/2}$. We conclude $|s_n(x)| \leq e^{-x^2/2}$ whenever $|x - x_0| \geq \delta$.

Since $f(x) \geq -|f(x)|$ for any x, the third fact implies $s_n(x)f(x) \geq -|f(x)|e^{-x^2/2}$ whenever $|x - x_0| > \delta$. Similarly, the second fact implies $s_n(x)f(x) > C^n \cdot c \cdot e^{-x^2/2}$ whenever $|x - x_0| < \delta/2$. Finally, the first fact implies $s_n(x)f(x) > e^{-x^2/2}c > -|f(x)|e^{-x^2/2}$ whenever $\delta/2 < |x - x_0| < \delta$, since $c > -|f(x)|$ on that interval.

The function $s_m(x)$ is a linear combination of $x^0 e^{-x^2/2}, x^1 e^{-x^2/2}, \ldots, x^{4m} e^{-x^2/2}$ for any positive integer m, and so we are assuming $0 = \langle s_n, f \rangle$ for each n. But the inner product equals

$$\int_{-\infty}^{x_0-\delta} s_n(x)f(x)\,dx + \int_{x_0-\delta}^{x_0-\delta/2} s_n(x)f(x)\,dx + \int_{x_0-\delta/2}^{x_0+\delta/2} s_n(x)f(x)\,dx$$
$$+ \int_{x_0+\delta/2}^{x_0+\delta} s_n(x)f(x)\,dx + \int_{x_0+\delta}^{\infty} s_n(x)f(x)\,dx.$$

From our inequalities, the sum is greater than or equal to

$$-\int_S |f(x)|e^{-x^2/2}\,dx + \int_{x_0-\delta/2}^{x_0+\delta/2} C^n c\, e^{-x^2/2}\,dx,$$

where $S = \{x : x \geq x_0 + \delta/2 \text{ or } x \leq x_0 - \delta/2\}$. Since $-|f(x)|e^{-x^2/2}$ is negative, the first integral is greater than or equal to $-\int_{-\infty}^{\infty} |f(x)|e^{-x^2/2}\,dx$, which is bounded below by $-M\sqrt{2\pi}$, where we are assuming the bounded function f satisfies $|f(x)| \leq M$ and are using the fact that $\int_{-\infty}^{\infty} e^{-x^2/2}\,dx = \sqrt{2\pi}$. In short, the first term satisfies $-\int_S |f(x)|e^{-x^2/2}\,dx \geq -K_1$ for a finite constant K_1.

Furthermore, since $e^{-x^2/2} > e^{-(|x_0|+\delta/2)^2/2}$ whenever $x_0 - \delta/2 < x < x_0 + \delta/2$, the second term satisfies $\int_{x_0-\delta/2}^{x_0+\delta/2} C^n c\, e^{-x^2/2}\,dx > \int_{x_0-\delta/2}^{x_0+\delta/2} C^n c\, e^{-(|x_0|+\delta/2)^2/2}\,dx = C^n c\, e^{-(|x_0|+\delta/2)^2/2} \int_{x_0-\delta/2}^{x_0+\delta/2} dx = C^n c\, e^{-(|x_0|+\delta/2)^2/2} \delta$. In summary, $0 = \langle s_n, f \rangle > -K_1 + C^n c\, e^{-(|x_0|+\delta/2)^2/2} \cdot \delta$.

Since $C > 1$, we may choose n large enough so that $C^n > e^{(|x_0|+\delta/2)^2/2} K_1/(c\,\delta)$, which implies $\langle s_n, f \rangle > 0$ and contradicts the assumption $\langle s_n, f \rangle = 0$. This proves the lemma. ∎

We are now ready to prove the main result.

Proof of Theorem 3.4.3. The result follows from Lemma 3.4.3 if $f(x)$ is continuous and bounded. If $f \in L^2(\mathbb{R})$ is not continuous and bounded, we examine the function $g \in L^2(\mathbb{R})$ defined as $g(x) = f(x)e^{-x^2}$. Assuming g has a generalized Fourier expansion $g(x) \sim \sum_{n=0}^{\infty} a_n \phi_n(x)$, then, for any $n = 0, 1, 2, \ldots$,

$$a_n = \langle g, \phi_n \rangle = \int_{-\infty}^{\infty} g(x)\phi_n(x)\,dx = (\sqrt{\pi}\, 2^n n!)^{-1/2} \int_{-\infty}^{\infty} f(x) e^{-x^2} H_n(x) e^{-x^2/2}\,dx$$

$$= (\sqrt{\pi}\, 2^n n!)^{-1/2} \int_{-\infty}^{\infty} f(x) \sum_{k=0}^{\infty} \frac{(-1)^k x^{2k}}{k!} H_n(x) e^{-x^2/2}\,dx$$

$$= (\sqrt{\pi}\, 2^n n!)^{-1/2} \sum_{k=0}^{\infty} \frac{(-1)^k}{k!} \int_{-\infty}^{\infty} f(x)\, x^{2k} H_n(x) e^{-x^2/2}\,dx.$$

Since $x^{2k} H_n(x) e^{-x^2/2}$ is a finite linear combination of terms in the system $\{x^n e^{-x^2/2}\}_{n=0}^{\infty}$, the assumption $\langle f, \phi_n(x) \rangle = 0$ for $n = 0, 1, 2, \ldots$ results in each of the integrals $\int_{-\infty}^{\infty} f(x)\, x^{2k} H_n(x) e^{-x^2/2}\,dx$ equaling zero for each n and k. Hence $a_n = 0$ for $n = 0, 1, 2, \ldots$.

Now we can apply Lemma 3.4.1. Assuming $g(x) = f(x)e^{-x^2} \sim \sum_{n=0}^{\infty} a_n \phi_n(x)$, we have, for each n:

$$a_n = \int_{-\infty}^{\infty} g(x)\phi_n(x)\,dx = (\sqrt{\pi}\, 2^n n!)^{-1/2} \int_{-\infty}^{\infty} f(x) e^{-x^2/2} H_n(x)\, e^{-x^2}\,dx$$

$$= \sqrt{2n+2} \int_{-\infty}^{\infty} F(x)\, H_{n+1}(x)\, e^{-x^2}\,dx,$$

where F, defined as $F(x) = \int_{-\infty}^{x} f(t)\, e^{-t^2/2}\,dt$, is bounded (by the Schwarz inequality[13]) and continuous, and $e^{-x^2/2} F(x)$ has limit 0 as $x \to \pm\infty$.

Hence, assuming the generalized Fourier expansion for $e^{-x^2/2} F(x)$ is $e^{-x^2/2} F(x) \sim \sum_{n=0}^{\infty} b_n \phi_n(x)$, then $0 = a_n = \sqrt{2n+2}\, b_{n+1}$. That is, the continuous and bounded function

---

[13] The Schwarz inequality is Theorem 3.2.1. The application here chooses $g(x) = e^{-x^2/2}$ (and $f = f$).

3.4 Application: Quantum Mechanics

G defined as $G(x) = e^{-x^2/2}F(x) - b_0$ has all its Fourier coefficients equal to zero, and so Lemma 3.4.3 implies $G(x) \equiv 0$. This means $F(x) = b_0 e^{x^2/2}$. Differentiating, $f(x)e^{-x^2/2} = F'(x) = b_0 x e^{x^2/2}$ a.e., or $f(x) = b_0 x e^{x^2}$ a.e. But $f \in L^2(\mathbb{R})$, and so $b_0 = 0$. Hence $f(x) = 0$ a.e., and the result follows. ∎

Theorem 3.4.3 begins to indicate why the L^2 spaces form a proper setting in which to study quantum mechanics. The acceptable solutions to the classical harmonic oscillator's Schrödinger equation $\psi''(x) + (2p + 1 - x^2)\psi(x) = 0$ happen exactly when $p = n$ and are of the form $\psi_n(x) = H_n(x)e^{-x^2/2}$. They are *eigenfunction* solutions, and the corresponding values of p, equal to the nonnegative integers n, are the *eigenvalues*. The eigenfunction solutions are in $L^2(\mathbb{R})$ and form a basis for the Hilbert space. That implies, as described in Section 3.3, that $f \in L^2(\mathbb{R})$ has a corresponding generalized Fourier series—an infinite series using the collection of Hermite functions $\{(\sqrt{\pi}2^n n!)^{-1/2} H_n(x) e^{-x^2/2}\}$, which we have normalized (they form an orthonormal basis). The generalized Fourier series for a function $f \in L^2(\mathbb{R})$ is $f(x) \sim \sum_{n=0}^{\infty} c_n (\sqrt{\pi}2^n n!)^{-1/2} H_n(x) e^{-x^2/2}$, where $c_n = \langle f, (\sqrt{\pi}2^n n!)^{-1/2} H_n(x) e^{-x^2/2}\rangle$ is the generalized Fourier coefficient formed from the $L^2(\mathbb{R})$ inner product.

This chapter has discussed function spaces, describing L^p as a Banach space of functions and focusing on the additional inner product structure that exists in L^2. In each case, a Lebesgue integral $\int |f|^p$ equips the space with a norm. We have explored particular cases of $L^2(I)$, where I might not be the entire real line but some interval subset, generating connections (via the corresponding basis elements) to the classical mathematics of orthogonal polynomials and to quantum mechanics. The Hilbert spaces have complete orthonormal bases that can be generated by an application of the Gram-Schmidt process.

The next chapter will define the integral with respect to different measures—no longer only Lebesgue measure. Each measure can be used to examine Banach and Hilbert spaces L^p, defined using integrals on them, in the same way we have discussed L^p spaces in this chapter using integrals on Lebesgue measure. The alternative scenarios are more general. Their norms are constructed as integrals of $|f|^p$ but now with a Lebesgue integral defined in terms of different measures. The generality of the function spaces opens an immense new set of considerations, which Chapter 5 explores as it examines operations on functions. Complicated operations on functions in one Hilbert space will often turn out to be equivalent to simple operations on functions in another. The Lebesgue integral typically forms the bridge needed to make the movements between Hilbert spaces work. Knowing how to make them (so to unravel the complexity of operations on functions) will be a major goal. Structures we have learned about in this chapter (such as Hilbert space isomorphisms) will serve as tools in the investigations. In summary, we have begun to develop in this chapter a framework (using L^p spaces, Hilbert space isomorphisms, etc.) that will soon be generalized, eventually empowering us to analyze successfully many operations on functions.

Solutions to Questions

3.4.1 Using the product rule and the earlier calculation for H_2, $H_3(u) = -e^{u^2} \frac{d}{du}[(4u^2 - 2)e^{-u^2}] = 8u^3 - 12u$. Similarly, $H_3(u) = -e^{u^2} \frac{d}{du}[(8u^3 - 12u)e^{-u^2}] = 16u^4 - 48u^2 + 12$. The other three follow similarly; the results are $H_4(u) = 16u^4 - 48u^2 + 12$, $H_5(u) = 32u^5 - 160u^3 + 120u$, and $H_6(u) = 64u^6 - 480u^4 + 720u^2 - 120$.

3.4.2 Using integration by parts with $U = 4x^2$ and $V'(x) = 2xe^{-x^2}$ and then an appropriate u-substitution, $\int_{-\infty}^{\infty}(8x^3 - 4x)e^{-x^2}\,dx = \int_{-\infty}^{\infty} 8x^3 e^{-x^2}\,dx - \int_{-\infty}^{\infty} 4xe^{-x^2}\,dx = \int_{-\infty}^{\infty} 8xe^{-x^2}\,dx - 0 = 0 - 0 = 0$. Similarly, using integration by parts, $\int_{-\infty}^{\infty} 4x^2 e^{-x^2}\,dx = \int_{-\infty}^{\infty} 2e^{-x^2}\,dx = 2\sqrt{\pi}$, and $\int_{-\infty}^{\infty}(4x^2 - 2)^2 e^{-x^2}\,dx = \int_{-\infty}^{\infty}(16x^4 - 16x^2 + 4)e^{-x^2}\,dx = (12 - 8 + 4)\sqrt{\pi} = 8\sqrt{\pi} = 2!2^2\sqrt{\pi}$.

Reading Questions for Section 3.4

1. What is the Schrödinger equation? Describe each of the variables in your answer.

2. What equation, called the Hermite equation, does the Schrödinger equation transform itself into under a change of variable and a product structure transformation, in the case of a classical harmonic oscillator?

3. What constraints, due to standard physical conditions, are traditionally imposed upon the Schrödinger equation's wave functions?

4. What are the solutions to the Hermite equation, according to the standard constraints that are imposed upon them?

5. For what L^2 space does the set of solutions form an orthogonal basis?

6. How can a function f in $L^2(\mathbb{R})$ be expanded as an infinite series in terms of the Hermite functions $H_n(x)e^{-x^2/2}$, $n = 0, 1, 2, \ldots$?

Exercises for Section 3.4

Exercises 1–6 are general investigations into the Hermite polynomials.

1. Continue identifying the Hermite polynomials by examining the infinite series solution for $g_1(u)$ and $g_2(u)$ with $p = 2n$. What are $H_4(u)$, $H_5(u)$, $H_6(u)$, $H_7(u)$, and $H_8(u)$, up to a constant multiple?

2. Show explicitly, by taking derivatives and through substitution, that the degree one Hermite polynomial $H_1(u) = 2u$ forms a solution to $g''(u) - 2u \cdot g'(u) + 2pg(u) = 0$. What is the value for p?

3. Show if $\psi_a(u)$ and $\psi_b(u)$ are two solutions to $g''(u) - 2u \cdot g'(u) + 2pg(u) = 0$, then so is $c_a \psi_a(u) + c_b \psi_b(u)$ for constants c_a and c_b.

4. All solutions to $g''(u) - 2u \cdot g'(u) + 2pg(u) = 0$ are of the form $c_1 g_1(u) + c_2 g_2(u)$, where $c_1, c_2 \in \mathbb{R}$ and g_1 and g_2 are as defined in Section 3.4.1. When $p = n$, is $H_n(u)$ the only polynomial $p(u)$, up to a constant, to form a solution $\psi(u) = p(u) \cdot e^{-u^2/2}$ that satisfies the corresponding Schrödinger equation?

5. On the same coordinate plane, graph the polynomial functions defined by $f_n(x) = H_n(x)/n^2$ for $n = 1, 2, 3, 4, 5$. (Here, the scaling factor n^2 allows the graphs to fit on the same plot.)

6. The Hermite differential equation with $p = 5$ is $g''(x) - 2xg'(x) + 10g(x) = 0$. Determine the solution that satisfies the initial conditions $g(0) = 1$ and $g(1) = 0$.

Exercises 7–8 work with various forms of the Schrödinger equation.

3.4 Application: Quantum Mechanics

7. Substitute the change of variable $u = 2\pi \sqrt{fm/h} \cdot x$ into the Schrödinger equation
$$\psi''(x) + (8\pi^2 m/h^2)(E - k[x(t)]^2/2)\psi(x) = 0$$
to prove it is transformed into $\psi''(u) + (2p + 1 - u^2)\psi(u) = 0$, where p is the constant $p = E/(hf) - 1/2$.

8. Assume $\lim_{x \to \infty} g(x)e^{-x^2/2} = 0$ and $\lim_{x \to -\infty} g(x)e^{-x^2/2} = 0$ for a function g that is not a polynomial. Explain why the representation $g(x)e^{-x^2/2} = \sum_{n=0}^{\infty} a_{2n}x^{2n} / \sum_{n=0}^{\infty} b_{2n}x^{2n} = \dfrac{a_0 + a_2 x^2 + a_4 x^4 + \cdots + a_{2n}x^{2n} + \cdots}{b_0 + b_2 x^2 + b_4 x^4 + \cdots + b_{2n}x^{2n} + \cdots}$ means there cannot exist a value $N \in \mathbb{N}$ such that $a_{2n} > b_{2n}$ for $n \geq N$.

Exercises 9–12 develop the generating function for the Hermite polynomials.

9. Evaluate $f(t) = e^{2tx - t^2}$ at $t = 0$. Compare to $H_0(x)$.

10. Repeat the last exercise, but now evaluate the first derivative with respect to t at 0, finding $f'(0)$. Compare to $H_1(x)$.

11. For $f(t) = e^{2tx - t^2}$, evaluate the nth derivative with respect to t at $t = 0$, finding $f^n(0)/n!$. Compare with $H_n(x)$ for $n = 0, 1, 2, 3, 4, 5$.

12. Use your results from the previous exercise to determine the Maclaurin series for $f(t) = e^{2tx - t^2}$.

Exercises 13–18 prove facts about the Hermite polynomials.

13. Prove $H_n(x)$ is an even function—in other words, that $H_n(-x) = H_n(x)$—for n even.

14. Prove $H_n(x)$ is an odd function—in other words, that $H_n(-x) = -H_n(x)$—for n odd.

15. Use the fact (see Exercise 12) that $e^{2tx - t^2} = \sum_{n=0}^{\infty} \dfrac{H_n(x)}{n!} t^n$ to say why
$$H_n(x) = \dfrac{\partial^n}{\partial t^n} e^{2xt - t^2}\bigg|_{t=0} = e^{x^2} \cdot \dfrac{\partial^n}{\partial t^n} e^{-(x-t)^2}\bigg|_{t=0}.$$
Then use a change of variable $z = x - t$ to prove Rodrigues' formula; namely, that $H_n(x) = (-1)^n e^{x^2} \dfrac{d^n}{dx^n} e^{-x^2}$.

16. The last exercise proved the following theorem: if $e^{2tx - t^2} = \sum_{n=0}^{\infty} \dfrac{H_n(x)}{n!} t^n$, then $H_n(x) = (-1)^n e^{x^2} \dfrac{d^n}{dx^n} e^{-x^2}$. Prove the converse is true.

17. Prove $H_n'(x) = 2n H_{n-1}(x)$. (Hint: without loss of generality assume $n = 2k$ is even, use the expression derived before Theorem 3.4.1, which is $H_{2k}(x) = 1 - (4k/2!)x^2 + \cdots - (4^k 2k(2k-2) \cdots 2/(4k)!)x^{2k}$, and differentiate both sides.)

18. Prove the recurrence relation for Hermite polynomials: $H_{n+1}(x) = 2x H_n(x) - 2n H_{n-1}(x)$. (Hint: differentiate both sides of Rodrigues' formula and apply the result in the previous exercise.)

The proof of Theorem 3.4.2 used the fact that the leading coefficient of $H_n(x)$ is $c_n = 2^n$. Exercises 19–21 describe why this formula holds.

19. Show the leading coefficients for the Hermite polynomials $H_0(x)$ and $H_1(x)$ are of the form 2^0 and 2^1, respectively.

20. Show the leading coefficient of $H_n(x)$ is $2c_{n-1}$, where c_{n-1} is the leading coefficient for $H_{n-1}(x)$. (*Hint*: use either $H'_n(x) = 2nH_{n-1}(x)$ or $H_{n+1}(x) = 2xH_n(x) - 2nH_{n-1}(x)$.)

21. We know $\int_{-\infty}^{\infty} e^{-x^2} \cdot d^n(H_m(x))/dx^n \, dx = n!c_n \int_{-\infty}^{\infty} e^{-x^2} \, dx$, where c_n is the leading coefficient of $H_n(x)$. Use the last two exercises to prove $c_n = 2^n$.

Advanced Exercises. Exercises 22–30 prove the fact used in Theorem 3.4.2 and Question 3.4.2; namely, $\int_{-\infty}^{\infty} e^{-x^2} \, dx = \sqrt{\pi}$. The proof relies on an ability to integrate functions $f(x, y)$ of two variables (functions defined on $\mathbb{R}^2$), which we develop in Exercises 22–26, as well as Fubini's theorem, which we present and use in Exercises 27–30. It requires us to define a step function f on $\mathbb{R}^2$, which starts by examining characteristic functions. As for the one-dimensional case, a characteristic function $\mathcal{X}_S(x, y)$ on a set S in $\mathbb{R}^2$ is defined to equal 1 if the point (x, y) is in S and 0 if (x, y) is not in S. On $\mathbb{R}^2$, a bounded interval I is defined as a rectangle in $\mathbb{R}^2$; it is of the form $I = \{(x, y) : a < x < b, c < y < d\}$. I is open, but any of the inequalities in this description of I can be replaced by strict inequalities to get a closed or semi-open interval I. A characteristic function on it is

$$\mathcal{X}_I(x) = \begin{cases} 1 & \text{if } x \in (a, b) \text{ and } y \in (c, d) \\ 0 & \text{if } x \notin (a, b) \text{ or } y \notin (c, d), \end{cases}$$

but any of the finite endpoints $a, b, c,$ or d may also be included in the interval (according to whether the corresponding inequalities defining I are strict or not).

22. Give four examples of characteristic functions on a bounded interval of $\mathbb{R}^2$, where at least one of your intervals is open and one is closed. Graph each of your four interval sets on an xy-coordinate axis.

23. Using three-dimensional $x, y,$ and z axes (with the positive z axis pointing upward), graph the four characteristic functions $z = \mathcal{X}_I$ you defined in the last exercise. (The height of the function above the interval I should be 1, and so each of the functions looks like a solid plane at height 0, except a rectangular interval I is cut out of the plane and raised to height 1.)

24. A step function $f(x, y)$ on $\mathbb{R}^2$ is of the form $f(x, y) = \sum_{j=1}^{n} c_j \cdot \mathcal{X}_{I_j}(x, y)$, where c_j is a real constant and I_j is a bounded interval in $\mathbb{R}^2$. For the characteristic functions $z = \mathcal{X}_{I_j}$, $j = 1, 2, 3, 4$ you graphed in the last exercise, use constants $c_1 = 1$, $c_2 = 2$, $c_3 = -1$, and $c_4 = -2$ to formulate the step function $f(x, y) = \sum_{j=1}^{4} c_j \cdot \mathcal{X}_{I_j}(x, y) = \mathcal{X}_{I_1} + 2\mathcal{X}_{I_2} - 1\mathcal{X}_{I_3} - 2\mathcal{X}_{I_4}$. Then graph it in three dimensions.

3.4 Application: Quantum Mechanics

25. The measure of an interval $I = \{(x, y) : a < x < b, c < y < d\}$ is $m(I) = (b - a)(d - c)$. Then the Lebesgue integral of a step function $f(x, y) = \sum_{j=1}^{n} c_j \cdot \mathcal{X}_{I_j}(x, y)$ is $\int f = \sum_{j=1}^{n} c_j \cdot m(I_j)$. As for functions on one variable (see Theorem 1.3.1), the definition is consistent. Calculate $\int f$ for the function you formulated in the previous exercise. The integral is also written $\int_{\mathbb{R}^2} f(x, y) \, d(x, y)$.

26. Two-dimensional sets of measure zero are defined in terms of intervals just as they were for one-dimensional sets (see Definition 1.2.2). Also, for any (x, y), the term $\lim_{n \to \infty} f_n(x, y) = f(x, y)$ means the sequence $\{f_n(x, y)\}$ approaches the value $f(x, y)$ in the following sense: Given $\varepsilon > 0$, there exists a positive real value N such that $|f_n(x, y) - f(x, y)| < \varepsilon$ whenever $n > N$. In this case, we say $f(x, y)$ is the (pointwise) limit of the sequence $\{f_n(x, y)\}$. With these concepts in place, we examine any function $f(x, y)$ that has a nondecreasing sequence $\{\phi_n(x, y)\}$ of step functions that converges pointwise almost everywhere to a function $f(x, y)$. Then $\int f \equiv \lim_{n \to \infty} \int \phi_n$. When f has a finite integral, we say f belongs to the space $L^0(\mathbb{R}^2)$. Use this definition to show $\int_{\mathbb{R}^2} (x + y) \cdot \mathcal{X}_I \, d(x, y) = 1$, where the characteristic function $\mathcal{X}_I$ is taken over the interval $I = \{(x, y) : 0 \leq x \leq 1, 0 \leq y \leq 1\}$. Conclude $f(x, y) = x + y$ is in $L^0(I)$.

27. The space $L^1(\mathbb{R}^2)$ of Lebesgue integrable functions on $\mathbb{R}^2$ consists of functions f of the form $f = g - h$, where g and h are in $L^0(\mathbb{R}^2)$. Their integrals are defined as $\int f = \int g - \int h$. As for functions of one variable (see Corollary 1.5.1), the definition is consistent. Prove: For functions f and u in $L^1(\mathbb{R}^2)$ and constants α and β, $\alpha f + \beta u$ is in $L^1(\mathbb{R}^2)$ and $\int \alpha f + \beta u = \alpha \int f + \beta \int u$.

28. The *polar transformation* is a change of variables $T(r, \theta) = (x, y)$ over $\mathbb{R}^2$ with $x = r \cos \theta$ and $y = r \sin \theta$, transforming a point (x, y) represented in the rectangular coordinate system into (r, θ). Here r is the distance from the point to the origin and θ the angle (between 0 and 2π) formed by the positive x axis and the ray emanating from the origin that goes through the point. A multivariate calculus class introduces this transformation for the Riemann integral. The substitution turns out to work similarly and in broad ways for the Lebesgue integral: for $f \in L^1(\mathbb{R}^2)$, $\int_{\mathbb{R}^2} f(x, y) \, d(x, y) = \int_{\mathbb{R}^2} f(r, \theta) \, r \, d(r, \theta)$. Note the addition of the multiplier r, which is the Jacobian and sometimes called the transformation's weight function. When integrating $f(x, y)$ over a measurable subset U of $\mathbb{R}^2$, the transformation therefore works similarly; in this case the transformed integral is over the subset V, where $T(V) = U$. Using this polar transformation, show $\int_U e^{-(x^2+y^2)} \, d(x, y) = \int_V e^{-r^2} \, r \, d(r, \theta)$. Furthermore, if $U = \{(x, y) : 0 < x, y < \infty\}$, say why $V = \{(r, \theta) : 0 < r < \infty \text{ and } 0 < \theta < \pi/2\}$.

29. Use the last exercise to show $\int_U e^{-(x^2+y^2)} \, d(x, y) = \pi/4$ where $U = \{(x, y) : 0 < x, y < \infty\}$.

30. *Fubini's theorem* is a famous and useful result that says, for $f \in \mathbb{R}^2$, $\int_{\mathbb{R}^2} f(x, y) \, d(x, y) = \int_{\mathbb{R}} (\int_{\mathbb{R}} f(x, y) \, dx) dy$. That is, we can calculate the integral by integrating one variable at a time (in either order), and the function $F(y) = \int_{\mathbb{R}} f(x, y) \, dx$ is in $L^1(\mathbb{R})$. Use Fubini's theorem and the result of the last exercise to prove $\int_0^\infty e^{-x^2} \, dx = \sqrt{\pi}/2$. Conclude that $\int_{-\infty}^\infty e^{-x^2} \, dx = \sqrt{\pi}$.

Notes for Chapter 3

Banach spaces (normed linear spaces) are named after the brilliant Polish (later Soviet, as Poland was incorporated into the U.S.S.R.) mathematician Stefan Banach, considered one of the founders of functional analysis. One of Banach's most famous theorems, which he discovered and published with coauthor Alfred Tarski, is the Banach-Tarski paradox. It has to be one of the most astounding results in the history of mathematics, especially since it is so nonintuitive that it seems impossible. Here is the general version of the result that Banach and Tarski published in 1924 [9]:

Take any bounded subsets A and B of a Euclidean space of at least three dimensions, where A and B both have non-empty interior. Then you can partition A and B into the same finite number n of disjoint subsets, $A = A_1 \cup A_2 \cup \cdots \cup A_n$ and $B = B_1 \cup B_2 \cup \cdots \cup B_n$, so the sets A_i and B_i are congruent for each i between 1 and n.

A simple illustration (and the one most often used in the general public) of this result's preposterously amazing meaning comes from defining A to be a sphere with radius 1 in 3-space (so A is the unit ball) and B to be two copies of A (so B is made up of two unit balls). The theorem says we can partition a single unit ball A into a finite number of pieces and then rearrange those pieces (renaming them B_1 through B_n) so they now come together to form two unit balls—they exactly form a partition of B.

Huh? There must be some funny business going on, but there's not. The Banach-Tarski paradox really does describe a reality present in the mathematics of high-dimensional Euclidean space. Or maybe there is—in the sense of one of the cruxes of this particular three-dimensional construction: the sets forming the partition can be nonmeasurable. (This chapter described nonmeasurable sets in one-dimensional space, and there are similar nonmeasurable sets in higher dimensions.) In fact, for the unit ball description, at least one of the sets forming the partition has to be nonmeasurable. The nonmeasurable sets don't have a volume (the three-dimensional equivalent of a set's length or measure) in any straightforward way, and so you are repiecing together subsets that don't have a well-understood notion of volume. Still, whether or not they have a nicely defined volume, they are the same sets—exactly—that piece together to form one unit ball, and then are repositioned remarkably to piece together and form two unit balls!

Banach's creative genius developed new mathematics connected with every topic presented in this chapter (and, it could be argued, with every topic in this book). For example, he was the first to demonstrate an orthonormal basis set (complete) for an L^2 space but not complete in the Banach space L^1. His 1932 text *Théorie des opérations linéaires* has been broadly noted, in publications spanning 40 years, for its clarity and expository force (see [115] and [5]).

Banach survived the conflicts of World War II that led to his native city of Lwów, Poland, being overtaken first by the Soviets (in 1939), then by the Germans (1941), then back to the Soviets (1945). But he survived just barely. His leadership position as president of the Polish Mathematical Society and his connections through his founding in 1929 and editing (with Hugo Steinhaus, who was one of Banach's Ph.D. advisors) the journal *Studia Mathematica*, along with a naturally caring and outgoing personality, meant he established partnerships and friendships with numerous Soviet mathematicians. Until the German invasion, he was thus allowed to continue his work at Lwów University and was appointed dean of the Faculty. He was arrested by the Gestapo but released. On July 3, 1941, the German S.S. Nachtigall battalion unit rounded up 40 Polish Lwów professors, scholars, and other intellectuals and murdered them. In his account, the mathematical historian Ioan James writes in [62] that the commando unit "appeared, ... did its work and then disappeared again. The atrocity apparently had no official Wehrmacht sanction, but was carried out by the direct order of [Heinrich] Himmler, who was visiting the city at the time." Several of Banach's students were killed by the Gestapo, including the brilliant Juliusz Schauder, who was posthumously awarded the international Metaxas Prize [115]. It is fair to say Nazi cruelty touched in some way the life of every Polish mathematician. For example, Alfred Tarski's father, mother, and brother were murdered by the Nazis in 1942.

David Hilbert (after whom the term Hilbert space was named) was a brilliant mathematician who was a product of Königsberg (then in Prussia and now part of Russia, known for the famous Königsberg bridge problem that Euler solved) and a leader (as the chairman of the mathematics department) for the great mathematical institute at Göttingen University. While Hilbert was chair at Göttingen, it was widely regarded as the premier site for mathematical research. There were many reasons Hilbert's mathematical influence was widespread. He had great abilities to solve fundamental mathematical problems in many subfields, including mathematical logic (where his efforts to resolve the logical foundation basis of mathematics became known as Hilbert's Problem and were developed by 1939 powerfully in his two volumes on the topic with coauthor Paul Bernays [58] and [59]). He also

Figure 3.6. David Hilbert.

had (perhaps more than any mathematician throughout history) a sophisticated understanding of the continuing development and progress of mathematics as it was unfolding during his lifetime, combined with a fundamental ability to communicate it to others in the mathematics community. Evidence of this ability comes from his historical address at the 1900 International Congress of Mathematicians, held in Paris, where he gave a list of unsolved problems. (He defined eight of them in the speech and developed the full list of "Hilbert's 23 Problems" in an article later that year.) The announcement declared Hilbert's sense of the problems as the most significant at the time. That famous 1900 speech presented challenges. Mathematicians responded by solving eleven of them and substantially resolving eight more in the 20th century. Hence four are left unresolved at this time. The list's eighth problem is the most famous open problem in mathematics: the Riemann Hypothesis. It is connected to the distribution of primes and claims the real part of each nontrivial zero of the Riemann Zeta Function equals $1/2$. Fermat's Last Theorem, which was not solved until 1995 by Andrew Wiles, was not on Hilbert's list. (Perhaps he omitted it because one of his goals was to present problems that he felt helped determine the foundation on which mathematics was produced, and Fermat's last theorem is arguably not of that type.) Hilbert had a special talent for teaching mathematics and producing an environment and community that encouraged and resulted in research successes, which helped to develop fully his students' abilities. The Mathematics Genealogy Project lists Hilbert with 75 Ph.D. students that include such notables as Richard Courant, Erich Hecke, Hugo Steinhaus, and Hermann Weyl. His book on mathematical physics, coauthored with Courant, is in print to this day [30]. The interested reader will want to see [96], Constance Reid's exceptionally good biography.

The Gram-Schmidt process is named after Jorgen Pedersen Gram (Danish) and Erhard Schmidt (German, and a student of David Hilbert). Gram's work came first, when by 1879 he had presented the paper *On series expansions determined by the methods of least squares* that contained the elements of the process and was later published in 1883. Besides the Gram-Schmidt process, Gram is best known for the "Gram determinant," a value associated with a matrix of numbers called a Gram (or Gramian) matrix. The Gram matrix (used, for example, in differential equations) is constructed from a sequence of vectors $\{\vec{v}_1, \vec{v}_2, \ldots, \vec{v}_n\}$ in a Hilbert space: the entry in the ith row and jth column of the $n \times n$ matrix is $\langle \vec{v}_i, \vec{v}_j \rangle$. If the determinant of the matrix is nonzero, then the vectors are linearly independent.

Section 3.4 has mentioned Charles Hermite as the French mathematician behind the development of Hermite polynomials. A brilliant mathematician, Hermite is also famous for proving in 1873 that e is a transcendental number (which means it is not a zero of a polynomial with rational coefficients), as well as for his proof in 1858 that a fifth-degree polynomial (which Niels Abel had earlier established as not generally solvable using only basic arithmetic operations and roots) *could* be solved using a type of function known as an elliptic function. Hermite is also recognized as being a marvelous teacher; his students include Émile Picard, Émile Borel, and Henri Poincaré. The other orthogonal polynomials described in this chapter are similarly named after notable 19th-century mathematicians. Edmond Laguerre, another French mathematician, developed the Laguerre polynomials as solutions $L(x)$ to the differential equation $xL''(x) + (1-x)L'(x) + nL(x) = 0$ where $n \in \mathbb{N}$. He also developed a powerful algorithm to find (through successive approximations) zeros of a general polynomial. His algorithm is known as Laguerre's method and uses derivative properties of logarithms, which the method applies to the polynomial in a clever way. It works well because it is almost guaranteed to find a root no matter what value is chosen for the initial guess. This characteristic is in contrast to other more famous root-finding algorithms such as the Newton-Raphson method, which is often taught in calculus courses and is effective because of speedy convergence, but which can fail to converge if the initial value is outside of a convergence region. Adrien Legendre determined the Legendre polynomials P_n in the 1780s as solutions $P(x)$ to $\frac{d}{dx}\left[(1-x^2)\frac{d}{dx}P(x)\right] + n(n+1)P(x) = 0$ where $n \in \mathbb{N}$. The polynomials arise as coefficients in a famous Taylor series expansion; namely, $(1 - 2xt + t^2)^{-1/2} = \sum_{n=0}^{\infty} P_n(x)t^n$. As explained in the last section's exercises, such an expression is a *generating function* for the polynomials since it generates them via Taylor's formula for the coefficients, involving the evaluation of nth derivative with respect to t at $t = 0$. Legendre's numerous notable mathematical discoveries (which earned him recognition as one of 72 scholars whose names are engraved on the Eiffel Tower) include an 1830 proof of Fermat's Last Theorem for $n = 5$ (which says there are no positive integers a, b, and c such that $a^5 + b^5 = c^5$). An edited translation of his seminal textbook on Euclidean geometry, first published in 1794, is still available (see [82]).

The classical harmonic oscillator is but one example of a broad expanse of quantum mechanics, which historically developed to explain the physics of electromagnetic radiation in a hollow cavity, especially when Wilhelm Wien noted such blackbody radiation produced an energy output $U(\nu)$ that did not follow the classical Rayleigh-Jeans law (which described the density as an approximate bell-shaped distribution). Instead, Wien found the energy density to behave approximately as $\nu^3 e^{-h\nu/kT}$, where h and k are constants and T is temperature. The constant k is the famous Boltzmann's constant and, in the later development of quantum mechanics, Wien's constant h showed up consistently and was termed the "quantum of action." A highly recommended exposition on quantum theory is the classic text by David Bohm [13], which includes the historical development. Bohm was the first to describe quantum mechanics as evolving out of classical mechanics (including electromagnetics and Fourier analysis). He explains the theory, including a thorough explanation of mathematical foundations, as motivated from an applied and experimental standpoint rather than a purely axiomatic one. It is quite detailed and thorough. For example, it obtains the mean value of the potential energy for the harmonic oscillator [13, p. 308] as exactly half the value of the total energy, and the so-called probability density of the wave packet (which describes a particle's position) as $P(y) = e^{-(y-y_0 \cos \omega t)^2}$, where t is time, ω is the angular frequency of oscillation, and y_0 describes the particle's initial center of location. Other acclaimed authors of classical texts on quantum theory include Richard Feynman [41] and Paul Dirac [35].

4
Measure Theory

Chapter 1 showed the dependence of the Lebesgue integral on the concept of measure: to understand how to integrate, we first need to define the Lebesgue measure of an interval and describe which sets have Lebesgue measure zero. Measure is important. We can define the Lebesgue measure of any set S (not just intervals), as long as the characteristic function of S is a measurable function (in the sense of Section 3.1). The full story of set measure is measure theory, and is interesting in its own right—it gives a way to approach with new eyes key properties of sets. The story is also interesting historically. Henri Lebesgue defined the Lebesgue integral (and hence L^1) by figuring out how to calculate the measure of sets. A property of his measure, *countable additivity* (explained in this chapter), was the breakthrough that made his definition work. Prior to his description of measure, mathematicians had tried but failed to produce measures that were countably additive. The French mathematician Camille Jordan had described what he termed the "content" of a set S. If we label his content $m(S)$, then Jordan was able to show $m(S \cup T) = m(S) + m(T)$ for two disjoint (nonoverlapping) sets, and more generally that $m(S_1 \cup S_2 \cup \cdots \cup S_n) = m(S_1) + m(S_2) + \cdots + m(S_n)$ for a finite number of disjoint sets. But Lebesgue's description of set measure had the additivity property for a countable number of sets, finite or infinite. That was the missing piece to the puzzle of how to define the integral, allowing Lebesgue to use his definition of measure to produce his integral for functions in L^1.

We've already succeeded in finding the Lebesgue integral of such functions, so why should we look at the Lebesgue measure of general types of sets? The full study is quite advanced (and hence is difficult), and that was the reason we avoided it as much as possible as we constructed the Lebesgue integral in Chapter 1.

The good news is we've turned Lebesgue's investigative procedure on its head. His method had to describe the measure of sets first, and then base his integral on it. The Daniell-Riesz approach first defines the integral of functions in L^1, and now we can use that to describe measure. The measure of a set follows immediately, and is straightforward (and easy). A discussion of the properties of Lebesgue measure naturally follows, and our framework allows us to state many of the properties—some with proof and some without. The interested

Figure 4.1. Camille Jordan at age 77.

reader is introduced to a fascinating branch of mathematics—measure theory—and there are many wonderful books that should now be more accessible for further reading and study. In fact, students usually find Lebesgue's procedure more understandable after they have worked through the introductory material we present, and Lebesgue's methodology is still, appropriately, considered necessary for the most general framework to study measure theory. Courses on measure theory are common at the graduate mathematics level. This chapter hopefully will motivate further study.

Lebesgue's measure is not the only one that has the countable additivity property. Hence the natural question: what happens if we define a measure (on intervals) differently than Lebesgue but in a way that produces, once again, countable additivity? Would each measure we define produce a Lebesgue-type integral, similar to the way that Lebesgue measure produced the Lebesgue integral? The answer, well worth celebrating, is "yes," and Section 4.2 will show how it works. The study of Lebesgue measure (as it is connected with the Lebesgue integral as defined in Chapter 1) will then provide a foundation for a more general concept of the integral, one that is defined for other measures. Chapter 1's systematic method applies to other measures to produce an integral defined for that measure. Each measure produces a new integral, and all of them are constructed from the same systematic process. This generalization will have profound effects, generalizing the theory of the Lebesgue measure integral. To state just one example, a complete $L^1(\mu)$ space of functions integrable in terms of any one of these measures μ (not just Lebesgue measure) will have many applications.

4.1 Lebesgue Measure

This section describes the Lebesgue measure of a measurable set S. Both definitions—of a measurable set and of its Lebesgue measure—arise from the definition of the Lebesgue integral as presented in Chapter 1. Because of the connection to the integral, properties of the Lebesgue integral illuminate properties of the Lebesgue measure. The next definition provides the basic framework.

Definition 4.1.1 *A set S in $\mathbb{R}$ is Lebesgue measurable when its characteristic function $\mathcal{X}_S$ is measurable in the sense of Definition 3.1.1. In this case, we define the Lebesgue measure of S, denoted $m(S)$, as*

$$m(S) = \int \mathcal{X}_S$$

whenever $\mathcal{X}_S \in L^1$. If $\mathcal{X}_S$ is not Lebesgue integrable, then we define $m(S) = \infty$.

We already know not every set is Lebesgue measurable (the Vitali set presented in Section 3.1 is an example). We call such sets nonmeasurable. A full study of Lebesgue measure theory often begins by contrasting nonmeasurable sets in $\mathbb{R}$ with those that are measurable. Then the question is how to gather together measurable sets into *sigma-algebra* collections. The term "sigma-algebra" is typically written as σ-algebra. We'll define what this term means later. At this point, we'll say that a standard study of measure theory uses such a σ-algebra as the playing field—once a collection of measurable sets forming a σ-algebra is established, then a measure theorist can see how its sets interact with one another under intersections and unions. For example, the σ-algebra structure is set up so that countable unions or intersections

4.1 Lebesgue Measure

of its sets also belong to the σ-algebra. This chapter studies a standard one, where bounded intervals are always measurable. It is easy to see that Definition 4.1.1 generalizes what we already know about Lebesgue measure. Namely, if S is a bounded interval, say (a, b), then $\mathcal{X}_{(a,b)}$ is measurable (by Theorem 3.1.1); in fact it is integrable. Hence Definition 4.1.1 applies and says $m((a, b)) = \int \mathcal{X}_{(a,b)} = b - a$. This result agrees with Definition 1.2.1's determination of the Lebesgue measure of a bounded interval as its length.

The smallest collection of sets forming a σ-algebra that contains every open interval (a, b) of $\mathbb{R}$ is called a *Borel algebra* (or a Borel field). Borel fields are useful σ-algebras; we will use their structure as we construct measures in the next section. We'll see the Borel algebra playing field provides a sleek and useful structure on which to build many interesting measures. After you see how the constructions work, you'll naturally be interested in more general constructions. They can get wild and crazy, but, in many positive ways, many measure-theoretic structures are tame enough to develop broadly general theorems that show how the measures act. To mention one such topic, many general measures have derivatives (said to be taken with respect to Lebesgue measure) that can be used in applications such as probability theory. We will touch on some of those topics in the last section of this chapter, but we intentionally skirt a broad discussion of measure theory—its waters are too deep for this initial investigation. Our hope, instead, is that this chapter's brief description will interest and motivate a more complete study beyond the boundaries of this text.

4.1.1 Properties of Lebesgue Measure

Definition 4.1.1 agrees with our earlier description of measure on intervals. Similarly, a set S with measure zero in the sense of Section 1.2 will have measure zero in the sense of Definition 4.1.1: $m(S) = \int \mathcal{X}_S = 0$. That fact will follow in Theorem 4.1.3, after we prove a few more fundamental results about Lebesgue measure.

Theorem 4.1.1 *If S and T are Lebesgue measurable sets in $\mathbb{R}$, then so are $S \cup T$ and $S \cap T$, and*

$$m(S \cup T) = m(S) + m(T) - m(S \cap T).$$

Proof. Since for a set A, $\mathcal{X}_A(x) = \begin{cases} 1 & \text{if } x \in A \\ 0 & \text{if } x \notin A, \end{cases}$ we get $\max\{\mathcal{X}_S, \mathcal{X}_T\} = \mathcal{X}_{S \cup T}$. (To verify this, let $x \in S \cup T$ and show the function values on each side of the equation match. Then repeat for $x \notin S \cup T$. See Exercise 11.) Similarly, $\min\{\mathcal{X}_S, \mathcal{X}_T\} = \mathcal{X}_{S \cap T}$. If f and g are measurable, then so are the functions $\max\{f, g\}$ and $\min\{f, g\}$ (the proof of this fact involves calculation and is outlined in this section's exercises). Hence $\max\{\mathcal{X}_S, \mathcal{X}_T\} = \mathcal{X}_{S \cup T}$ and $\min\{\mathcal{X}_S, \mathcal{X}_T\} = \mathcal{X}_{S \cap T}$ are measurable functions, and so $S \cup T$ and $S \cap T$ are by definition Lebesgue measurable sets.

The fact that $m(S \cup T) = m(S) + m(T) - m(S \cap T)$ then follows quickly from integrating both sides of $\mathcal{X}_{S \cup T} + \mathcal{X}_{S \cap T} = \mathcal{X}_S + \mathcal{X}_T$, which is proved via a straightforward calculation. (See Exercise 12.) ∎

We point out several facts. For a measurable set A, since $\mathcal{X}_A \geq 0$, we have $m(A) = \int \mathcal{X}_A \geq 0$. Then, from Theorem 4.1 and the fact that $m(S \cap T) \geq 0$, we get $m(S \cup T) \leq m(S) + m(T)$ for any measurable sets S and T. The same fact holds for a finite number or measurable sets

$S_1, S_2, \ldots, S_n$. In other words, $m(S_1 \cup S_2 \cup \cdots \cup S_n) \leq \sum_{j=1}^{n} m(S_j)$ always holds. This follows from applying Theorem 4.1 repeatedly, proving the result for three steps, then for four steps, etc., by taking the union one additional successive set at a time (and verifying the inequality each time) until all n sets are gathered into the union of sets. (This method of proof is commonly known as induction.)

We can in fact say more. When the sets are disjoint, then $m(S \cap T) = m(\emptyset) = 0$. Hence $m(S \cup T) = m(S) + m(T)$ for disjoint sets S and T. This is true not only for two sets but for any countable number of sets, including a countably infinite collection of sets.

Theorem 4.1.2 *Suppose $S = \bigcup_{n=1}^{\infty} S_n$, where S_n is a measurable set. Then S is measurable with*

$$m(S) \leq \sum_{n=1}^{\infty} m(S_n).$$

If in addition the sets S_n are disjoint, so $S_j \cap S_k = \emptyset$ when $j \neq k$, then

$$m(S) = \sum_{n=1}^{\infty} m(S_n).$$

Proof. Collecting together a finite number n of the first sets in the collection, we let

$$A_n = S_1 \cup S_2 \cup \cdots \cup S_n.$$

Then the sequence of functions $\{\mathcal{X}_{A_n}\}$ is nondecreasing because $A_1 \subseteq A_2 \subseteq A_3 \ldots$. Moreover, the sequence $\{\mathcal{X}_{A_n}\}$ converges (pointwise everywhere) to $\mathcal{X}_S$. Now if A_n has $m(A_n) = \int \mathcal{X}_{A_n} = \infty$, or if $\lim_{n \to \infty} m(A_n) = \infty$, then by definition $m(S) = \infty$, and the result follows. Otherwise, the monotone convergence theorem and then Theorem 4.1.1 apply:

$$m(S) = \int \mathcal{X}_S = \lim_{n \to \infty} \int \mathcal{X}_{A_n} = \lim_{n \to \infty} m(A_n) \leq \lim_{n \to \infty} \sum_{k=1}^{n} m(S_k) = \sum_{k=1}^{\infty} m(S_n),$$

which proves the theorem's first statement. The inequality in the last calculation is replaced with equality when the sets S_n are disjoint, which finishes the proof. ∎

Theorem 4.1.3 *A set S has measure zero in the sense of Definition 1.2.2 if and only if it has measure zero in the sense of Definition 4.1.1. That is, the two definitions are equivalent.*

Proof. We start by assuming S has measure zero in the sense of Definition 1.2.2. Then $\mathcal{X}_S$ is measurable because it equals the zero function almost everywhere and hence is integrable. Definition 1.2.2 states S can be covered with a sequence of open intervals $I_1, I_2, I_3, \ldots$ whose bounded total measure $\sum_{n=1}^{\infty} m(I_n) < \varepsilon$ for any $\varepsilon > 0$. So $S \subseteq \cup I_n$, and (apply Theorem 4.1.2 twice) $m(S) \leq m(\cup I_n) \leq \sum_{n=1}^{\infty} m(I_n) < \varepsilon$. Since this is true for any positive ε, it must be the case that $m(S) = 0$ in the sense of Definition 4.1.1.

4.1 Lebesgue Measure

To prove the converse, assume S has measure zero in the sense of Definition 4.1.1, so $m(S) = \int \mathcal{X}_S = 0$. By Corollary 2.2.1, $\mathcal{X}_S = 0$ almost everywhere. But that means S is null (in the sense of Definition 1.2.2), since the places where $\mathcal{X}_S \neq 0$ are precisely the elements of S. ∎

Question 4.1.1 Does Definition 4.1.1 say a finite set $S = \{s_1, s_1, \ldots, s_n\}$ has measure zero? Why? ∎

4.1.2 Consideration of Other Ways to Define Measure

The remaining material of this section will show how difficult the behavior of real-valued sets, as described by measure theory, can be. To illustrate its messes, we note Lebesgue historically thought about measure theory in terms of four defining (or at least desirable) properties. We would naturally want:

1. *Universalness*: the measure of any real set to be properly defined.
2. *Countable Additivity*: the measure of the union of a countable number of disjoint sets to equal the summation of the measure of each.
3. *Length Agreement*: for any interval I, the measure of I to equal its length.
4. *Translation Invariance*: the measure of any set E to equal, for any given real c, the measure of the translated set $E + c = \{x + c : x \in E\}$.

Some of these properties hold for Lebesgue measure, but not all of them. We have seen that the first does not—the Vitali set is an example of a nonmeasurable set. Could we have defined Lebesgue measure differently, somehow, so that all the properties would have held?

It's an abstract question: could we have started similarly and defined the measure of an interval as its length (so the measure would possess length agreement), but then defined the measure of arbitrary sets in some other way, so the other three properties would also have held? Unfortunately, the answer is "no." Though there are many ways—different from Lebesgue measure—to define the measure of sets, and though they are often interesting and useful, it turns out it's impossible to satisfy all four fundamental properties.

To see why, consider the following construction using some abstract measure m on real-valued sets (see [105, p. 64]). Here we are thinking of m not as Lebesgue measure but some other construction. As we are wondering if it is possible for m to satisfy all four properties simultaneously, it is reasonable to assume m satisfies countable additivity and translation invariance. Can m simultaneously satisfy both of the other two properties? To show the answer is no, we construct a nonmeasurable set S, so the universalness property is not satisfied. (The reader may wish to compare this construction with the one in Example 3.1.1.) This construction is managed well if you think of it as a five-step process, outlined here.

Step 1: Define a relationship on the numbers in the interval $[0, 1)$: two real numbers a and b in $[0, 1)$ are related if $a - b$ is a rational number. So all rational numbers in $[0, 1)$ are related. For example, $\pi/4 = .735898\ldots$ is related to $.624787\ldots$, since their difference is the rational number $.1111\ldots = 1/9$. We put each collection of numbers that are related to each other into an equivalence class. For example, the rational numbers in $[0, 1)$ form one equivalence class under the relation. All the numbers related to $\pi/4$ form another.

Step 2: Choose one number from each equivalence class and put them together into a set S.

Step 3: Write the rational numbers in $[0, 1)$ as a list $\{0, r_1, r_2, r_3, \ldots\}$. We put the number 0 as the first value (so $0 = r_0$) in the list. Now form, using each r_n in this list, the infinite collection of sets $S_n = \{s + r_n : s \in S\}$, $n = 0, 1, 2, \ldots$. Notice $S = S_0$. Also, the sets S_n are disjoint from one another. For if any number $x \in S_j \cap S_k$, then $x = s + r_j$ and $x = t + r_k$ for some values s and t in S. But $s - t = (x - r_j) - (x - r_k) = r_k - r_j$ is a rational number, and so s is related to t—they belong to the same equivalence class. But we formed S so it has only one element from each equivalence class, and so $0 = s - t = r_k - r_j$, and we get $r_j = r_k$. The only way that can happen is if $S_j = S_k$, which proves the sets S_n are pairwise disjoint.

Step 4: Realize $[0, 1) = \bigcup_{n=0}^{\infty} S_n$. For if s is any value in $[0, 1)$, then s must be related to some element t in S. But that means $s - t = r_n$ for some rational r_n, which places s in S_n. Hence $[0, 1) \subseteq \bigcup_{n=0}^{\infty} S_n$. By construction, we have $\bigcup_{n=0}^{\infty} S_n \subseteq [0, 1)$. The two are therefore equal. Furthermore, since the sets S_n are disjoint, countable additivity (one of the two properties we assume holds) implies $m([0, 1)) = \sum_{n=0}^{\infty} m(S_n)$.

Step 5: Realize $m(S_n) = m(S)$ for each of the sets S_n. For each $S_n = \{s + r_n : s \in S\}$ is a translation of S by the fixed value r_n. Since we assume translation invariance, the two sets have the same measure. Combining that fact with the conclusive statement in Step 4, we get $m([0, 1)) = \sum_{n=0}^{\infty} m(S_n) = \sum_{n=0}^{\infty} m(S)$. If S is measurable, then only two possibilities exist: $m(S) = 0$, and then the last equation implies $m([0, 1)) = 0$; or $m(S) > 0$, and then the last equation implies $m([0, 1)) = \infty$. Both possibilities violate the property of length agreement.

The conclusion is either S is nonmeasurable, and then the property of universalness fails, or $m([0, 1)) \neq 1$, and then the property of length agreement fails. There can unfortunately be no measure m that simultaneously satisfies all four of the basic desirable properties.

Lebesgue measure is a natural measure to define, since it possesses length agreement, and so the measure of an interval is its length. That consistency with the size of interval sets then translates well. For example, how large (in Lebesgue measure) is the size of two disjoint intervals? Why, naturally, it's the sum of their lengths. This consistency exists for other measurable sets. In this way, Lebesgue measure is a natural generalization of the measurement of length.

We have alluded to the possibility of defining measure in different ways, and it turns out we can do that easily. The other measures will generally no longer work like Lebesgue measure—they can, for example, measure intervals without any regard to their lengths—but they can be useful, as we will see later in this chapter. This section concludes with one example, called a point mass measure. It's a preview of coming attractions.

Example 4.1.1 A unit point mass measure m_p is defined for a real number (the "point") p. For a set S, if $p \in S$, then $m_p(S) = 1$, and if $p \notin S$, then $m_p(S) = 0$. The measure of every set S is defined, and so m_p satisfies the property of universalness.

For example, let $p = 1/2$. Depending upon whether or not $1/2$ is in the set, its point mass measure at $1/2$ is either 0 or 1. So $m_{1/2}(\mathbb{R}) = 1$, $m_{1/2}([0, 1]) = 1$, $m_{1/2}([-10, 10]) = 1$, $m_{1/2}([3, 4]) = 0$, and $m_{1/2}(C) = 0$ for the Cantor set C because $1/2 \notin C$. You can see $m_{1/2}$ does not satisfy the property of length agreement. ∎

Question 4.1.2 Which of the four fundamental properties (universalness, countable additivity, length agreement, and translation invariance) does a general unit point mass measure m_p satisfy? Prove each property holds or give a counterexample to its proof. ∎

The study of measure theory is a broad and interesting field in its own right. The fact that the four fundamental properties can never be satisfied with any measure shows just how intricate and complex are the real numbers—sets of real numbers resist being measured in basic ways. But the wonderful news is the study of measure theory is not only interesting, it is extraordinarily useful. Constructions of different measures allow mathematicians to form different types of Lebesgue integrals. Each measure produces a Lebesgue integral in terms of that measure. Just as we formed the Lebesgue integral out of Lebesgue measure, so can we form other integrals out of other measures. And just as we have seen the Lebesgue integral to be useful, for example, in understanding functions via L^2 spaces and other function spaces, so will we see these other integrals are useful in understanding functions. For example, functions can be put together into function spaces formed by integrals on other measures. Studying their properties can tell us properties of the functions. That investigation is what the remainder of this text is about, and it blossoms out of a formulation of the Lebesgue integral with respect to other measures.

Solutions to Questions

4.1.1 Since S is finite, $\mathcal{X}_S(x) = 0$ almost everywhere (except at $x = s_1, s_2, \ldots, s_n$). By Definition 4.1.1, $m(S) = \int \mathcal{X}_S = \int 0 = 0$.

4.1.2 Universalness holds for m_p because $m_p(S) = 1$ for a set S containing p and $m_p(S) = 0$ for a set S not containing p. Countable additivity holds because, for a disjoint collection of sets $S_1, S_2, \ldots$, the point p is in at most one of them, and then there are two cases.
Case 1: $p \in S_j$ for some S_j in the collection. Then $m_p(\cup S_n) = 1$ because $p \in \cup S_n$, and $\sum m_p(S_n) = \sum_{n \neq j} m_p(S_n) + m_p(S_j) = 0 + 1 = 1$ because p is not in any of the sets except for S_j.
Case 2: $p \notin S_j$ for any S_j in the collection. Then $m_p(\cup S_n) = 0$ because the union cannot contain p, and $\sum m_p(S_n) = 0$ because none of the S_n sets contains p. Neither length agreement nor translation invariance hold: for length agreement, $m_p([a, b]) = 1$ when $p \in [a, b]$, which is generally not the length of $[a, b]$; and for translation invariance, if $p \in [a, b]$, then, for example, $m_p([a, b]) = 1 \neq 0 = m_p([b + 1, 2b - a + 1])$.

Reading Questions for Section 4.1

1. Is every real set measurable?

2. What is the Lebesgue measure of a measurable set A?

3. For real sets S and T, is the measure of $S \cup T$ always equal to $m(S) + m(T)$?

4. If S and T are disjoint real measurable sets, what can you say about the measure of $S \cup T$?

5. Is $S \cup T$ always measurable for measurable S and T?

6. Describe countable additivity. Does Lebesgue measure have it?

7. Give two different but equivalent definitions of measure zero sets.

8. Is it possible for a measure of real-valued sets to satisfy all four of the properties of universalness, countable additivity, length agreement, and translation invariance?

Exercises for Section 4.1

In Exercises 1–10, determine if the set is measurable in the sense of Definition 4.1.1, giving a reason for your answer. Find the Lebesgue measure of each measurable set, using either Definition 4.1.1 or a theorem presented in this section.

1. $A = [0, 1) \cup [3, 4.5) \cup (10, 12)$

2. $B = [0, 1) \cup [.5, 4.5) \cup (5, 7] \cup (10, 12)$

3. $C =$ the Cantor set

4. $D =$ the subset of $[0, 1]$ constructed exactly as the Cantor set, except that the middle interval removed at the kth step has length $c \cdot 3^{-k}$ for c between 0 and 1

5. $E = \bigcup_{n=1}^{\infty} (1 - 1/2^{n-1}, 1 - 1/2^n)$

6. $F = \bigcup_{n=1}^{\infty} (K - K/2^{n-1}, K - K/2^n)$, where K is a real number

7. $G = \bigcup_{n=1}^{100,000} (n, n+1)$

8. $H = \bigcup_{n=-\infty}^{\infty} (n, n+1]$

9. $I = \bigcap_{n=1}^{\infty} (1/2^n, 1)$

10. $J = \bigcap_{n=1}^{\infty} (5 + 1/2^n, 6)$

Exercises 11 and 12 ask you to prove facts used in proofs of theorems in this section.

11. Prove $\max\{\mathcal{X}_S, \mathcal{X}_T\} = \mathcal{X}_{S \cup T}$. (*Hint:* see the commentary in the proof of Theorem 4.1.1.)

12. Prove $\mathcal{X}_{S \cup T} + \mathcal{X}_{S \cap T} = \mathcal{X}_S + \mathcal{X}_T$. (*Hint:* evaluate each side of the equation for the cases $x \in S$ and $x \in T$, $x \in S$ and $x \notin T$, $x \notin S$ and $x \in T$, and $x \notin S$ and $x \notin T$.)

Exercises 13–20 ask you to consider properties of (Lebesgue) measurable sets.

13. If $A \cup B$ is measurable, is it always the case that A and B are measurable? Why or why not?

14. If $A \cap B$ is measurable, is it always the case that A and B are measurable? Why or why not?

15. If $A = [a_1, a_2]$ and $B = [b_1, b_2]$ are closed intervals, when is it the case that $m(A \cup B) = m(A) + m(B)$?

16. Do measurable sets A and B have to be disjoint in order for $m(A \cup B) = m(A) + m(B)$? Either prove it true or give a counterexample.

17. If $A = [a_1, a_2]$ and $B = [b_1, b_2]$ are closed intervals, when is it the case that $m(A \cap B) = m(A)$?

18. Is it possible for $m(A \cap B) = m(A) + m(B)$ for measurable sets A and B? Either prove it not possible or give an example.

19. Is it possible for $m(A \cap B) > m(A) + m(B)$ for measurable sets A and B? Either prove it not possible or give an example.

20. Describe the set $S = \{x : \cos x \geq \sqrt{2}/2 \text{ and } 0 \leq x \leq 2\pi\}$ in terms of intervals. Is S measurable, and, if so, what is its Lebesgue measure?

Exercises 21–26 form a proof of the following: if f and g are (Lebesgue) measurable, then so is the function $\max\{f, g\}$.

21. For real numbers f and g, prove $\max\{f, g\} = .5(f + g) + .5|f - g|$. (*Hint:* consider the two cases where $f \geq g$, and where $g > f$.)

22. Prove: If f is a measurable function in the sense of Definition 3.1.1, then $|f|$ is a measurable function. (*Hint:* for a nonnegative step function ϕ, $\text{mid}\{-\phi, f, \phi\}$ is in L^1 when $|\text{mid}\{-\phi, f, \phi\}|$ is. See Theorem 3.1.1.)

23. Prove: If f and g are measurable functions in the sense of Definition 3.1.1, then $f + g$ is a measurable function. (*Hint:* for a nonnegative step function ϕ, define $f_n = \text{mid}\{-n \cdot \phi, f, n \cdot \phi\}$ and $g_n = \text{mid}\{-n \cdot \phi, g, n \cdot \phi\}$, and note f_n and g_n are both integrable (hence measurable). Then show the sequence of integrable functions $\text{mid}\{-\phi, f_n + g_n, \phi\}$ is dominated by ϕ and converges to $\text{mid}\{-\phi, f + g, \phi\}$. Finally, apply the Lebesgue dominated convergence theorem.)

24. Prove: If f and g are measurable functions in the sense of Definition 3.1.1, then $f - g$ is a measurable function.

25. Prove: If f is measurable, then $.5f$ is measurable.

26. Apply the results of Exercises 21–25 to prove: if f and g are measurable, then $\max\{f, g\}$ is measurable.

4.2 Lebesgue Integrals with Respect to Other Measures

We can define a Lebesgue-type integral with respect to measures that are different from Lebesgue measure. They (one for each measure) generally produce a different result when you integrate functions. This happens because the key defining feature, which is the measure of an interval, is different for each different measure. Fortunately, the other measures produce integrals that have much of the same underlying structure as the integral using Lebesgue measure. Hence many of the Lebesgue measure integral's properties, such as the Lebesgue dominated convergence theorem, hold for Lebesgue integrals with respect to other measures of broad type. Also, Lebesgue integrals with respect to the other measures can be used to define function spaces, such as L^p spaces, in the same way as for the Lebesgue measure's integral. The function spaces turn out to have a vast usefulness in the study of orthogonal polynomials, bounded linear functionals, and subfields in physics such as harmonic analysis. The measures we construct that correspond to the integrals are of special type—Borel measures. They can be constructed from how they define the measure of intervals.

For simplicity, in this section we will discuss only measures on sets of real numbers. It is no more complicated to study, for example, $\mathbb{R}^n$—and that framework would allow us, for example, to look at measures on sets of complex numbers or other higher-dimensional sets. The general notions of measures on those metric spaces can be understood well in the simpler framework of measures on $\mathbb{R}$, and so that's what we'll examine.

In Chapter 1's description of the Lebesgue integral, it was important to define first the Lebesgue measure of intervals. That Lebesgue measure corresponded to $m([a, b]) = b - a$, so $m(I)$ was the length of I for a bounded interval I. Similarly, for our development of measures (those different from Lebesgue measure) and the integrals that correspond to them, it will be important to define first the measure on an interval. Intervals of real numbers can be considered as a foundation for a large collection of sets—the sigma-algebra of Borel sets first mentioned in Section 4.1. How does this work? To form a σ-algebra of Borel sets $\mathcal{B}$, first put all the intervals of $\mathbb{R}$ (including $\mathbb{R}$, single-point intervals of the form $\{x\}$, other infinite intervals, and the empty set) into $\mathcal{B}$. Then add into $\mathcal{B}$ all of the sets formed as countable unions, or countable intersections, or complements of these intervals. Then add in all of the sets that are countable unions, intersections, or complements of anything already in $\mathcal{B}$. Continue, forming $\mathcal{B}$ as closed with respect to countable unions, countable intersections, and complements. In this way, a σ-algebra is a collection of sets $\mathcal{B}$ where every union or intersection of a countable collection of sets in $\mathcal{B}$, along with any complement, is again in $\mathcal{B}$.

A measure μ defined naturally on such a σ-algebra $\mathcal{B}$ is a Borel measure. Here's how it works for Borel measures on $\mathbb{R}$: a measure is technically a function $\mu : \mathcal{B} \to [0, \infty]$, defined on all of $\mathcal{B}$, with three main properties. First, it is defined for each interval of $\mathbb{R}$, with the measure of the empty set $\mu(\emptyset)$ being zero and (so the measure is not trivial or always infinite) at least one interval having finite nonzero measure. Second, it is countably additive in the same sense we defined in Section 4.1, so $\mu(\cup B_n) = \sum \mu(B_n)$ for a countable number of disjoint sets B_n. (The fact that we are working inside a σ-algebra assures us—whenever each B_n is a member of $\mathcal{B}$—that $\cup B_n$ is also a member of $\mathcal{B}$.) Third, we insist μ is defined so it is finite on any closed and bounded set, which means μ is defined so it is finite on every closed and bounded interval. Much more general frameworks for measures are sometimes considered, but we outline these rules for our introductory description so the measures we construct are easy to understand.

The Daniell-Riesz method will use the fact that μ is nicely defined on all intervals to define an integral with respect to it, labeled $\int f \, d\mu$ for an integrable function f. As for any integral on real-valued functions, it takes a function f and outputs a real number. A theorem originally attributable to Riesz guarantees $\int \phi \, d\mu$ is well-defined on step functions ϕ and has standard properties such as linearity. (See Theorem 13.4.2 in [120, p. 126] or [120, p. 159] for the proof, which is similar in structure to our presentation in Chapter 1 and that we omit.) In a manner similar to the way we defined the Lebesgue integral for functions in L^1, the integral can then be extended beyond step functions to a space of functions determined to be integrable with respect to μ. We can mimic Definition 1.3.1 to define an integral in terms of step functions. The full description is a three-step process, which the next definition outlines.

Definition 4.2.1 **The Lebesgue Integral with Respect to a Borel Measure μ:** *Take a Borel measure μ, which has a measure $\mu(I)$ for any interval I. Then:*

(i) *The integral with respect to μ of the step function $\phi(x) = \sum_{j=1}^{n} c_j \cdot \mathcal{X}_{I_j}(x)$ is*

$$\int_{-\infty}^{\infty} \phi(x) \, d\mu(x) = \sum_{j=1}^{n} c_j \cdot \mu(I_j).$$

This integral is also denoted $\int \phi \, d\mu$.

4.2 Lebesgue Integrals with Respect to Other Measures

(ii) If a nondecreasing sequence of step functions $\{\phi_n\}$ converges μ-almost everywhere (except possibly on a set S with $\mu(S) = 0$) to a function f, then the integral of f with respect to μ is

$$\int f \, d\mu = \lim_{n \to \infty} \int \phi_n(x) \, d\mu(x).$$

If the integral is finite, then f is said to be in the collection of functions $L^0(\mu)$.

(iii) Whenever a function f can be written as $f = g - h$, where $g, h \in L^0(\mu)$, then

$$\int f \, d\mu = \int g \, d\mu - \int h \, d\mu.$$

Such functions f are said to be in the function space $L^1(\mu)$.

There are many Borel measures μ that generate such integrals. We describe three here, each one representative of a general type of measure. The first is an absolutely continuous measure, the second is based on a point mass, and the third a singular continuous measure. (A famous theorem by Lebesgue, the Lebesgue decomposition theorem, states any Borel measure can be decomposed as the sum of three measures, one of each of these types.)

An *absolutely continuous measure* (with respect to Lebesgue measure), which we call ν, produces an integral of the form $\int g \, d\nu = \int g(x) \cdot w(x) \, dx$, where the function w is measurable and takes on only nonnegative values, and dx refers to Lebesgue measure. The function w is sometimes called a weight function or a density function (a term that comes from probability theory—where absolutely continuous measures have strong applications). The function w is the Radon-Nikodym derivative of the measure ν (for historical reasons explained below). It turns out $\nu(S) = 0$ exactly when S has Lebesgue measure zero. (This will easily follow from the construction.) Here's an example, where we use the three-step format of Definition 4.2.1 to construct the integral with respect to a measure ν.

Example 4.2.1 **Absolutely Continuous Measure:** We choose a Radon-Nikodym derivative weight function $w(x) = e^{-x^2}/\sqrt{\pi}$, and develop the corresponding integral as in Definition 4.2.1. Before proceeding, we verify w is in $L^2(\mathbb{R})$ (w is measurable because it is continuous). The square of the L^2 norm is

$$\|w\|_2^2 = \int_{-\infty}^{\infty} e^{-2x^2}/\pi \, dx = \int_{-\infty}^{\infty} e^{-u^2}/(\sqrt{2}\pi) \, du = 1/\sqrt{2\pi}.$$

Now proceed to develop the integral with respect to the absolutely continuous measure for w. Use w to define the measure of an interval, say $I = (a, b)$, where a and b can be allowed to equal ∞ or $-\infty$. That measure, $\nu(I)$, is defined as the Lebesgue integral of the characteristic function $\mathcal{X}_I$ multiplied by w:

$$\nu(I) \equiv \int_{-\infty}^{\infty} \mathcal{X}_I(x) \cdot e^{-x^2}/\sqrt{\pi} \, dx.$$

Any absolutely continuous measure having a given weight function w works this way. In practice, the presence of the characteristic function restricts the values over which the integral is taken: here, $\nu(I) = \int_a^b e^{-x^2}/\sqrt{\pi} \, dx$. As an example, $\nu([0, \infty)) = \int_0^{\infty} e^{-x^2}/\sqrt{\pi} \, dx$, which turns out to be $\nu([0, \infty)) = \frac{1}{2}$.

Apply Definition 4.2.1. First, for a step function $\phi(x) = \sum_{j=1}^{n} c_j \cdot \mathcal{X}_{I_j}(x)$, we have $\int \phi \, dv = \sum_{j=1}^{n} c_j \cdot v(I_j)$. Applying the formula for $v(I)$, $\int \phi \, dv = \sum_{j=1}^{n} c_j \cdot [\int_{-\infty}^{\infty} \mathcal{X}_{I_j}(x) \cdot e^{-x^2}/\sqrt{\pi} \, dx]$. Interchanging the finite sum and the integral sign, $\int \phi \, dv = \int_{-\infty}^{\infty} (\sum_{j=1}^{n} c_j \cdot \mathcal{X}_{I_j}(x)) \cdot e^{-x^2}/\sqrt{\pi} \, dx = \int_{-\infty}^{\infty} \phi(x) \cdot e^{-x^2}/\sqrt{\pi} \, dx$. The main point is that the integral of the step function with respect to v turns out to be the Lebesgue integral of the step function multiplied against the weight function w. This integral exists (so it is well-defined), because $\int_{-\infty}^{\infty} \phi(x) \cdot e^{-x^2}/\sqrt{\pi} \, dx \leq \int_{-\infty}^{\infty} \phi(x)/\sqrt{\pi} \, dx < \infty$ and the function $\phi(x) \cdot e^{-x^2}/\sqrt{\pi}$ is continuous almost everywhere.

Second, given a nondecreasing sequence of step functions ϕ_n that converges a.e. (in the sense of v, which is the same as in the sense of Lebesgue measure) to a function f, the integral of f is

$$\int f \, dv = \lim_{n \to \infty} \int \phi_n \, dv.$$

By the monotone convergence theorem,

$$\int f \, dv = \lim_{n \to \infty} \int_{-\infty}^{\infty} \phi_n(x) \cdot e^{-x^2}/\sqrt{\pi} \, dx$$
$$= \int_{-\infty}^{\infty} \lim_{n \to \infty} \phi_n(x) \cdot e^{-x^2}/\sqrt{\pi} \, dx = \int f(x) \cdot e^{-x^2}/\sqrt{\pi} \, dx.$$

In short, the integral is understood as the Lebesgue integral of f multiplied by the weight $e^{-x^2}/\sqrt{\pi}$. We say such functions f, producing a finite integral, are in $L^0(v)$.

Third, when f can be written as $f = g - h$, for g and h in $L^0(v)$, then $f \in L^1(v)$ and $\int f \, dv = \int g \, dv - \int h \, dv$. From the last paragraph, we see $\int f \, dv = \int f(x) \cdot e^{-x^2}/\sqrt{\pi} \, dx$ for such functions.

The function $w(x) = e^{-x^2}/\sqrt{\pi}$ (which is known as a bell-shaped curve) is associated with the Hermite polynomials, as discussed in Section 3.4. ∎

Question 4.2.1 Using the weight function $w(x) = e^{-x^2}/\sqrt{\pi}$ to form the absolutely continuous measure v so that $dv(x) = w(x)dx$, find $\int x\mathcal{X}_{(0,\infty)}(x) \, dv(x)$ and $\int x^2 \, dv(x)$. ∎

A famous theorem, the Radon-Nikodym theorem, gives conditions on when a measure v results in an integral produced by the absolutely continuous integral's form (where $dv(x) = w(x)dx$ for some nonnegative measurable weight function w). In this case (for the framework of measures we are examining in this introduction), the condition characterizing when a measure v is absolutely continuous is that $v(A) = 0$ for every set A having Lebesgue measure $m(A) = 0$. The theorem is named for Johann Radon, who first proved it in 1913 for the case of the underlying domain space $\mathbb{R}^k$. Otton Nikodym then proved the more general result in 1930. His general theorem relates any two measures, not just a measure compared to Lebesgue measure. See [105, p. 238].

The second example defines a *point mass measure* μ concentrated at a value p to be a Borel measure μ_p such that $\mu_p(I) = 1$ for an interval I that contains p, and $\mu_p(I) = 0$ for an interval I that does not contain p. Example 4.1.1 already introduced a point mass measure;

4.2 Lebesgue Integrals with Respect to Other Measures

the next examples shows how it generates an integral. The point p, on which the measure is concentrated, is the *eigenvalue* of the measure.

Example 4.2.2 Point Mass Measure: Define a point mass measure μ as above, where $\mu_p(I) = 1$ for a interval I that contains a given point p, and $\mu_p(I) = 0$ for an interval I not containing p. This measure turns out to produce $\mu_p(S) = 0$ precisely when S does not contain p. The measure generates an integral of a function f with respect to μ_p that, essentially, evaluates f at p. For if $\phi = \sum_{j=1}^{n} c_j \cdot \mathcal{X}_{I_j}(x)$ is a step function (without loss of generality, we may assume the intervals I_j are disjoint), then $\int_{-\infty}^{\infty} \phi(x)\, d\mu(x) = \sum_{j=1}^{n} c_j \cdot \mu_p(I_j) = \sum_{\{j: p \in I_j\}} c_j = \phi(p)$. Working through the rest of the three-step process of Definition 4.2.1, we get $\int_{-\infty}^{\infty} f(x)\, d\mu_p(x) = f(p)$ for $f \in L^1(\mu_p)$.

This example generalizes to a point mass measure where the measure of a set equals either 0 or a predetermined mass c (the measure μ_p just presented has $c = 1$). So, in general, we define a point mass measure μ concentrated at an eigenvalue p and having mass c as $\mu_p(I) = c$ for an interval I that contains p, and $\mu_p(I) = 0$ for an interval I that does not contain p. It is then straightforward to show $\int_{-\infty}^{\infty} f(x)\, d\mu_p(x) = c \cdot f(p)$ for $f \in L^1(\mu_p)$. Sums of individual point mass measures (with multiple eigenvalues) are also important; the exercises include several examples. ∎

Question 4.2.2 Define the measure μ on a real set S to be $\mu(S) = 5$ if $1.5 \in S$ and $\mu(S) = 0$ if $1.5 \notin S$. Determine $\int x\, d\mu(x)$ and $\int x^2\, d\mu(x)$. ∎

The third example is of a *continuous singular measure*—typically considered the most complex of the three types. It is constructed from a function F that is singular with respect to Lebesgue measure, which means $F'(x) = 0$ a.e. Such functions seem peculiar: we will want to base the construction of a continuous singular measure on a nondecreasing singular function, but how can functions that increase have zero derivative almost everywhere? An example is the famous *Cantor function*, which is constant (and hence has zero derivative) on the complement of the Cantor set C defined in Question 1.2.3. All of the function's growth is therefore restricted to the Cantor set, which we showed in Section 1.2 has Lebesgue measure zero.

Here's one way to define the Cantor function $F : [0, 1] \to [0, 1]$ for $x \in [0, 1]$. Write x in its ternary (base 3) expansion as in Question 1.2.3. For technical reasons that make this definition work, do not use a 1, when possible, in the expansion (for example, when representing $x = 1/3$ we choose $0.02222\ldots[3]$ instead of $0.10000\ldots[3]$). Next, replace the first 1 in the expansion with a 2 and everything after it with 0. Finally, replace any 2 in the resulting expansion with a 1, and interpret the result as a binary number. That number is $F(x)$.

For example, $1/2 = 0.111111\ldots[3]$ in base 3 (there is no way not to use 1s in the expansion). We replace the first 1 with a 2 and everything after it with a 0, obtaining $0.2000\ldots$. Now we replace the 2 with a 1, obtaining $0.1000\ldots$. Interpreting that expansion in base 2, we get the resulting function value: $F(1/2) = 1/2 + 0/2^2 + 0/2^3 + \cdots = 1/2$.

We finish this section with a quick presentation of the singular continuous measure that is generated by the Cantor function F. Its construction is characteristic of how other singular continuous measures work, based on nondecreasing singular function. We also allude to the three-step process of Definition 4.2.1 as to how to construct the integral with respect to the measure.

Example 4.2.3 **Singular Continuous Measure:** To construct a singular continuous measure μ_F, start with a nondecreasing singular function such as the Cantor function, and define the corresponding measure of an interval I to be $\mu_F(I) = F(b) - F(a)$, where a and b are the interval's endpoints. Using the Cantor function as F, we have, for example, $\mu_F([0, 1]) = F(1) - F(0) = 1$, and the measure of any interval containing $[0, 1]$ is also 1. For other examples, note $\mu_F([0, \frac{1}{2}]) = F(\frac{1}{2}) - F(0) = \frac{1}{2}$, and $\mu_F([0, \frac{3}{18}]) = F(\frac{3}{18}) - F(0) = \frac{1}{4}$. The three-step process of Definition 4.2.1 develops the integral $\int f \, d\mu_F$. It is a peculiar measure with many interesting properties. For example, it is supported on the Cantor set C (which has Lebesgue measure zero), because $\mu_F(S) = 0$ for $S = \{x : x \notin C\}$. We call the measure μ_F the Cantor function measure, and we explore it further in this section's exercises.

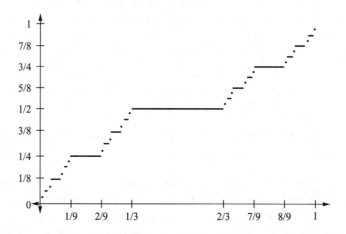

Figure 4.2. The Cantor Function is often fancifully called the *Devil's staircase*.

Using the calculations of $\mu_F(I)$ for the intervals I in the last paragraph and seen in Figure 4.2, we see, for example, $\int 1 \, d\mu_F = \int \mathcal{X}_{[0,1]} \, d\mu_F = 1$, $\int \mathcal{X}_{[0,1/2]} \, d\mu_F = 1/2$, and $\int \mathcal{X}_{[0,3/18]} \, d\mu_F = 1/4$. Additionally, $\int \mathcal{X}_{[a,b]} \, d\mu_F = 1$ for an interval $[a, b]$ that contains $[0, 1]$. In this way, the integral with respect to μ_F is well-determined on intervals, and the first step in Definition 4.2.1 nicely defines the integrals of step functions. The other two steps then follow accordingly to define the integral with respect to μ_F. ■

Question 4.2.3 Define the singularly continuous measure μ_F using the Cantor function F as in the last example. Find $\int_0^1 x \, d\mu_F(x)$. ■

The construction in Definition 4.2.1 is the basis for the Daniell-Riesz approach, creating an integral with respect to a Borel measure as structured under the framework of measure described here. More general measures exist, and this method is not always meaningful. An example is a measure μ that counts the number of elements in a set, for instance, $\mu(\{1, 3, 5\}) = 3$, but then $\mu(I) = \infty$ for every interval I with more than one element. Using Definition 4.2.1 on this measure doesn't produce an interesting integral. (The integral of any nontrivial step function would be infinite.)

4.2 Lebesgue Integrals with Respect to Other Measures

Still, the Daniell-Riesz approach of Definition 4.2.1 is quite powerful and useful. The exercises at the end of this section let you grapple with some complicated measures, showing you how the approach allows you to consider some interestingly complex integrals with respect to various measures. Then in the rest of this chapter you'll explore how the measures, with the integrals they create, construct interesting measure spaces that can be applied to fields such as probability theory. In this way, integrals with respect to Borel measures are incredibly useful and enjoyable to work with.

Solutions to Questions

4.2.1 Using u-substitution and the monotone convergence theorem, $\int x \mathcal{X}_{(0,\infty)}(x) \, dv(x) = \frac{1}{\sqrt{\pi}} \int_0^\infty x e^{-x^2} \, dx = \frac{1}{2\sqrt{\pi}} \lim_{n \to \infty} -e^{-x^2}\big|_0^n = \frac{1}{2\sqrt{\pi}}$. Using integration by parts and LDCT, $\int x^2 \, dv(x) = \frac{1}{\sqrt{\pi}} \int_{-\infty}^\infty x^2 e^{-x^2} \, dx = 1/2$.

4.2.2 The measure is a point mass measure with $c = 5$ and $p = 1.5$. Therefore $\int x \, d\mu(x) = 5 \cdot (1.5) = 7.5$ and $\int x^2 \, d\mu(x) = 5 \cdot (1.5)^2 = 11.25$.

4.2.3 Define step functions ϕ_n on 2^n equal-sized subintervals of $[0, 1)$ as $\phi_n(x) = (k-1)/2^n$ for x in the kth subinterval, where $k = 1, 2, \ldots, 2^n$. Note $\int \phi_2 \, d\mu_F = (1/2) \cdot \mu_F([1/2, 1]) = (F(1) - F(1/2))/2 = 1/4$, $\int \phi_3 \, d\mu_F = (1/4) \cdot \mu_F([1/4, 1/2]) + (2/4) \cdot \mu_F([1/2, 3/4]) + (3/4) \cdot \mu_F([3/4, 1]) = (1/4)(1/2 - 1/4) + 1/2(3/4 - 1/2) + (3/4)(1 - 3/4) = (1/4^2)(1 + 2 + 3) = 3/8$, and in general $\int \phi_n \, d\mu_F = (1/2^{2(n-1)}) \cdot (1 + 2 + \cdots + (2^{n-1} - 1)) = \frac{2^{n-1}(2^{n-1}-1)}{2^{2n-1}}$. Then $\int_0^1 x \, d\mu_F(x) = \lim_{n \to \infty} \int \phi_n \, d\mu_F = \lim_{n \to \infty} \frac{2^{n-1}(2^{n-1}-1)}{2^{2n-1}} = \lim_{n \to \infty} \frac{2^{2n-1}-2^n}{2 \cdot 2^{2n-1}} = 1/2$.

Reading Questions for Section 4.2

1. What does the term measure mean as we are using it in this section?

2. Describe the three-step process that constructs an integral with respect to a given measure μ.

3. Describe what is meant by an absolutely continuous measure.

4. Give an example of an integral created using Definition 4.2.1 and using an absolutely continuous measure v. For your example, what is $\int x \, dv$?

5. Describe what is meant by a point mass measure.

6. Give an example of an integral created using Definition 4.2.1 and using a point mass measure μ_p. For your example, what is $\int x \, d\mu_p$?

7. Describe what is meant by a singular continuous measure.

8. Give an example of an integral created using Definition 4.2.1 and using a singular continuous measure μ. For your example, what is $\int \mathcal{X}_{[0,1]} \, d\mu$?

Exercises for Section 4.2

Exercises 1–10 deal with integrals with respect to absolutely continuous measures.

1. Calculate $\int_0^\infty x \cdot e^{-x^2} \, dx$.

2. Find $\int_{-\infty}^\infty x \cdot w(x) \, dx$, where the weight function is $w(x) = e^{-x^2} \cdot \mathcal{X}_{(0,\infty)}$.

3. Describe why Lebesgue measure is absolutely continuous: what is the weight function w that expresses the Lebesgue integral of a function f in L^1 as $\int_{-\infty}^{\infty} f(x) w(x)\, dx$?

4. Define the $L^2(v)$ norm of an integrable (with respect to v) function f to be $\left(\int |f|^2\, dv\right)^{1/2}$. Determine the $L^2(v)$ norm of $f(x) = \sqrt{x}$, for the absolutely continuous measure $v = w(x)\, dx$ having weight function $w(x) = e^{-x^2}$.

5. Find $\int x^2 \cdot w(x)\, dx$, where $v = w(x)\, dx$ has weight function $w(x) = \mathcal{X}_{[0,1]}(x)$. (In probability theory, the function w is the uniform density function on the interval $[0, 1]$.)

6. Given a weight function $w(x) = \mathcal{X}_{[a,b]}(x)$ that is a characteristic function of a closed bounded interval, determine $\int_{-\infty}^{\infty} x^n \cdot w(x)\, dx$ for a nonnegative integer n.

7. For the weight function $w(x) = e^{-x} \cdot \mathcal{X}_{(0,\infty)}(x)$, what is $\int \mathcal{X}_{[a,b]} w \cdot dx$ for a finite interval $[a, b]$?

8. For the weight function $w(x) = e^{-x} \cdot \mathcal{X}_{(0,\infty)}(x)$, what is $\int_{-\infty}^{\infty} x^n\, w(x) \cdot dx$ for $n = 0, 1, 2, \ldots$?

9. For the weight function $w(x) = x$, what is $\int \mathcal{X}_{[a,b]} w \cdot dx$ for a closed bounded interval $[a, b]$?

10. For the weight function $w(x) = x \cdot \mathcal{X}_{(0,1)}(x)$, what is $\int_{-\infty}^{\infty} x^n\, w(x) \cdot dx$ for $n = 0, 1, 2, \ldots$?

Exercises 11–20 deal with integrals with respect to point mass measures.

11. For an interval I, calculate $\int \mathcal{X}_I d\mu$, where μ is the point mass $\mu = 3\mu_{1/2}$. Hence, for an interval I, $\mu(I) = 3\mu_{1/2}(I) = 3$ if $1/2 \in I$, and $\mu(I) = 0$ if $1/2 \notin I$.

12. For a real-valued function f whose domain includes $x = 1/2$, calculate $\int f d\mu$, where μ is the point mass $\mu = 3\mu_{1/2}$ described in the previous exercise.

13. For a function f whose domain includes $x = 1/2$ and for the measure μ described in the previous exercise, define the $L^2(\mu)$ norm of f to be $\left(\int |f|^2\, d\mu\right)^{1/2}$. Calculate the $L^2(\mu)$ norm of $f(x) = x$.

14. For the point mass measure μ used in the previous exercise and for any $n = 0, 1, 2, \ldots$, calculate the $L^2(\mu)$ norm of $f(x) = x^n$.

15. Define μ as the sum of unit point mass measures $\mu = \mu_1 + \mu_2$, so for a set S,

$$\mu(S) = \begin{cases} 2 & \text{if 1 and 2 are both in } S \\ 1 & \text{if exactly one of 1 or 2 is in } S \\ 0 & \text{otherwise.} \end{cases}$$

For an interval I, calculate $\int \mathcal{X}_I d\mu$.

16. For a real-valued function f whose domain includes $x = 1$ and $x = 2$, calculate $\int f d\mu$, where μ is the sum of the unit point masses described in the previous exercise.

17. For a function f whose domain includes $x = 1$ and $x = 2$ and for the measure μ described in the previous exercise, define the $L^2(\mu)$ norm of f to be $\left(\int |f|^2\, d\mu\right)^{1/2}$. Calculate the $L^2(\mu)$ norm of $f(x) = x$.

4.2 Lebesgue Integrals with Respect to Other Measures

18. For the measure μ defined as the sum of unit point mass measures and used in the previous exercise, and calculate the $L^2(\mu)$ norm of $f(x) = x^n$ for $n = 0, 1, 2, \ldots$.

19. Let $\mu = \mu_{1/2} + \mu_{1/4} + \cdots + \mu_{1/2^n} + \cdots = \sum_{n=1}^{\infty} \mu_{1/2^n}$, where, for an interval I, each component in the infinite sum has $\mu_{1/2^n}(I) = 1/2^n$ if $1/2^n \in I$ and $\mu_{1/2^n}(I) = 0$ if $1/2^n \notin I$. Find $\mu([0, 1])$.

20. For the countable sum μ of point mass measures defined in the previous exercise, calculate $\int f d\mu$ where $f(x) = x$. Then repeat for $f(x) = x^2$.

Exercises 21–32 deal with the Cantor function's singular continuous measure μ_F and its integrals.

21. Determine $\mu_F([0, 1/3])$, $\mu_F([0, 1/9])$, and $\mu_F([0, 1/27])$.

22. Calculate $\mu_F([0, (1/3)^n])$ for $n = 0, 1, 2, \ldots$.

23. Determine $\int \mathcal{X}_{[0,1/3]} \, d\mu_F$.

24. For $n = 0, 1, 2, \ldots$, calculate $\int \mathcal{X}_{[0,(1/3)^n]} \, d\mu_F$.

25. Find $\int x^2 \, d\mu_F(x)$. (*Hint*: see Question 4.2.3.)

26. Find $\int (ax + bx^2) \, d\mu_F(x)$ for $a, b \in \mathbb{R}$.

27. Determine $\int \mathcal{X}_{[8/9,1]} \, d\mu_F$.

28. For $n = 0, 1, 2, \ldots$, calculate $\int \mathcal{X}_{[1-(1/3)^n, 1]} \, d\mu_F$.

29. For $n = 0, 1, 2, \ldots$ and real constants a and b, find $\int a\mathcal{X}_{[0,(1/3)^n]} + b\mathcal{X}_{[1-(1/3)^n, 1]} \, d\mu_F$.

30. Define the function $f(x) = 0$ if $x \in (1/3, 2/3)$ and $f(x) = 1$ if $x \in [0, 1/3] \cup [2/3, 1]$, and so that it is periodic with period 1 off of the interval $[0, 1]$ (see [89]). Explain why the function $\mathcal{X}(x) \equiv \prod_{n=0}^{\infty} f(3^n x)$ is the characteristic function of the Cantor set.

31. Using the function f defined in the last exercise, construct the function $F(x) = \lim_{k \to \infty} \int_0^x \prod_{n=0}^{k} \frac{3}{2} \cdot f(3^n t) \, dt$ (see [89]). Show $F(1/3)$ and $F(1/2)$ both equal $1/2$.

32. For the function F defined in the last exercise, show $F(1/9) = 1/4$. Can you explain why F is the Cantor function? What familiar measure is then defined on a subinterval I of $[0, 1]$ according to $\mu(I) = \lim_{k \to \infty} \int_I \prod_{n=0}^{k} \frac{3}{2} \cdot f(3^n t) \, dt$?

Exercises 33–38 construct another singular continuous measure, different from the Cantor function measure, originally due to Riesz.

33. We construct a function F on $[0, 1]$ with range $[0, 1]$ as the limit of a sequence of functions $\{F_n\}, n = 0, 1, 2, \ldots$. The first function F_0 in the sequence is the line segment that goes from the origin $(0, 0)$ to the point $(1, 1)$, so $F_0(x) = x$, where $x \in [0, 1]$. To create the sequence, choose $t \in (0, 1)$ and define it iteratively: to form the graph of F_{n+1}, take each line segment that goes from the point (x, y) to the point (a, b) in the graph of F_n and replace it with two line segments—one that goes from (x, y) to $((x + a)/2, [(1 - t)/2]y + [(1 + t)/2]b)$,

and one that goes from $((x+a)/2, [(1-t)/2]y + [(1+t)/2]b)$ to (a, b). The notation is messy, but you can see F_1's graph consists of two line segments: the first goes from the origin to $(1/2, (1+t)/2)$, and the second from $(1/2, (1+t)/2)$ to $(1, 1)$.

Describe the four line segments that make up the graph of F_2. Also, for $t \in (0, 1)$ that you choose, graph F_0, F_1, and F_2 on the same set of axes.

34. If t is chosen in the last exercise to equal $1/2$, describe the graphs of F_0, F_1, F_2, and F_3.

35. Let $F(x) = \lim_{n \to \infty} F_n(x)$ and define the measure ν on an interval $I = [c, d]$ contained in $[0, 1]$ as $\nu(I) = F(d) - F(c)$. It turns out ν is singular continuous. Choosing $t = 1/2$, what is $\nu([0, 1])$? What is $\nu([0, 1/2])$?

36. For t and the measure ν constructed as in the previous exercise, determine $\nu([0, 1])$ and $\nu([0, 1/2])$.

37. Choosing $t = 1/2$ to form the measure ν as in the previous exercise, what is $\int \mathcal{X}_{[0,1/2]} \, d\nu$?

38. For $t \in (0, 1)$ forming the measure ν, what is $\int \mathcal{X}_{[0,1/2]} \, d\nu$?

4.3 The Hilbert Space $L^2(\mu)$

Given a Borel measure μ as constructed in Section 4.2, a Hilbert space $L^2(\mu)$ is defined as the set of μ-measurable functions f with finite $L^2(\mu)$ norm. The collections of functions provide a setting where we can investigate properties of functions. We've already seen some of these explorations. For example, we have developed a theory of Fourier series for functions in L^2 spaces. The material in Section 4.4 on quantum mechanics used the Hilbert space structures inherent in the function space $L^2(\mu)$, where μ was the absolutely continuous measure with weight function $w(x) = e^{-x^2}$. This connection is not unusual in many applications. The Hilbert space properties provide good reasons why physicists, for example, look to $L^2(\mu)$ spaces as a setting in which they can study general theories of quantum mechanics.

The next definition formalizes what you probably already intuitively understood as to how these function spaces are constructed.

Definition 4.3.1 *A function is μ-measurable when $\mathrm{mid}\{-g, f, g\}$ is in $L^1(\mu)$ for every nonnegative $g \in L^1(\mu)$. The Hilbert space $L^2(\mu)$ is then the collection of μ-measurable functions f with finite $L^2(\mu)$ norm, which is defined as $\|f\| = \sqrt{\int |f|^2 \, d\mu}$.*

Each $L^2(\mu)$ space is a Hilbert space. The inner product is $\langle f, g \rangle = \int f \cdot g \, d\mu$. From the discussion in Section 3.3, we know two Hilbert spaces having the same finite or countably infinite dimension are equivalent: there is always a Hilbert space isomorphism between them. The isomorphism maps orthonormal basis elements to orthonormal basis elements, and then the isomorphism map of every linear combination of basis elements is determined. So in this sense, there is only one infinite separable (countably infinite-dimensional) Hilbert space.

In this way, each infinite-dimensional $L^2(\mu)$ space is isometrically isomorphic to every other one, and each is isomorphic to ℓ^2. Why, then, would we want to study different ones? And why would we care to distinguish between the structures and construction of basis elements

4.3 The Hilbert Space $L^2(\mu)$

for different ones? One answer is in terms of the study of linear maps on functions in the spaces; the situation is similar to the study of linear transformations that takes place in an undergraduate course on linear algebra. In a finite-dimensional vector space, such as $\mathbb{R}^n$, a linear transformation on vectors in the space is often best studied by viewing it not in terms of $\mathbb{R}^n$'s standard basis, but in terms of some other basis. (In the words of matrix algebra, the linear transformation defined as multiplication by the matrix T can often be diagonalized, making the transformation transparent in its diagonal form D. This is manifested as multiplication on vectors in a different basis.) The same thing happens in infinite dimensions: we can often study a linear transformation T on a Hilbert space $L^2(\mu)$ by thinking of it as acting isometrically on an isomorphic Hilbert space—some other $L^2(\nu)$ space—where T's actions become transparent as a much simpler "diagonal" transformation. We will discuss such investigations in Chapter 5.

In the remaining material in this section, we look at different examples of $L^2(\mu)$ spaces. The next example looks at the simplest of such spaces, which comes from a measure μ formed as a point mass.

Example 4.3.1 Let $\mu_{1/2}$ be the unit point mass measure at $p = 1/2$, so for a real set S, $\mu_{1/2}(S) = 1$ if $1/2 \in S$ and $\mu_{1/2}(S) = 0$ if $1/2 \notin S$. As we have already discussed in Section 4.2, every set S is measurable, and $\int f \, d\mu_{1/2} = f(1/2)$ for functions f that include the point $1/2$ in their domain. In this way, the function space $L^2(\mu_{1/2})$ is the collection of functions f for which $f(1/2)$ is well-defined and finite. For such functions, the $L^2(\mu_{1/2})$ norm is $\|f\| = \sqrt{\int |f|^2 \, d\mu_{1/2}} = |f(1/2)|$. For $f, g \in L^2(\mu_{1/2})$, the inner product is $\langle f, g \rangle = \int f \cdot g \, d\mu_{1/2} = f(1/2) \cdot g(1/2)$.

At first glance $L^2(\mu_{1/2})$ might appear complex and multidimensional. After all, it includes a huge number of functions: any function whose domain contains $1/2$ is a member of it. But remember such L^2 spaces are defined in terms of the integral, which does not distinguish between two functions equal almost everywhere. As long as $L^2(\mu_{1/2})$ functions f and g satisfy $f(1/2) = g(1/2)$, then they agree μ-almost everywhere. We can therefore think of functions that differ, or are μ-distinct, as those that differ at $p = 1/2$. A set S that does not include $1/2$ has measure zero.

That determines $L^2(\mu_{1/2})$ as a one-dimensional Hilbert space. We deem its functions as different only according to the measure $\mu_{1/2}$, and so they are described simply: they are determined as the different possible (finite) real values $f(1/2)$ that could exist, which are the real values $\mathbb{R}$. In this way, they are each equal μ-almost everywhere to a real constant function $f(x) = c$. Since the constants together make up $\mathbb{R}$, which is one-dimensional, so is $L^2(\mu_{1/2})$ one-dimensional. Put another way, every μ-distinct function f in $L^2(\mu_{1/2})$ is a real multiple of the single basis element, the (constant) function $b(x) = 1$. We can think of f as equal μ-almost everywhere to the constant function $c = f(1/2) = f(1/2) \cdot 1$, which would then say $f(x) = c \cdot b(x)$, where $c = f(1/2) = \int f \, d\mu = \int f \cdot 1 \, d\mu = \langle f, 1 \rangle = \langle f(x), b(x) \rangle$. This last equation develops a Fourier series expansion for a function $f \in L^2(\mu_{1/2})$: it is a one-dimensional expansion with Fourier coefficient c described in terms of the single basis element as $c = \langle f(x), b(x) \rangle$. ∎

Question 4.3.1 Describe the collection of functions in $L^2(\mu)$ when μ is defined as the sum of two unit point mass measures $\mu = \mu_{1/2} + \mu_0$. Here, $\mu_{1/2}$ is defined as in the previous example, and μ_0 is defined the same way, except the point mass is at $p = 0$.

1. What real-valued functions are in $L^2(\mu)$? (*Hint:* such a function must be well-defined at 0.)
2. Why is the space $L^2(\mu)$ two-dimensional? (*Hint:* describe it as the space of linear functions $f(x) = c_1(1/2 - x) + c_2 x$.)
3. Why do the functions $1 - 2x$ and $2x$ form an orthonormal basis for $L^2(\mu)$? (Note the normalizations of the elements.)
4. How can you describe any function f in $L^2(\mu)$ in a Fourier series expansion of the form $f = \langle f(x), 1 - 2x \rangle \cdot (1 - 2x) + \langle f(x), 2x \rangle \cdot 2x$? ∎

When the weight function for an absolutely continuous measure μ is a characteristic function of an interval I, so $w(x) = \mathcal{X}_I$, the corresponding $L^2(\mu)$ space is the space $L^2(I)$. For example, when the weight function is $w(x) = \mathcal{X}_{(-1,1)}$, the corresponding $L^2(\mu)$ space is $L^2(-1, 1)$, which we studied previously. For example, in the exercises of Section 3.3, we saw the Legendre polynomials form a basis for it. It is therefore infinite-dimensional, and a function f in $L^2(-1, 1)$ can be written in terms of a Fourier series expansion $f(x) = \sum_{n=0}^{\infty} \langle f, L_n \rangle \cdot L_n(x)$, where $L_n(x)$ is the normalized nth degree Legendre polynomial.

Question 4.3.2 What if the weight function for an absolutely continuous measure μ is the characteristic function of $(0, \infty)$?

1. What is the corresponding $L^2(\mu)$ space we have already studied in Section 3.3?
2. What set of polynomials, constructed by the Gram-Schmidt process, forms a basis for this $L^2(\mu)$ space?
3. How can you describe a function f in $L^2(\mu)$ in a Fourier series expansion, using them? ∎

A famous theorem of Lebesgue, the Lebesgue decomposition theorem, says a Borel measure μ can be decomposed into three components described as point masses, an absolutely continuous part, and a singularly continuous piece. For example, we could define the measure μ as the sum of an absolutely continuous measure ν corresponding to a weight function, say $w(x) = \sqrt{1 - x^2}$ for $x \in [-1, 1]$, plus a single point mass measure μ_p, say at the point $p = \frac{4x_0^2 + 1}{4x_0}$ for a value $x_0 > 1/2$ and having a given mass c. We could write $\mu = \nu + \mu_p$. The corresponding $L^2(\mu)$ space consists then of functions both in $L^2(\nu)$ and in $L^2(\mu_p)$. They are those ν-measurable with finite $L^2(\nu)$ norm, which is $\|f\| = (\int_{-1}^{1} |f|^2 \sqrt{1 - x^2} \, dx)^{1/2}$, and defined and with finite function values at $x = p$.

Solutions to Questions

4.3.1 (1) A function whose domain includes 0 and 1/2 is in $L^2(\mu)$. (2) Two functions $f, g \in L^2(\mu)$ are equal μ-almost everywhere so long as $f(0) = g(0)$ and $f(1/2) = g(1/2)$. Hence every function $f(x) \in L^2(\mu)$ satisfies $f(x) = c_1 \cdot (1/2 - x) + c_2 \cdot x$ μ-almost everywhere for some (unique) choice of c_1 and c_2. In fact, $c_1 = 2f(0)$ and $c_2 = 2f(1/2)$. (3) Assume $\langle f(x), 1 - 2x \rangle = 0$ and $\langle f(x), 2x \rangle = 0$. We prove $f = 0$ a.e. The first condition says $0 = \int f(x) \cdot (1 - 2x) \, d\mu = f(0) \cdot 1 + f(1/2) \cdot 0 = f(0)$, and the second condition says $0 = \int f(x) \cdot (2x) \, d\mu = f(0) \cdot 0 + f(1/2) \cdot 1 = f(1/2)$. In other words, $f = 0$ μ-almost everywhere. The basis elements are orthogonal because $\int (1 - 2x) \cdot 2x \, d\mu = 0 + 0$ when evaluation at 0 and 1/2 occur. They are orthonormal because $\|1 - 2x\|^2 = \int (1 - 2x)^2 \, d\mu = 1^2 + 0^2 =$

4.3 The Hilbert Space $L^2(\mu)$

1 and $\|2x\|^2 = \int (2x)^2 \, d\mu = 0^2 + 1^2 = 1.$ (4) In terms of the generalized Fourier series expansion, $\langle f(x), 1 - 2x \rangle \cdot (1 - 2x) + \langle f(x), 2x \rangle \cdot 2x = \int f(x) \cdot (1 - 2x) \, d\mu \cdot (1 - 2x) + \int f(x) \cdot (2x) \, d\mu \cdot 2x = f(0) \cdot (1 - 2x) + f(1/2) \cdot 2x.$ But that function equals $f(x)$ μ-almost everywhere because they agree at 0 and at $1/2$: $f(0) \cdot (1 - 2 \cdot 0) + f(1/2) \cdot 2 \cdot 0 = f(0)$ and $f(0) \cdot (1 - 2 \cdot (1/2)) + f(1/2) \cdot 2 \cdot (1/2) = f(1/2).$

4.3.2 (1) $L^2(\mu) = L^2(0, \infty)$ (2) The generalized Laguerre polynomials $\phi_n(x) = c_n L_n(x) e^{-x/2}$, where c_n is a normalizing constant and $n = 0, 1, 2, \ldots$, form the orthonormal basis for the space. See Example 3.3.2. (3) If $f \in L^2(\mu)$, then $f(x) \sim \sum_{n=0}^{\infty} \langle f, \phi_n \rangle \phi_n(x).$

Reading Questions for Section 4.3

1. For a Borel measure μ, what does it mean for a function to be μ-measurable?

2. For a Borel measure μ, what does it mean for a function to be in $L^2(\mu)$?

3. Give an example of a finite-dimensional function space $L^2(\mu)$.

4. What functions form a basis for $L^2(\mu)$ if μ is absolutely continuous with weight function $w(x) = \mathcal{X}_{(-1,1)}$?

5. What does the Lebesgue decomposition theorem say?

6. Give an example of a measure μ that is a combination of at least two of the three fundamental types of measures: absolutely continuous, a (countable) sum of point masses, and singularly continuous.

7. Describe which functions are in the L^2 space for the measure μ you constructed as your answer to the last question.

Exercises for Section 4.3

Exercises 1–10 explore L^2 spaces defined on a collection of point mass measures.

1. Let $\mu_5(x)$ be the unit point mass measure at $p = 5$. Determine which real sets S are measurable, and determine the value of $\int f \, d\mu_5$ for a function f whose domain includes $x = 5$. For such functions, determine the $L^2(\mu_5)$ norm. Also evaluate the $L^2(\mu_5)$ inner product $\langle f, g \rangle$ for functions f and g whose domains both include $x = 5$.

2. What is the dimension of the Hilbert space $L^2(\mu_5)$ in the previous exercise?

3. Let $\mu(x) = \mu_5(x) + \mu_{10}(x) + \mu_{15}(x)$ be the sum of three unit point mass measures at $p = 5, 10,$ and 15, respectively. Determine which real sets S are measurable. Characterize those functions f that are μ-measurable and determine the value of $\int f \, d\mu$. Determine the $L^2(\mu)$ norm and evaluate the inner product $\langle f, g \rangle$ for μ-measurable functions f and g.

4. Let μ be a point mass measure at $p = 5$ with weight 2, so, for a function f whose domain includes $x = 5$, $\int f \, d\mu = 2 \cdot f(5)$. For such functions, determine the $L^2(\mu)$ norm. Also evaluate the $L^2(\mu)$ inner product $\langle f, g \rangle$ for functions f and g whose domains include $x = 5$.

5. Let $\mu = \mu_5 + \mu_{10} + \mu_{15}$ be the sum of three point mass measures at $p = 5, 10,$ and 15, respectively, where μ_5 has mass 1, μ_{10} has mass 2, and μ_{15} has mass 3, so, for a μ-measurable

function f, $\int f\,d\mu = 1 \cdot f(5) + 2 \cdot f(10) + 3 \cdot f(15)$. For such functions, determine the $L^2(\mu)$ norm. Also evaluate the $L^2(\mu)$ inner product $\langle f, g\rangle$ for μ-measurable functions f and g.

6. For $n \in \mathbb{N}$, let $\mu = \mu_1 + \mu_2 + \mu_3 + \ldots + \mu_n$ be the finite sum of point mass measures at $p = 1, 2, 3, \ldots, n$, respectively, where the mass at $k = 1, 2, 3, \ldots, n$ is k. Therefore, for a μ-measurable function f, $\int f\,d\mu = 1 \cdot f(1) + 2 \cdot f(2) + 3 \cdot f(3) + \cdots + n \cdot f(n)$. Characterize the μ-measurable functions and determine their $L^2(\mu)$ norm. Also evaluate the $L^2(\mu)$ inner product $\langle f, g\rangle$ for μ-measurable functions f and g.

7. Let $\mu = \mu_1 + \mu_2 + \mu_3 + \cdots + \mu_n + \cdots$ be the countable sum of point mass measures at $p = 1, 2, 3, \ldots$, respectively, where the mass at n is $1/2^n$. Therefore, for any μ-measurable function f, $\int f\,d\mu = 1 \cdot f(1) + (1/2) \cdot f(2) + (1/4) \cdot f(3) + \cdots + (1/2^n)f(n) + \cdots$. Characterize the μ-measurable functions and determine their corresponding $L^2(\mu)$ norm. Evaluate the $L^2(\mu)$ inner product $\langle f, g\rangle$ for a μ-measurable functions f and g. The measure μ is sometimes called a *geometric measure* because its total mass, taken as the sum of the masses at each eigenvalue, is determined as a geometric series.

8. For a measurable set S, suppose $\mu(S) = m(S) + \mu_\pi(S)$ is the sum of its Lebesgue measure m on $\mathbb{R}$ and the point mass measure μ_π at $p = \pi$ with mass $\sqrt{2}$. Find $\int f\,d\mu$, where f is the characteristic function $f = \mathcal{X}_{[0,10]}$. Also calculate $\int_{-2\pi}^{2\pi} \mathcal{X}_{[2,4]}(x) + xe^{-x^2}\,d\mu$.

9. What are the μ-measurable functions for the measure $\mu = dx + \mu_\pi$ defined in the previous exercise?

10. Suppose $\mu = dx + \mu_\pi$ is the sum of the Lebesgue measure dx on $[-\pi, \pi]$ and the point mass measure μ_π at $p = \pi$ with mass $\sqrt{2}$. Find the $L^2(\mu)$ inner product $\langle \cos x, \sin x\rangle$.

Exercises 11–20 explore L^2 spaces defined on a collection of absolutely continuous measures.

11. Suppose the weight function w for an absolutely continuous measure μ is a constant, so $w(x) = k$, where $k \in \mathbb{R}$. Show a Lebesgue-measurable (as in Definition 3.1.1) function f has $L^2(\mu)$ norm equal to $\sqrt{k} \cdot \|f\|$, where the norm refers to the standard L^2 norm defined in Section 3.2. Furthermore, say why the square of the $L^2(\mu)$ norm is equal to $\int |f|^2 \cdot k\,dx$.

12. Let $w(x) = \frac{1}{\pi}\mathcal{X}_{(-\pi,\pi)}(x)$ be the weight function w for an absolutely continuous measure μ, so $d\mu = \frac{1}{\pi}\mathcal{X}_{(-\pi,\pi)}(x)dx$ (here dx refers to Lebesgue measure). Show the $L^2(\mu)$ norm satisfies $\|\cos nx\| = 1$ for any positive integer n.

13. Let $w(x) = \frac{1}{\pi}\mathcal{X}_{(-\pi,\pi)}(x)$ be the weight function w for an absolutely continuous measure μ, so $d\mu = \frac{1}{\pi}\mathcal{X}_{(-\pi,\pi)}(x)dx$ (here dx refers to the Lebesgue measure differential). Show the $L^2(\mu)$ norm satisfies $\|\sin nx\| = 1$ for any positive integer n.

14. The preceding two exercises set up the classical Fourier series expansion for a function $f \in L^2(\mu)$, where $d\mu = \frac{1}{\pi}\mathcal{X}_{(-\pi,\pi)}(x)dx$ (here dx refers to Lebesgue measure). That series is $f(x) = a_0 + \sum_{n=1}^{\infty}(a_n \cos nx + b_n \sin nx)$ where $a_n = \int f \cdot \cos nx\,d\mu$ for $n \neq 0$, $b_n = \int f \cdot \sin nx\,d\mu$, and $a_0 = (1/2)\int f\,d\mu$ (the factor $1/2$ follows from the required normalization). Use the formulas to determine the classical Fourier series expansion for the function $f(x) = x$, $-\pi < x < \pi$.

15. Determine the $L^2(\mu)$ norm $\|x\|$, where μ is the absolutely continuous function given in the previous exercise. Determine the $L^2(\mu)$ inner products $\langle x, \cos nx\rangle$ and $\langle x, \sin nx\rangle$, where $n = 1, 2, 3, \ldots$.

16. Repeat the last exercise, but replace the function x with the function $\cos^2 x$.

17. Let $w(x) = e^{-x^2}$ be the weight function w for an absolutely continuous measure μ, so $d\mu = e^{-x^2}\,dx$ (here dx refers to the Lebesgue measure differential). Determine the $L^2(\mu)$ norm $\|H_n(x)\|$, where $H_n(x)$ is the nth Hermite polynomial defined in Section 3.4. (*Hint:* see Theorem 3.4.2.)

18. Let $w(x) = e^{-x}$, where $x \geq 0$, be the weight function w for an absolutely continuous measure μ, so $d\mu = \mathcal{X}_{[0,\infty)}(x)e^{-x}\,dx$ (here dx refers to the Lebesgue measure differential). Determine the $L^2(\mu)$ norm $\|L_n(x)\|$, where $L_n(x)$ is the nth Laguerre polynomial defined in Section 3.3.

19. For a Lebesgue-measurable set A with finite Lebesgue measure $m(A)$, let $w(x) = \mathcal{X}_A(x)$ and define the absolutely continuous measure μ so $d\mu = w(x)\,dx$ (here dx refers to the Lebesgue measure differential). Determine the $L^2(\mu)$ norm $\|c\|$, where c is a real constant function.

20. Suppose we are given a sequence of Lebesgue-measurable sets $A_1, A_2, A_3, \ldots$, nested so $A_1 \subset A_2 \subset A_3 \ldots$, with finite Lebesgue measure. Let $w(x) = \mathcal{X}_{(\bigcap_{n=1}^{\infty} A_n)}(x)$ and define the absolutely continuous measure μ so $d\mu = w(x)\,dx$ (here dx refers to the Lebesgue measure differential). Determine the $L^2(\mu)$ norm $\|c\|$, where c is a real constant function.

4.4 Application: Probability

Probability is the mathematical measurement of chance. It calculates the odds of a given event occurring. All of us have an intuitive understanding of the chance of an event occurring; for example, we all know the odds of a fair coin tossing heads or tails should be 50-50. The mathematics of probability theory is a systematic approach to the study of chance. To be credible, it has to match the intuitive understanding, but it also has to go beyond: it has to handle complex situations too complicated for us to think about intuitively. In this section, we develop a mathematical model for probability theory that satisfies the goals. We also develop one that can uniformly model any probability question that (reasonably) arises from real-world concerns.

Any mathematical study of probability begins with an understanding of the objects for which it is defined. This definition is framed in terms of what we call an experiment (also known as a random experiment). We take liberties and define the term "experiment" nontechnically—we don't want our explanation to be overly burdensome, and at times we err on the side of the intuitive development of probability theory. This is one of those times.

Definition 4.4.1 *An* experiment *is an activity that can result in many different outcomes, where the term* outcome *is a label for the most fundamental type of result of the experiment. We also define the* sample space S *of an experiment as the set of all possible outcomes. An* event *is a subset of the sample space, and so it is a collection of outcomes.*

Mathematicians often think of probability theory in terms of random variables, which also play a fundamental role in statistics.

Definition 4.4.2 *Given an experiment and a sample space S of all possible outcomes, a random variable X assigns a real number x to each outcome $s \in S$. We sometimes write $X(s) = x$, a notation that lends itself to thinking of the random variable X as a function that maps a domain set of outcomes into a range set of real numbers.*

There is a formal model to understand how random variables behave, based on the concept of a probability space, which is a triplet that makes explicit the items connected with a random variable (specifically, those items in Definition 4.4.2). For the most general model, the triplet can be thought of as $(S, \mathbb{F}, \mu)$. Here, S is the sample space. The symbol $\mathbb{F}$ represents the set of all subsets of S. $\mathbb{F}$ is required to form a σ-algebra of subsets (similar to the way we described σ-algebras in Section 4.1). Finally, μ is a type of measure called a *probability distribution*. For our purposes and for our interest in studying real random variables, we will typically identify the sample space S in terms of a set of real numbers $\mathbb{R}$—then S will be a subset of the reals. The σ-algebra $\mathbb{F}$ can always be thought of in our studies as the Borel field of μ-measurable sets. The probability distribution μ can then be taken in this text's investigations as a specific type of Borel measure—exactly the type we defined in Section 4.2. These items then describe a real-valued random variable with Borel distribution.

Definition 4.4.3 *A real-valued random variable X with Borel distribution is a random variable defined on a probability space $(S, \mathbb{F}, \mu)$, where $S \subseteq \mathbb{R}$ is the sample space of values X assigns. $\mathbb{F}$ is required to be a collection of sets containing elements of S (always including the empty set and S itself) and for which any complement of a set in $\mathbb{F}$ is again in $\mathbb{F}$ and every countable union of sets in $\mathbb{F}$ is again in $\mathbb{F}$. Furthermore, we require μ to be a Borel measure in the sense of Section 4.2, for which every set in $\mathbb{F}$ is μ-measurable and for which $\int \mathcal{X}_S \, d\mu = 1$.*

Definition 4.4.3 connects a real-valued random variable X with Borel distribution to a probability space $(S, \mathbb{F}, \mu)$. Specifically, the sample space for X appears as S. In other words, S is the set of real numbers X might assign. The probabilities associated with X are thereby governed according to the probability measure μ. In particular, the probability that X assigns a number in a set A of $\mathbb{F}$ is

$$P[X \in A] = \int \mathcal{X}_A \, d\mu.\text{[1]}$$

The probabilistic model is designed to let us understand mathematically how real-world random variables behave. A couple of standard examples can illustrate how the model works in its abstract structure and for real-world calculations.

Example 4.4.1 Suppose a gambler rolls a fair die, so the probabilities associated with the six rolls (1 through 6) are $1/6$. Define X to equal the roll of the die. Then S can be identified as

---

[1] We could have described this probabilistic model in terms of a different sample space—the one that has outcomes on which X assigns values (instead of the one defined in terms of the values X assigns). The alternate description is the connection most textbooks use. Describing the sample space as the values X assigns allows us to think of the sample space as a subset of $\mathbb{R}$ and the outcomes as the σ-algebra of μ-measurable subsets of $\mathbb{R}$ from the values X assigns.

4.4 Application: Probability

$S = \{1, 2, 3, 4, 5, 6\}$, the σ-algebra $\mathbb{F}$ consists of subsets of S (there are $2^6 = 64$ of them), and μ is the sum of six point mass measures located at a value of S and having equal point masses of $1/6$. Probabilities about X then follow from the probabilistic model using the integral with the measure μ.

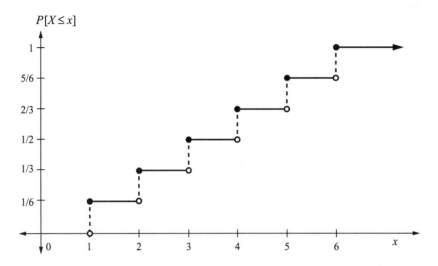

Figure 4.3. The cumulative distribution function for Example 4.4.1.

For example, $P[X = 1]$, which is the same as the probability the gambler rolls a 1 with the die, is $P[X = 1] = \int \mathcal{X}_{\{1\}} \, d\mu$, where μ has mass $1/6$ at $p = 1$. Hence $P[X = 1] = 1/6$.

In the same way, $P[X = x] = 1/6$ for $x = 1, 2, 3, 4, 5, 6$. This fact is often described in introductory statistics courses as the probability distribution function for the random variable X. A random variable's *cumulative probability distribution function* is $P[X \leq x]$, and in this case is $P[X \leq x] = n/6$ for real x, where n equals the number of values in $S = \{1, 2, 3, 4, 5, 6\}$ less than or equal to x. The cumulative probability distribution's values can be calculated directly from the probabilistic model. For example,

$$P[X \leq 5.5] = \int \mathcal{X}_{\{x: \, x \in S \text{ and } x \leq 5.5\}} \, d\mu$$
$$= \int \mathcal{X}_{\{1,2,3,4,5\}} \, d\mu = 1/6 + 1/6 + 1/6 + 1/6 + 1/6 = 5/6.$$

Figure 4.3 graphs the cumulative probability distribution function $F(x) = P[X \leq x]$. ∎

Question 4.4.1 Suppose a blackjack player deals two cards from a standard 52-card deck. If an ace can be counted as 11 points, a face card is 10 points, and any other card matches its rank, what is the probability that the cards sum to 21 points? ∎

The last example is described in elementary textbooks as an example of a finite (discrete) random variable. Because X assigns only a finite number of values, its probabilistic model is described simply—in terms of its distribution function $P[X = x]$. The next example is dependent upon an absolutely continuous probability measure. The elementary textbook description has to model it in a way that looks different from the distribution function description. Indeed, random variables with absolutely continuous probability measures always have distribution functions that give no information, since $P[X = x] = 0$ for them. The next example illustrates why this is the case. It also shows the probabilistic model (where $P[X \in A] = \int \mathcal{X}_A \, d\mu$) provides a unified approach (and a general one) to model real-valued random variables with Borel distribution of any type, not just finitely discrete or continuous.

Example 4.4.2 The height X (in feet) of autumn purple white ash trees in Jackson Forest is assumed to be distributed according to the probabilistic model: $P[X \in A] = \int \mathcal{X}_A \, w(x) \, dx$, where $w(x) = \frac{1}{4\sqrt{2\pi}} e^{-\frac{(x-36)^2}{32}}$. The function w is the *density* function. In this case, w has a bell-shaped curve and is a so-called normal density function. In terms of our definitions, the experiment to whose outcomes X assigns a number is to "randomly select an ash tree from Jackson Forest." The sets $\mathbb{F}$ corresponding to X as described in Definition 4.4.3 comprise the Borel field of μ-measurable sets, where the corresponding measure $\mu(x) = w(x) \cdot dx$ works as a probability measure. The reason it satisfies the properties of a probability measure is that $\frac{1}{4\sqrt{2\pi}} \int_{-\infty}^{\infty} e^{-\frac{(x-36)^2}{32}} \, dx = 1$. (The proof of this follows from the result of Exercise 3.4.30.) A graph of the density function $w(x)$ is in Figure 4.4.

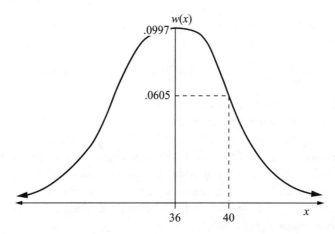

Figure 4.4. The normal density function $w(x) = \frac{1}{4\sqrt{2\pi}} e^{-\frac{(x-36)^2}{32}}$ for Example 4.4.2.

This probabilistic model for the behavior of the heights of the ash trees then determines probabilities of real-life concerns. For example, the proportion of the ash trees in the forest that have height between 30 feet and 40 feet would be $P[30 \leq X \leq 40]$. According to the model, that value is

$$P[30 \leq X \leq 40] = \int \mathcal{X}_{[30,40]} \, w(x) \, dx = \frac{1}{4\sqrt{2\pi}} \int_{30}^{40} e^{-\frac{(x-36)^2}{32}} \, dx.$$

Numerical integration methods calculate this value to be approximately .7745. (For example, the TI-83 calculator makes this calculation as "normalcdf(30,40,36,4).") So if one ash tree in the forest is selected at random, the probability its height will be between 30 and 40 feet is approximately .7745. About 77.45% of the ash trees in the forest are between 30 and 40 feet tall. ■

4.4 Application: Probability

Question 4.4.2 The scores X on a mathematics graduate entrance examination follow an approximately normal distribution that determines $P[X \in A] = \int \mathcal{X}_A\, w(x)\, dx$, where $w(x) = \frac{1}{100\sqrt{2\pi}} e^{-\frac{(x-500)^2}{20{,}000}}$. Determine the approximate proportion of exam scores that are above 650. ∎

We have developed many facts about the probabilistic model. It presents a unified approach to model a real-valued random variable with Borel distribution. In fact, any probability measure (of the Borel type we have described here) can define such a random variable X; it only needs to corresponds to X according to $P[X \in A] = \int \mathcal{X}_A\, d\mu$. Hence the collection of random variables we can now consider is equivalent to the collection of the probability Borel measures we can define. The two examples of random variables this section has given correspond to finite and continuous random variables, respectfully, but the exercises allow you to explore other types of random variables (those with other types of probability measures).

Solutions to Questions

4.4.1 We define the random variable X to be the sum of the points for the two cards. Using combinatorics, there are $\binom{52}{2} = 52!/(2! \cdot 50!) = 1326$ two-card deals. To get 21 points in the two cards, the dealer must deal an ace (which happens in four possible ways) and either a face card or ten (which happens in 16 ways). Hence the number of two-card deals resulting in 21 points is $4 \cdot 16 = 64$. The point mass assigned to the random variable outcome $X = 21$ is therefore $p = 64/1326 = 32/663$. In terms of the probabilistic model, where μ is the measure for X, $P[X = 21] = \int \mathcal{X}_{\{x:\, x \in S \text{ and } x=21\}}\, d\mu = \int \mathcal{X}_{\{21\}}\, d\mu = 32/663$.

4.4.2 The proportion is $P[X > 650] = \int_{650}^{\infty} \frac{1}{100\sqrt{2\pi}} e^{-\frac{(x-500)^2}{20{,}000}}\, dx \approx .0668$ (numerically from normalcdf(650,100000,500,100) on a TI-83 calculator).

Reading Questions for Section 4.4

1. What does this section mean by the term "experiment"?

2. What is a sample space of an experiment?

3. What is an outcome of an experiment? An event?

4. What action does a random variable take for a given well-defined experiment with a sample space S?

5. What is a probability space $(S, \mathbb{F}, \mu)$?

6. On what object is a real-valued random variable with Borel distribution defined? Explain the elements of the object in detail.

7. Given a real-valued random variable with Borel distribution X, how do we calculate mathematically the probability that X assigns a number in a set A of $\mathbb{F}$?

8. Give an example of a finite random variable and describe its cumulative probability distribution function.

9. Give an example of a random variable with probability measure μ that is absolutely continuous.

Exercises for Section 4.4

Exercises 1–10 describe random variables with Borel distributions. Identify the objects from Definition 4.4.3. In particular, list S, $\mathbb{F}$, and μ.

1. Let $X = 1$ if a fair coin tosses heads and $X = 0$ if tails.

2. Let Y be the number of heads tossed with three coins. The sample space has eight elements, and, for example, the outcome HTH is different from the outcome HHT. Both have $Y = 2$.

3. Let P be the number appearing on a single roll of an eight-sided octagonal die, where the die is fair (so each side comes up equally often), and the whole numbers from 1 to 8 appear on the die's faces.

4. Let R be the total points appearing on a roll of two fair six-sided dice. The two outcomes, for example, of rolling a three, which are "1 , 2" and "2 , 1" (both have $R = 3$) are different. There are $6 \cdot 6 = 36$ two-dice rolls.

5. Let X be the number of kings dealt in a well-shuffled two-card deal. Using combinatoric formulas, it turns out there are 1326 hands. The chance of dealing two kings is $6/1326 = 1/221$, of dealing one king is $192/1326 = 32/221$, and of dealing zero kings is $1128/1326 = 188/221$.

6. Let Y be the number of heads (H) tossed in 10 tosses of a fair coin. Combinatoric formulas say there are 2^{10} coin tosses and there are $_{10}C_r = r!/(10! \cdot (10-r)!)$ tosses that result in exactly r heads tossed for $r = 0, 1, 2, \ldots, 10$. Each happens equally often with a fair coin.

7. Let N be the number of heads (H) tossed in n tosses of a fair coin, where $n \geq 1$. See the previous exercise for the case $n = 10$.

8. Let U be a continuous random variable that assigns values x in the set $[0, 1]$ with equal weight, so the density function is $f(x) = \mathcal{X}_{[0,1]}$. ($U$ is the uniform random variable on $[0, 1]$.)

9. Modify the last exercise to let Y be a continuous random variable that assigns values y in the set $[a, b]$ with equal weight, where $a, b \in \mathbb{R}$ are constants with $a < b$.

10. Toss a coin and then roll a die. Let $W =$ the sum of the number of heads tossed and the number rolled. Compare your answer with the results of Exercise 1 and Example 4.4.1.

In Exercises 11–20, determine the cumulative distribution function for each given random variable.

11. Let $X = 1$ if a coin tosses heads and $X = 0$ if tails. (See Exercise 1.)

12. Let N be the number of heads (H) tossed in n tosses of a fair coin, where $n \geq 1$. (See Exercise 7.)

13. Let Y be the continuous random variable that assigns values y in the set $[a, b]$ with equal weight. (See Exercise 9.)

14. Let S be the sum of the number of points appearing on a roll of three fair six-sided dice.

15. Let K be the number of kings randomly dealt in a two-card (blackjack) hand. (See Exercise 5.)

16. Let Q be the number of queens randomly dealt in a three-card hand.

17. Spin a roulette wheel twice. The wheel has numbered slots from 0 to 36. Let X be the number of spins resulting in any of the values from 1 to 12. (Such a result is an "outside bet win.")

18. Let W be the continuous (exponential) random variable that assigns values x in the set $[0, \infty)$ according to the density function e^{-x}.

19. When tossing a coin and rolling a die, let W be the sum of the number of heads tossed and the number rolled. (See Exercise 10.)

20. Toss two coins, and then have a machine (which is highly irregular and works unreliably) fill a one-gallon paint can with a volume x of paint. Suppose any value x between 0 and 1 gallon is the filled amount of paint randomly attained with equal weight. Let S be (the number of heads tossed $+ x$).

21. Let Y be an insurance company's net profit, which is dependent on the damage amount X for a given policyholder, and where the probability measure for X is absolutely continuous with density, in x thousands of dollars, $f(x) = x \cdot e^{-x}$ for $x > 0$ (and 0 otherwise).[2] Assume policyholders are charged a premium of $1,000 and have a deductible of $1,500 applied to the damage amount payment. If the resulting payout to policyholders is capped at $5,000, find the cumulative distribution function for Y.

Notes for Chapter 4

Many other methods can be used to develop integrals with respect to other measures on the real line. Since it corresponds to the methodology used to develop the Lebesgue measure's integral in Chapter 1, we have chosen Percy John Daniell's method (which first appeared in 1917 in [34]) to develop the material in this chapter. Raised in England from the time he was six years old, Daniell studied first at Cambridge and then under Hilbert and Max Born at Göttingen. In 1914, he emigrated to America to accept an appointment as professor of mathematics at what is now Rice University in Houston. He worked there until 1923. His seminal paper on the integral is broadly considered to have developed an integration method that is maximally abstract, as reflected by this quote from the paper:

> The idea of an integral has been extended by Radon, Young, Riesz and others so as to include integration with respect to a function of bounded variation. These theories are based on the fundamental properties of sets of points in a space of a finite number of dimensions.... In this paper a theory is developed which is independent of the nature of the elements. They may be points in a space of a denumerable number of dimensions or curves in general or classes of events so far as the theory is concerned.

---

[2] This is a reasonable assumption about the density. Thanks are due to Ken Constantine for his input on this exercise.

His later work also generalized the derivative to correspond to the integrals (considered at the time a more difficult problem), a statistical result that is considered the first paper in what statisticians now call "order statistics," and extended applications of frequency spectra to provide automatic control of equipment, such as radar, in need of band pass filter systems (see [112]).

Figure 4.5. The accomplished Greek mathematician Constantin Carathéodory (seated), around 1910 in Germany, with the Hungarian mathematician Lipót Fejér (who was the dissertation director for John von Neumann, Paul Erdős, and George Pólya).

Another method to develop integrals, more commonly cited and more broadly understood, is the Lebesgue-Stieltjes method. Though Lebesgue-Stieltjes integrals are typically described differently from integrals using the Daniell method, they turn out to be the same on spaces $L^1(\mu)$ that contain all step functions. A third method, which is the classical method taught in many graduate mathematics programs and due originally to Carathéodory (see [23]), starts with the concept of a general measure and then shows how the integral is built on it. Such a measure can often be constructed out of a nonempty collection $\mathcal{C}$ of real subsets that satisfy the condition if S and T are in $\mathcal{C}$, then $S \cap T$ and $S \setminus T$ can be expressed as finite disjoint unions of sets in $\mathcal{C}$. When that is the case, then (see [120]) that Carathéodory treatment's resulting integral is equivalent to the Daniell integral. We hope to convey there exist many methods to develop integrals with respect to measures, and mathematicians have gotten to know each of them by different names. Indeed, they can develop differently over different set domains. But to a large degree—in the sense there is a huge number of settings for most investigations typical in function theory—these methods are equivalent, or at least they overlap, and the Daniell method works in this way quite powerfully. Weir's second volume on integration theory [120] gives a further development of this topic and on the different integration methods. Similarly, we refer the interested reader to Rudin [107] and Royden [105].

Many famous 20th-century mathematicians contributed to the development of probability in terms of measure theory. One of the most prominent was Laurent Schwartz (not to be confused

Figure 4.6. Laurent Schwartz in 1975.

with Hermann Schwarz of the Schwarz inequality). Laurent Schwartz was a French mathematician of Jewish descent from the Alsace area who managed to escape Nazi detection by using the alias Laurent-Marie Sélimartin and going into hiding during World War II. Schwartz is credited with developing a comprehensive theory of distributions—including functions nondifferentiable and discontinuous. He extended some of the Russian mathematician Sergei Sobolev's results, which defined a generalized function, in the 1930s. In 1950, Schwartz received the Fields Medal (the most prestigious award in mathematics) for this work. The Fields Medal is considered roughly equivalent to a Nobel Prize, as there is no Nobel Prize in mathematics. Other Fields Medalists are: Lars Ahlfors and Jesse Douglas (in 1936); Atle Selberg (with Schwartz in 1950); Kunihiko Kodaira and Jean-Pierre Serre (in 1954); Klaus Roth and René Thom (in 1958); Lars Hörmander and John Milnor (in 1962); Michael Atiyah, Paul Cohen, Alexander Grothendieck, and Stephen Smale (in 1966); Alan Baker, Heisuke Hironaka, Sergei Novikov, and John G. Thompson (in 1970); Enrico Bombieri and David Mumford (in 1974); Pierre Deligne, Charles Fefferman, Grigory Margulis, and Daniel Quillen (in 1978); Alain Connes, William Thurston, and Shing-Tung Yau (in 1982); Simon Donaldson, Gerd Faltings, and Michael Freedman (in 1986); Vladimir Drinfel'd, Vaughan Jones, Shigefumi Mori, and Edward Witten (in 1990); Efim Zelmanov, Pierre-Louis Lions, Jean Bourgain, and Jean-Christophe Yoccoz (in 1994); Richard Borcherds, Timothy Gowers, Maxim Kontsevich, and Curtis T. McMullen (in 1998); Laurent Lafforgue and Vladimir Voevodsky (in 2002); Andrei Okounkov, Grigori Perelman, Terence Tao, and Wendelin Werner (in 2006); Elon Lindenstrauss, Ngô Bào Châu, Stanislav Smirnov, and Cédric Villani (in 2010); and Artur Avila, Manjul Bhargava, Martin Hairer, and (the first female medalist) Maryam Mirzakhani (in 2014).

5
Hilbert Space Operators

Anyone who studies functions eventually looks at ways in which they can be manipulated. Actions on functions can often be thought of as maps themselves. For example, calculus describes the way to take a differentiable function f and calculate its derivative f', a process that can be thought of as a mapping from f to f'. In fact, the mapping (or transformation) $D : f \to f'$ is linear—we all know it satisfies the linearity property $D(af + bg) = aD(f) + bD(g)$ for constants a and b and differentiable functions f and g. This algebraic description of linearity describes a combination of two familiar rules: the derivative of the sum is the sum of the derivatives and the derivative of a constant times a function is the constant times the derivative of the function. Another example of a familiar map on functions is the integral. Any integrable function $f(x)$, say, on $[0, \infty)$ can be mapped to the antiderivative function $F(x) = \int_0^x f(t)\,dt$. The linearity of the integral guarantees that $T : f \to F$ is also linear.

These mappings are *operators*. We study operators that take elements of one vector space (the domain space) to the elements of another (the target space). We get a *linear operator* when the linearity property is satisfied. You can think of operators as acting on functions, or on any type of object in a vector space. Each element of the space on which the operator is defined is mapped to exactly one element in the target space. For example, if you have taken a course in linear algebra, then you have studied operators acting on finite-dimensional vectors—multiplication of a vector by an $n \times n$ matrix is a linear operator acting on an n-dimensional vector space. An operator on a function space works similarly in many ways. It takes elements in one space of functions to elements in another (possibly different) space of functions. In this chapter, we introduce some types of operators and give insights on how they are studied. We'll describe what it means for a linear operator to be bounded (over the function space on which it acts). We'll provide examples of bounded linear operators on Hilbert spaces, starting with the case L^2. One type of operator is the shift operator. It appears often in applications, acts in constructive ways, and can even provide a model for the way operators behave. Section 5.3 defines it. The last section describes a way many operators can be well understood: instead of thinking of the operator as acting on its own domain Hilbert space, we can think of it as a different but equivalent operator acting on some other Hilbert space—where the action on that space is simple. That process is embodied in the spectral theorem.

What does the study of operators have to do with the Lebesgue integral? A lot. First, operators are often naturally studied on the function spaces L^2, whose formation and structures depend on the Lebesgue integral. Second, we will see operators can look (and can be) very complicated as they are viewed as acting on objects in their domain space. But for many of them, we can set up an equivalence with an operator acting more simply on another, equivalent, domain space. The equivalent space is often $L^2(\mu)$ for some Borel measure μ. When the equivalence is possible, it is the Lebesgue integral with respect to the measure μ that has to be used to describe the simpler action on the equivalent space. Spectral theorem equivalence is a major application of the Lebesgue integral, as it can be applied to discover many facts about operators, and hence many facts about functions. We'll take up the spectral theorem in Section 5.4.

Historically, these mathematical ideas were developed throughout the 1900s, and they continue to be an active area of research. Some of the greatest function theorists of all time were part of this development, including such mathematicians as G. H. Hardy (Great Britain), David Hilbert (Germany), Stephan Banach (Poland), Hermann Weyl (Germany and United States), John von Neumann (Germany and United States), Frigyes Riesz (Hungary), Arne Beurling (Sweden), Walter Rudin (United States), and Paul Halmos (United States). Topics in operator theory are advanced and often saved for graduate study. They are also typically considered over spaces of functions defined on the field of complex numbers rather than on the reals, as complex-valued functions provide a richness that reflects the complexity inherent in many operators. Our description avoids complex functions because we want to keep the discussion broadly accessible. This chapter's goal is not to present the material in its complete theoretical development and format. Instead, it introduces definitions, examples, and statements and uses of theorems, along with a few proofs of interesting results.

5.1 Bounded Linear Operators on L^2

Operators on a Hilbert space $\mathcal{H}$ are objects that map the elements of $\mathcal{H}$ to another Hilbert space, say $\mathcal{G}$ (where perhaps $\mathcal{H}$ and $\mathcal{G}$ are the same space)[1]. So if the operator is T, then we describe the map from one space to the other as

$$T : \mathcal{H} \to \mathcal{G}.$$

If f is an element in $\mathcal{H}$, then we write $T(f)$ as its image in $\mathcal{G}$. We will often relax the need for parentheses and write Tf for $T(f)$.

This chapter studies bounded linear operators defined on a general Hilbert space $\mathcal{H}$. We will assume $\mathcal{H}$ is separable: it has a countable basis. This section examines the situation where $\mathcal{H}$ is an $L^2(\mu)$ space. Then an operator on $L^2(\mu)$ maps the functions in $L^2(\mu)$ to the objects in another Hilbert space $\mathcal{G}$, where perhaps $\mathcal{G} = L^2(\mu)$. In general, we say the operator T is linear when, for $f, g \in \mathcal{H}$ and scalar constants a and b,

$$T(af + bg) = aT(f) + bT(g).$$

The operator T is bounded when there exists a constant $K > 0$ such that, for every f in $\mathcal{H}$,

$$\|T(f)\|_{\mathcal{G}} \leq K \cdot \|f\|_{\mathcal{H}}.$$

---

[1] In the 1960s, Louis de Branges and James Rovnyak examined a certain type of functional Hilbert space, which they described as "square-summable" (see [16]). Their space of square-summable power series is a vector space over the complex numbers that admits an inner product. It is naturally seen to be equivalent to ℓ^2. Studying operators on square-summable power series continues to form a productive area of investigative research.

5.1 Bounded Linear Operators on L^2

The smallest such K, described in terms of the infimum over all such constants, is the norm of the operator, or the *operator norm*, and so a bounded operator is one with finite operator norm. The value of the norm for a given operator T is written $\|T\|$, and it is calculated as

$$\|T\| = \sup\{\|Tf\|_{\mathcal{G}}\},$$

where the supremum of the $\mathcal{G}$-norm values is taken over all elements f in $\mathcal{H}$ with $\|f\|_{\mathcal{H}} \leq 1$.

We begin by looking at an example of a bounded linear operator on an L^2 space.

Example 5.1.1 Choose a function $r(x)$ in an L^2 space. Then, for f in it, define the operator $T_r : L^2 \to L^2$ by

$$T_r(f) = \langle f, r \rangle \cdot r(x).$$

You can see that every function in L^2 is mapped to a constant multiple of $r(x)$. Hence (considering L^2 as a vector space of functions), the range of T_r (the set of output values) consists of the one-dimensional span of the function r. (The one-dimensional span of r is as defined in any linear algebra course as the set of functions g of the form $g(x) = c \cdot r(x)$ for a constant c. In this case, the output values of T_r are exactly that type, with $c = \langle f, r \rangle$.) Because its range is one-dimensional, the operator T_r is a *rank-one operator*. It turns out T_r is a bounded linear operator, and we can show why each of the two properties hold.

Linearity: T_r is linear because, for L^2 functions f and g and constants a and b,

$$T_r(af + bg) = \langle af + bg, r \rangle \cdot r(x) = a\langle f, r \rangle \cdot r(x) + b\langle g, r \rangle \cdot r(x) = aT_r(f) + bT_r(g).$$

Boundedness: We can see T_r is bounded by examining the L^2-norm of any $f \in L^2$:

$$\|T_r(f)\| = \|\langle f, r \rangle \cdot r(x)\| = |\langle f, r \rangle| \cdot \|r(x)\|.$$

By Schwarz's inequality, $|\langle f, r \rangle| \leq \|f\| \cdot \|r\|$, and so

$$\|T_r(f)\| \leq (\|f\| \cdot \|r\|) \cdot \|r\| = \|r\|^2 \cdot \|f\|.$$

Boundedness is therefore satisfied by setting $K = \|r\|^2$. This value is the operator norm $\|T_r\|$, since $\sup\{\|T_r(f)\|\} = \|r\|^2$ is attained when $\|f\| = 1$. ∎

Question 5.1.1 For the function space $L^2(-\pi, \pi)$, define $T_{\cos x}(f) = \langle f, \cos x \rangle \cdot \cos x$. What is $T_{\cos x}(f)$ when $f(x) = x^2$? Determine the operator norm $\|T_{\cos x}\|$ and show $\|T_{\cos x}(x^2)\| \leq \|T_{\cos x}\| \cdot \|x^2\|$. ∎

For an example of a rank-one operator, choose a function r in $L^2(\mathbb{R})$, such as $r(x) = e^{-x^2/2}$. Then $T_r f = \langle f(x), e^{-x^2/2} \rangle \cdot e^{-x^2/2} = \int_{-\infty}^{\infty} f(x) e^{-x^2/2} \, dx \cdot e^{-x^2/2}$, and the norm of $T_{e^{-x^2/2}}$ is $\sqrt{\pi}$, since $\|r\|^2 = \int_{-\infty}^{\infty} |e^{-x^2/2}|^2 \, dx = \int_{-\infty}^{\infty} e^{-x^2} \, dx = \sqrt{\pi}$.

Another classic example of a bounded linear operator is an *integral operator*, a label used whenever the operator maps an element of L^2 to another function determined by integration in the L^2 space. A basic structure of an integral operator T, as it acts on an element f in the L^2 space, is

$$T(f)(x) = \int k(x, y) f(y) \, dy.$$

The function k is the kernel of the operator. The operator's output $T(f)(x)$ is a function of x. (We often suppress the variable notation and write the output function as $T(f)$.) The linearity of T follows from the linearity of the integral. Mathematicians began studying in earnest such operators in the first decade of the 20th century, and they motivated some of the earliest investigations into Hilbert space theory. The next example describes one such operator, which Erhard Schmidt[2] investigated.

Example 5.1.2 Let $k(x, y)$ be a (real-valued) continuous function defined for real x and y in a finite interval (a, b), so $\int_a^b \int_a^b |k(x, y)|^2 \, dx \, dy = M < \infty$ for some constant M. For $f \in L^2(a, b)$ define the operator T by

$$T(f)(x) = \int_a^b k(x, y) \cdot f(y) \, dy.$$

Then T is linear because for constants c_1 and c_2 and functions $f, g \in L^2(a, b)$, we have $T(c_1 f + c_2 g) = \int_a^b k(x, y)[c_1 f(y) + c_2 g(y)] dy = c_1 T(f) + c_2 T(g)$, which follows from the linearity of the integral. The operator T turns out to be bounded, and $T : L^2(a, b) \to L^2(a, b)$. Both facts result from the Schwarz inequality (Theorem 3.2.1), which implies

$$\|T(f)\|^2 = \int_a^b \left| \int_a^b k(x, y) \cdot f(y) \, dy \right|^2 dx \leq \int_a^b \left(\int_a^b |k(x, y)|^2 \, dy \right) \cdot \left(\int_a^b |f(y)|^2 \, dy \right) dx$$
$$= M \cdot \|f\|^2.$$

Taking the supremum over $f \in L^2(a, b)$ having $\|f\| = 1$, we have $\|T\| = \sup\{\|T(f)\|\} = \sqrt{M}$, which proves T is bounded.

The operator T is an example of a broad class of Hilbert-Schmidt integral operators. In the early 1900s, Schmidt and Hilbert examined them in specific cases; for example, when the kernel k is symmetric so $k(x, y) = k(y, x)$ for $x, y \in (a, b)$. Schmidt (see [109]) was able to determine there was, in this case, a countable collection of real numbers $\{\lambda_n\}_{n=0}^{\infty}$, the *eigenvalues*, and a corresponding sequence of functions e_n, $n = 0, 1, 2, \ldots$, the *eigenfunctions*, so the kernel could be expressed completely in terms of the eigenvalues and eigenfunctions:

$$k(x, y) = \sum_{n=0}^{\infty} \lambda_n e_n(x) e_n(y).$$

T could be completely described in a similarly remarkable way:

$$T(f)(x) = \sum_{n=0}^{\infty} \lambda_n \langle f, e_n \rangle e_n(x),$$

for $f \in L^2(a, b)$. The inner product is the $L^2(a, b)$ inner product, so $\langle f, e_n \rangle = \int_a^b f(y) \cdot e_n(y) \, dy$, and the infinite series converges in the L^2-norm (so $\lim_{N \to \infty} \| \sum_{n=0}^{N} \lambda_n \langle f, e_n \rangle e_n(x) - T(f)(x) \| = 0$). Later sections will say more about this type of representation of operators; it can be considered, historically, to be the first time a nontrivial class of operators on an infinite-dimensional Hilbert space was diagonalized. ∎

---

[2] Erhard Schmidt is the same German mathematician after whom the Gram-Schmidt process is named. He was one of the most prolific of 75 doctoral students of David Hilbert at Göttingen, defending his dissertation in 1905 and working throughout his career and into the 1950s at the University of Berlin, where he directed 32 Ph.D. students of his own.

5.1 Bounded Linear Operators on L^2

An operator similar in many ways to the Hilbert-Schmidt operator but more complicated is the Wiener-Hopf operator, defined on functions in $L^2(0, \infty)$. The next example describes it, and then we show how it can be thought of in other Hilbert-space settings. For example, it can be understood as an operator on ℓ^2.

Example 5.1.3 To define the Wiener-Hopf operator T with kernel function $k(x)$, choose a function $k \in L^2(\mathbb{R})$. Then T is based on k (and so we write T as T_k), and it maps $L^2(0, \infty)$ to $L^2(0, \infty)$. Here's how T_k is defined: for a function $f \in L^2(0, \infty)$,

$$T_k : f(x) \to f(x) + \int_0^\infty k(x-t) \cdot f(t) \, dt.$$

T_k is another example of an integral operator. It is a *convolution operator*, one that has a structure established from a dummy variable—here the variable t. The convolution $g * f$ is an operation (usually an integral or sum) applied to the product $g(x - t)f(t)$. You can see how this appears in T_k. Convolution operators are useful in computer science, the physical sciences, and applied mathematics.

For example, let $k(t) = -(1/2)e^{-|t|/2}$. Then for $f \in L^2(0, \infty)$,

$$T_k f(x) = f(x) - \frac{1}{2} \int_0^\infty e^{-|x-t|/2} \cdot f(t) \, dt. \qquad \blacksquare$$

The last example describes an operator, like many, defined on one Hilbert space (in this case $L^2(0, \infty)$), which has a more pleasing form if we reformulate it as one acting on a different Hilbert space. The reformulation often provides a visually pleasing description, and that's what happens for the Wiener-Hopf operator. The Wiener-Hopf kernel function $k(t)$ is related to a function $w(x)$ that develops the reformulation. To get w, first define a related function

$$W(x) = 1 + \int_{-\infty}^\infty e^{ixt} \cdot k(t) \, dt.$$

When, for example, $k(t) = -(1/2)e^{-|t|/2}$, $W(x) = 1 - \frac{1}{2}\int_{-\infty}^\infty e^{ixt} \cdot e^{-|x-t|/2} \, dt$, or

$$W(x) = \frac{x^2 - 1/4}{x^2 + 1/4}.$$

Now define $w(t)$ as

$$w(t) = W(x) \text{ where } t = \frac{x - i/2}{x + i/2}.$$

Continuing with the example $k(t) = -(1/2)e^{-|t|/2}$, where $W(x) = \frac{x^2-1/4}{x^2+1/4}$, we get $w(e^{it}) = \cos(t)$ (see [101, p. 61] for a complete description). The calculation is in Exercise 13 at the end of this section. We will sometimes write $w(t)$ for $w(e^{it})$ when the composition with the complex exponential is well-understood.

The correspondence between the functions k and w establishes a different way of looking at the operator T_k. Instead of viewing T_k as acting on $L^2(0, \infty)$, we can view it as an equivalent operator T_w that acts on ℓ^2. The equivalence is established through an equivalence of the Hilbert spaces $L^2(0, \infty)$ and ℓ^2 (via an isomorphic mapping of basis elements in the former Hilbert space to basis element in the latter) and an equivalence of operator action. The Wiener-Hopf operator T_k on $L^2(0, \infty)$ works the same as an operator T_w on ℓ^2. We will study bounded linear operators on ℓ^2 more completely in the next section, but this example serves as a preview, and it shows how to reformulate T_k into an equivalent operator on ℓ^2.

Why is the reformulation interesting and pleasing? One reason is that it provides a new way of viewing the operator's action: operators on ℓ^2 can be thought of as multiplication by an infinite-dimensional matrix. (We will soon describe how this multiplication works.) In this case, it turns out the multiplication operator T_w manifests itself in the matrix form

$$T_w \sim \begin{bmatrix} a_0 & a_1 & a_2 & a_3 & a_4 & \ldots \\ \overline{a_1} & a_0 & a_1 & a_2 & a_3 & \ldots \\ \overline{a_2} & \overline{a_1} & a_0 & a_1 & a_2 & \ldots \\ \overline{a_3} & \overline{a_2} & \overline{a_1} & a_0 & a_1 & \ldots \\ \overline{a_4} & \overline{a_3} & \overline{a_2} & \overline{a_1} & a_0 & \ldots \\ & & \ldots & & & \end{bmatrix}.$$

The bar on any of the matrix's terms refers to the complex conjugate, which is not needed if the terms are real. This matrix $T_w = [a_{ij}]$ is in Toeplitz form, which means that it is constant along its diagonals. The elements $a_0, a_1, a_2, \ldots$ of the matrix show up in a complex-valued series expansion of $w(t)$; the correspondence is $w(t) = \sum_{n=-\infty}^{\infty} a_n e^{int}$. In this expansion, the negatively indexed coefficients are linked to the positively indexed terms via $a_{-n} = \overline{a_n}$. For the example, where $w(t) = \cos t$, the expansion is determined by realizing $w(t) = \cos t = (1/2)e^{it} + (1/2)e^{-it}$. (This is seen from adding $e^{it} = \cos t + i \sin t$ and $e^{-it} = \cos t - i \sin t$. See Exercise 2.4.1.) For this choice of $w(t) = \cos t$, each a_n is therefore equal to 0 except $a_1 = a_{-1} = 1/2$. The matrix for $T_{\cos t}$ is

$$T_{\cos t} \sim \begin{bmatrix} 0 & 1/2 & 0 & 0 & \ldots \\ 1/2 & 0 & 1/2 & 0 & \ldots \\ 0 & 1/2 & 0 & 1/2 & \ldots \\ 0 & 0 & 1/2 & 0 & \ldots \\ & & \ldots & & \end{bmatrix}.$$

As a linear operator on ℓ^2, $T_{\cos t}$ acts as a matrix product—it's the matrix of $T_{\cos t}$ multiplied by an arbitrary column vector $\vec{v}$. The product is taken in the same way as a product of a finite-dimensional matrix and column—we multiply each row of the matrix by the column, where we add the product of corresponding entries. That product then produces each of the resulting column's entries:

$$T_{\cos t} \cdot \vec{v} = \begin{bmatrix} 0 & 1/2 & 0 & 0 & \ldots \\ 1/2 & 0 & 1/2 & 0 & \ldots \\ 0 & 1/2 & 0 & 1/2 & \ldots \\ 0 & 0 & 1/2 & 0 & \ldots \\ & & \ldots & & \end{bmatrix} \cdot \begin{bmatrix} v_0 \\ v_1 \\ v_2 \\ v_3 \\ \vdots \end{bmatrix} = \begin{bmatrix} v_1/2 \\ (v_0 + v_2)/2 \\ (v_1 + v_3)/2 \\ \vdots \end{bmatrix}.$$

Viewed in this way, a wiener-Hopf operator T on $L^2(0, \infty)$ is equivalent to an operator T_w on ℓ^2. We will describe what we mean by this equivalence in the next paragraph. The operator T_w is a Toeplitz operator and is useful, since it occurs naturally in many applied and theoretical investigations. It is named after the early 20th-century German mathematician Otto Toeplitz, who connected complex function series expansions to such infinite-dimensional patterned matrices.

5.1 Bounded Linear Operators on L^2

Question 5.1.2 Determine $T_w = T_{\sin t}$ in its matrix form, using the fact that $\sin t = (-i/2)e^{it} + (i/2)e^{-it}$. Then describe $T_{\sin t}$ as an operator on ℓ^2 by determining the output value for $T_{\sin t} \cdot \vec{v}$, where $\vec{v}$ is an element in ℓ^2. ∎

A mapping from one Hilbert space to another that forms the equivalence of operators is an isometry (or a Hilbert space isomorphism). The equivalence of Example 5.1.3's Weiner-Hopf operator T_k and the Toeplitz operator T_w is expressed in the preservation of inner products under the isometry (which we will illustrate in a moment). Section 3.3 introduced the concept of an isometry by looking at how the isometric isomorphism mapped orthonormal basis elements of one Hilbert space to orthonormal basis elements of another. In the same way as Hilbert spaces are described as being isometrically isomorphic, equivalent operators are also said to be isometrically isomorphic or "unitarily equivalent" to one another. So what isometry from $L^2(0, \infty)$ to ℓ^2 makes T_k and T_w equivalent? As you would expect, it is the mapping U that takes one set of basis elements to the other. In this case, U pairs basis elements by taking $f_{n-1}(t)$ to $\vec{e}_n$, $n = 1, 2, 3, \ldots$ where $f_n(t) = e^{-t/2} L_n(t)$ is built from the normalized Laguerre polynomial $L_n(t)$ (see Section 3.4 on how the functions form an orthonormal basis for $L^2(0, \infty)$),

and $\vec{e}_n = \begin{bmatrix} 0 \\ 0 \\ 0 \\ 1 \\ 0 \\ \vdots \end{bmatrix}$ is the column vector having 1 in the nth entry. So $U(f_{n-1}) = \vec{e}_n$ determines

a pairwise map between basis entries. The inverse map U^{-1} satisfies $U^{-1}(\vec{e}_n) = f_{n-1}$. The isometry U preserves inner products associated with the operators in the different function spaces, following an identity known as the Toeplitz isomorphism theorem and proved by the American mathematician Marvin Rosenblum in 1965 (see [101, p. 60]).

$$\langle T_k f_n, f_m \rangle_{L^2(0,\infty)} = a_{n-m} = \langle U T_k U^{-1} \vec{e}_n, \vec{e}_m \rangle_{\ell^2} = \langle T_w \vec{e}_n, \vec{e}_m \rangle_{\ell^2}, \text{ for } n, m = 1, 2, 3, \ldots.$$

In the identity, a_n refers to the diagonal constant in the matrix for T_w. Similar identities allow you to think of the Wiener-Hopf operator on other Hilbert spaces. In Section 5.4, we will explore the operator isometrically isomorphic to T_w on a particular Hilbert space H^2, a "Hardy space." In that case, the operator action is multiplication on an element $f \in H^2$ by $w(t)$ (though the operator adds a small extra twist—a projection—after the multiplication).

Solutions to Questions

5.1.1 $T_{\cos x}(f) = T_{\cos x}(x^2) = \langle x^2, \cos x \rangle \cdot \cos x = (\int_{-\pi}^{\pi} x^2 \cos x \, dx) \cdot \cos x = -4\pi \cos x$. From the last comment in Example 5.1.1, the operator norm is $\|T_{\cos x}\| = \|\cos x\|^2 = \int_{-\pi}^{\pi} \cos^2 x \, dx = \pi$. Since $\|x^2\|^2 = \int_{-\pi}^{\pi} x^4 \, dx = 2\pi^5/5$ and $\|T_{\cos x}(x^2)\|^2 = 16\pi^2 \int_{-\pi}^{\pi} \cos^2 x \, dx = 16\pi^3$, we get $\|T_{\cos x}(x^2)\| = 4\pi^{3/2} < 30 < \sqrt{2/5}\pi^{7/2} = \pi \cdot \sqrt{2\pi^5/5} = \|T_{\cos x}\| \cdot \|x^2\|$.

5.1.2 $T_{\sin t} \sim \begin{bmatrix} 0 & -i/2 & 0 & 0 & \cdots \\ i/2 & 0 & -i/2 & 0 & \cdots \\ 0 & i/2 & 0 & -i/2 & \cdots \\ 0 & 0 & i/2 & 0 & \cdots \\ & & \vdots & & \end{bmatrix}$ and $T_{\sin t} \cdot \vec{v} = \begin{bmatrix} 0 & -i/2 & 0 & 0 & \cdots \\ i/2 & 0 & -i/2 & 0 & \cdots \\ 0 & i/2 & 0 & -i/2 & \cdots \\ 0 & 0 & i/2 & 0 & \cdots \\ & & \vdots & & \end{bmatrix} \cdot \begin{bmatrix} v_0 \\ v_1 \\ v_2 \\ \vdots \end{bmatrix} = \begin{bmatrix} -iv_1/2 \\ i(v_0-v_2)/2 \\ i(v_1-v_3)/2 \\ \vdots \end{bmatrix}$.

Reading Questions for Section 5.1

1. What do we mean by an operator T on a given $L^2(\mu)$ space?

2. When is an operator linear? When is it bounded?

3. Give an example of a rank-one operator on $L^2(0, 1)$.

4. What is the formula for the map provided by a Wiener-Hopf operator T_k with kernel function $k(x)$ on $L^2(0, \infty)$?

5. What is a Toeplitz operator on ℓ^2?

6. When are two operators isometrically isomorphic?

Exercises for Section 5.1

Exercises 1–4 deal with the shift operator S on ℓ^2, which is defined for a vector $\vec{v} \in \ell^2$ by

$$S(\vec{v}) = S\left(\begin{bmatrix} v_1 \\ v_2 \\ v_3 \\ \vdots \end{bmatrix}\right) = \begin{bmatrix} 0 \\ v_1 \\ v_2 \\ \vdots \end{bmatrix}.$$

That is, it shifts each component of the vector down one spot and places a 0 in the first component. We take up the shift operator for an arbitrary Hilbert space in Section 5.3.

1. Determine $S(\vec{v})$, where $\vec{v} = \begin{bmatrix} 1 \\ 0 \\ 0 \\ \vdots \end{bmatrix}$.

2. Determine $S(\vec{v})$, where $\vec{v} = \begin{bmatrix} 0 \\ \vdots \\ 0 \\ 1 \\ 0 \\ \vdots \end{bmatrix}$, where the 1 appears in the nth coordinate spot.

3. Determine $S(\vec{v})$, where $\vec{v} = \begin{bmatrix} 1 \\ 1/2 \\ \vdots \\ 1/2^n \\ \vdots \end{bmatrix}$.

4. Find the ℓ^2-norm of $\vec{v}$ and $S(\vec{v})$, where S is the shift operator, and $\vec{v} = \begin{bmatrix} 1 \\ 1/2 \\ \vdots \\ 1/2^n \\ \vdots \end{bmatrix}$. (*Hint:* The norm of $\vec{v} \in \ell^2$ is defined by $\|\vec{v}\|^2 = \sum_{n=1}^{\infty} v_i^2$.)

5. Calculate the ℓ^2-norm of $\vec{v}$ and $S(\vec{v})$ for an ℓ^2 element $\vec{v} = \begin{bmatrix} v_1 \\ v_2 \\ v_3 \\ \vdots \end{bmatrix}$. Describe the relationship between them.

6. Use your result of the last exercise to compute the norm of the shift operator S on ℓ^2 and show S is bounded. Make sure you use the definition of the norm of an operator: the smallest K that satisfies, for every $\vec{v}$ in ℓ^2, $\|S(\vec{v})\|_{\ell^2} \leq K \cdot \|\vec{v}\|_{\ell^2}$.

Exercises 7–12 deal with the rank-one compact operator T on $L^2(0, 1)$ defined as $T(f) = \langle f, r \rangle \cdot r(x)$ for some $r \in L^2(0, 1)$.

7. Prove T is linear.

8. Suppose $r(x)$ is continuous on the interval $[0, 1]$, so it attains its maximum. Define M so $|r(x)| \leq M$ for $x \in [0, 1]$. Prove $\|T(f)\|_{L^2(0,1)} \leq M \cdot \|f\|_{L^2(0,1)}$.

9. Assume r is as in the previous exercise, and assume $M = |r(x_0)|$ for some $x_0 \in [0, 1]$. Define a sequence of functions $g_n(x) = \sqrt{n/2}$, if $x \in [x_0 - 1/n, x_0 + 1/n]$ (and 0 otherwise), $n = 1, 2, 3, \ldots$. Find the $L^2(0, 1)$-norm of g_n.

5.1 Bounded Linear Operators on L^2

10. Find an expression (in terms of the integral) for $\|T(g_n)\|_{L^2(0,1)}$, where g_n is as defined in the previous exercise.

11. Use the fact that, for f continuous at x_0, $(n/2)\int_{x_0-1/n}^{x_0+1/n} f(t)\,dt \to f(x_0)$ to evaluate your expression in the previous exercise to find $\lim_{n\to\infty} \|T(g_n)\|_{L^2(0,1)}$.

12. Use your results of the previous exercise and Exercise 8 to show $\|T\| = \max\{|r(x)| : 0 \le x \le 1\}$.

Exercises 13–22 examine Hilbert-Schmidt integral operators of the type in Example 5.1.2, where
$$T(f)(x) = \int_a^b k(x,y) \cdot f(y)\,dy.$$

13. Let $k(x,y) = x + y$ with $a = 0$ and $b = 1$. Determine the formula for $T(f)(x)$. Show $\int_a^b \int_a^b |k(x,y)|^2\,dx\,dy = M < \infty$ by finding M.

14. Let $k(x,y) = x + y$ be as in the previous exercise with $a = 0$ and $b = 1$. Determine $T(f)(x)$ when $f(y) = 1$ and when $f(y) = y^2$.

15. Let $k(x,y) = e^{x+y}$ with $a = 0$ and $b = 1$. Determine the formula for $T(f)(x)$. Show $\int_a^b \int_a^b |k(x,y)|^2\,dx\,dy = M < \infty$ by finding M.

16. Let $k(x,y) = e^{x+y}$ be as in the previous exercise with $a = 0$ and $b = 1$. Determine $T(f)(x)$ when $f(y) = 1$ and when $f(y) = y$.

17. Let $k(x,y) = x$ with $a = -1$ and $b = 1$. Determine the formula for $T(f)(x)$. Show $\int_a^b \int_a^b |k(x,y)|^2\,dx\,dy = M < \infty$ by finding M.

18. Let $k(x,y) = x$ be as in the previous exercise with $a = -1$ and $b = 1$. Determine $T(f)(x)$ when $f(y) = 1$ and when $f(y) = y^2$.

19. Let $k(x,y) = \sin(x+y)$ with $a = -\pi$ and $b = \pi$. Determine the formula for $T(f)(x)$. Show $\int_a^b \int_a^b |k(x,y)|^2\,dx\,dy = M < \infty$ by finding M.

20. Let $k(x,y) = \sin(x+y)$ be as in the previous exercise with $a = -\pi$ and $b = \pi$. Determine $T(f)(x)$ when $f(y) = 1$.

21. Let $k(x,y) = \sin(x+y)$ be as in the previous exercise with $a = -\pi$ and $b = \pi$. Determine $T(f)(x)$ when $f(y) = y$.

22. Let k be the characteristic function $k(x,y) = \mathcal{X}_{\{(x,y):\ x \le y\}}(x,y)$ with $a = 0$ and $b = 1$. For example, $k(.5, .6) = \mathcal{X}_{\{x \le y\}}(.5, .6) = 1$ since $.5 \le .6$, and $k(.7, .6) = \mathcal{X}_{\{x \le y\}}(.7, .6) = 0$ since $.7 > .6$. Determine $T(f)(x)$ when $f(y) = 1$ and when $f(y) = y$.

Exercises 23–30 examine Weiner-Hopf operators as Toeplitz operators on ℓ^2.

23. As in Example 5.1.2, if the Weiner-Hopf kernel function $k(t) = -(1/2)e^{-|t|/2}$ is transformed to $W(x) = 1 + \int_{-\infty}^{\infty} e^{ixt} \cdot K(t)\,dt$, then $W(x) = (4x^2 - 1)/(4x^2 + 1)$. Prove $w(e^{it}) = \cos t$, using the fact that $w(e^{it}) = \cos t$ means
$$w(t) = \cos(-i\log t) = (1/2)[e^{i(-i\log((2x-i)/(2x+i)))} + e^{-i(-i\log((2x-i)/(2x+i)))}].$$
Also use $i^2 = -1$, $e^{\log A} = A$ for any A in the log function's domain, and $(2x-i)/(2x+i) = 2(4x^2 - 1 - 4xi)/(4x^2 + 1)$. (The last follows from multiplication in both

the numerator and denominator by the complex conjugate term $2x - i$.) In other words, show the Weiner-Hopf operator acts on ℓ^2 as $T_{\cos t}$.

24. Let $k(t) = -(1/2)e^{-|t|/2}$ so the Wiener-Hopf operator on $L^2(0, \infty)$ is $T_k f = f(x) - \frac{1}{2}\int_0^\infty e^{-|x-t|/2} \cdot f(t)\, dt$. Now let $f(t) = e^{-t/2}$ and note $f \in L^2(0, \infty)$. Show $T_k f(x) = \frac{1}{2}(1 - x)e^{-x/2}$.

25. From Example 3.3.6, a complete orthonormal basis for $L^2(0, \infty)$ is $\{L_n(x)e^{-x/2}\}$, where $n = 0, 1, 2, \ldots$ and $L_n(x)$ is the nth normalized Laguerre polynomial defined in Example 3.3.3. Use the last exercise to prove T_k maps the first basis element (which is $L_0 e^{-x/2}$) to half of the second (which is $\frac{1}{2}L_1(x)e^{-x/2}$), when $k(t) = -(1/2)e^{-|t|/2}$.

26. Use the matrix representation $T_{\cos t} \sim \begin{bmatrix} 0 & 1/2 & 0 & 0 & \cdots \\ 1/2 & 0 & 1/2 & 0 & \cdots \\ 0 & 1/2 & 0 & 1/2 & \cdots \\ 0 & 0 & 1/2 & 0 & \cdots \\ & & \cdots & & \end{bmatrix}$ to determine $T_{\cos t}(\vec{e}_0)$, where $\vec{e}_0 = \begin{bmatrix} 1 \\ 0 \\ 0 \\ \vdots \end{bmatrix}$.

27. From Question 3.3.1, a complete orthonormal basis for ℓ^2 is $\{\vec{e}_n\}$, where $n = 0, 1, 2, \ldots$ and $\vec{e}_n$ is the column vector with a 1 in the nth spot. Use the last exercise to prove T_w maps the first basis element (which is $\vec{e}_0$) to half of the second (which is $\frac{1}{2}\vec{e}_1$), when $w(e^{it}) = \cos t$.

28. Compare the result of Exercise 27 with that of Exercise 25. Why did that result follow from the concluding paragraph of Section 5.1?

29. What is $T_{\cos t}(\vec{e}_1)$, where $\vec{e}_1$ is the second basis element of ℓ^2?

30. What is $T_k(e_1)$, where $k(t) = -(1/2)e^{-|t|/2}$, T_k is the Weiner-Hopf operator acting on $L^2(0, \infty)$, and $e_1 = L_1(x)e^{-x/2} = (1 - x)e^{-x/2}$ is the second basis element of $L^2(0, \infty)$? (*Hint:* compare to the solution of the previous exercise and use the fact that T_k acting on $L^2(0, \infty)$ is equivalent to $T_{\cos t}$ acting on ℓ^2.)

5.2 Bounded Linear Operators on General Hilbert Spaces

The last section introduced bounded linear operators on L^2 Hilbert spaces. This section describes examples of many operators on other Hilbert spaces. Those must of course be vector spaces, and we start by describing the simplest case—finite-dimensional ones. An n-dimensional Hilbert space may always be thought of as a space of column vectors of length n. In this case, any linear operator can be studied in a straightforward manner. In fact, that investigation forms the material in a linear algebra course. Linear algebra studies finite-dimensional matrices, and can more broadly be thought of as a study of finite-dimensional linear operators. Any such operator can be described as multiplication by a matrix. This section presents a few examples and theorems to study those operators. It then provides examples of linear operators in a variety of infinite-dimensional Hilbert spaces. The finite-dimensional results will inform us about what to expect on infinite dimensions.

Most of the Hilbert spaces we study will be spaces of functions, but they will include ℓ^2 for example, on which certain Toeplitz operators were examined in Section 5.1. They will also sometimes include functions defined on complex values. We have not yet dealt with

5.2 Bounded Linear Operators on General Hilbert Spaces

complex-valued functions, and this course does not assume a thorough or even comfortable understanding of them. The later examples will therefore sometimes omit significant detail, giving only rudimentary sketches of how the operators work. Still, operator theoretic problems are often studied using complex function theory, and some operators act on complex functions. The explanations can therefore be helpful even without full descriptions.

5.2.1 Operators on Finite-Dimensional Spaces

A set of column vectors of finite length n forms a Hilbert space with the dot product (the scalar product) forming the inner product. For example, the set of column vectors $\begin{bmatrix} v_0 \\ v_1 \\ v_2 \end{bmatrix}$ with real entries forms a three-dimensional[3] Hilbert space $\mathbb{R}^3$. The Hilbert space inner product is

$$\left\langle \begin{bmatrix} v_0 \\ v_1 \\ v_2 \end{bmatrix}, \begin{bmatrix} w_0 \\ w_1 \\ w_2 \end{bmatrix} \right\rangle = v_0 w_0 + v_1 w_1 + v_2 w_2.$$

Since $\|\vec{v}\|^2 = \langle \vec{v}, \vec{v} \rangle$, it follows that $\|\vec{v}\|^2 = v_0^2 + v_1^2 + v_2^2$.

Any linear operator (or linear transformation) T on such a space can be thought of as matrix multiplication. To multiply an $n \times n$ matrix T by a length n column vector $\vec{v}$, multiply each row of T by $\vec{v}$, where that multiplication is the sum of the products of corresponding entries. The values produced form the n component entries of the vector $T\vec{v}$.

Example 5.2.1 We illustrate the product $T\vec{v}$, where $T = \begin{bmatrix} 1 & 2 & 3 \\ 4 & 5 & 6 \\ 7 & 8 & 9 \end{bmatrix}$ and $\vec{v} = \begin{bmatrix} 10 \\ 11 \\ 12 \end{bmatrix}$. Multiplying the column by the first row of T, we get $\begin{bmatrix} 1 & 2 & 3 \end{bmatrix} \begin{bmatrix} 10 \\ 11 \\ 12 \end{bmatrix} = 1 \cdot 10 + 2 \cdot 11 + 3 \cdot 12 = 68$. Similarly, the second row of T times the vector produces $4 \cdot 10 + 5 \cdot 11 + 6 \cdot 12 = 167$. The third row of T times the vector is $7 \cdot 10 + 8 \cdot 11 + 9 \cdot 12 = 266$. Hence

$$T\vec{v} = \begin{bmatrix} 1 & 2 & 3 \\ 4 & 5 & 6 \\ 7 & 8 & 9 \end{bmatrix} \begin{bmatrix} 10 \\ 11 \\ 12 \end{bmatrix} = \begin{bmatrix} 68 \\ 167 \\ 266 \end{bmatrix}.$$

∎

Question 5.2.1 For the Toeplitz matrix $T = \begin{bmatrix} 1 & 2 & 3 \\ 4 & 1 & 2 \\ 5 & 4 & 1 \end{bmatrix}$ and the vector $\vec{v} = \begin{bmatrix} -1 \\ 6 \\ -2 \end{bmatrix}$, find the product $T\vec{v}$. ∎

Matrix multiplication times column vectors is important. We will also need to know how to multiply two square matrices, say A and B, of equal $n \times n$ size. To find $A \cdot B$, multiply each row of A by each column of B—sum the product of corresponding entries. The products give the entries in $A \cdot B$. Label them c_{ij}, where the subscripts indicate it is in row i and column j. Row i of A times column j of B produces c_{ij}.

For example, $\begin{bmatrix} 1 & 2 \\ 3 & 4 \end{bmatrix} \cdot \begin{bmatrix} 5 & 7 \\ 6 & 8 \end{bmatrix}$ has four row-times-column products: $1 \cdot 5 + 2 \cdot 6 = 17$, $1 \cdot 7 + 2 \cdot 8 = 23$, $3 \cdot 5 + 4 \cdot 6 = 39$, and $3 \cdot 7 + 4 \cdot 8 = 53$. Therefore $A \cdot B = \begin{bmatrix} 17 & 23 \\ 39 & 53 \end{bmatrix}$.

---

[3] The length of the vectors (the number of component entries) is the dimension of the space. The term dimension refers to the minimal number of vectors required to generate, by taking "linear combinations," all the vectors in the space. Such a set of generating vectors is a basis. For example, $\left\{ \begin{bmatrix} 1 \\ 0 \\ 0 \end{bmatrix}, \begin{bmatrix} 0 \\ 1 \\ 0 \end{bmatrix}, \begin{bmatrix} 0 \\ 0 \\ 1 \end{bmatrix} \right\}$ forms a basis for $\mathbb{R}^3$ because every vector in it can be written as a linear combination $\vec{v} = \sum c_i \vec{v}_i$ of them: $\begin{bmatrix} v_0 \\ v_1 \\ v_2 \end{bmatrix} = v_0 \begin{bmatrix} 1 \\ 0 \\ 0 \end{bmatrix} + v_1 \begin{bmatrix} 0 \\ 1 \\ 0 \end{bmatrix} + v_2 \begin{bmatrix} 0 \\ 0 \\ 1 \end{bmatrix}$. A full explanation is in any linear algebra textbook.

Multiplication of an n-dimensional vector by an $n \times n$ matrix is linear. The proof for a finite dimension n is illustrated here in the case of $n = 2$. Set $T = \begin{bmatrix} a & b \\ c & d \end{bmatrix}$, $\vec{u} = \begin{bmatrix} e \\ f \end{bmatrix}$, and $\vec{v} = \begin{bmatrix} g \\ h \end{bmatrix}$. Then for constants c_1 and c_2,

$$T(c_1 \vec{u} + c_2 \vec{v}) = \begin{bmatrix} a & b \\ c & d \end{bmatrix} (c_1 \begin{bmatrix} e \\ f \end{bmatrix} + c_2 \begin{bmatrix} g \\ h \end{bmatrix}) = \begin{bmatrix} a & b \\ c & d \end{bmatrix} \begin{bmatrix} c_1 e + c_2 g \\ c_1 f + c_2 h \end{bmatrix}$$
$$= \begin{bmatrix} a(c_1 e + c_2 g) + b(c_1 f + c_2 h) \\ c(c_1 e + c_2 g) + d(c_1 f + c_2 h) \end{bmatrix}.$$

Similarly,

$$c_1 T\vec{u} + c_2 T\vec{w} = c_1 \begin{bmatrix} a & b \\ c & d \end{bmatrix} \begin{bmatrix} e \\ f \end{bmatrix} + c_2 \begin{bmatrix} a & b \\ c & d \end{bmatrix} \begin{bmatrix} g \\ h \end{bmatrix} = \begin{bmatrix} c_1 ae + c_1 bf \\ c_1 ce + c_1 df \end{bmatrix} + \begin{bmatrix} c_2 ag + c_2 bh \\ c_2 cg + c_2 dh \end{bmatrix}$$
$$= \begin{bmatrix} a(c_1 e + c_2 g) + b(c_1 f + c_2 h) \\ c(c_1 e + c_2 g) + d(c_1 f + c_2 h) \end{bmatrix}.$$

Since the two equations have the same right-hand term, $T(c_1 \vec{u} + c_2 \vec{v}) = c_1 T\vec{u} + c_2 T\vec{w}$, and linearity follows. The general case is similar.

Thinking of a finite-dimensional matrix multiplication as a linear operator T, we might immediately wonder about its operator norm. The two-dimensional case is a good example: what is $\|T\|$ when $T\vec{v} \equiv \begin{bmatrix} a & b \\ c & d \end{bmatrix} \begin{bmatrix} v_0 \\ v_1 \end{bmatrix}$?

The answer comes from examining the defining equation for the operator norm: the norm is the largest value (taken as a supremum) of $\|T(\vec{v})\|$, for $\vec{v}$ with norm one in the two-dimensional Hilbert space. To calculate this value, even for the two-dimensional case, is messy, but when T is diagonal it is straightforward and gives insight. In that case,

$$\|T(\vec{v})\| = \left\| \begin{bmatrix} a & 0 \\ 0 & d \end{bmatrix} \begin{bmatrix} v_0 \\ v_1 \end{bmatrix} \right\| = \left\| \begin{bmatrix} av_0 \\ dv_1 \end{bmatrix} \right\| = \sqrt{(av_0)^2 + (dv_1)^2}$$

and the right-hand side is $K\sqrt{v_0^2 + v_1^2}$. Equating terms and squaring, $v_0^2 a^2 + v_1^2 d^2 = K^2(v_0^2 + v_1^2)$. Without loss of generality, we may assume $\|\vec{v}\|$ is a constant, say equal to 1 (otherwise the defining equation would simplify to this result). Hence $v_0^2 + v_1^2 = 1$, and the equation simplifies to $v_0^2 (a^2 - d^2) + d^2 = K^2$. There are two cases, depending on the sizes of the diagonal entries a and d. If $a \geq d$, K is maximized when $v_0 = 1$. Then $K = a$. The other case is $d > a$, and then K is maximized when $v_0 = 0$. Then $K = d$. In summary (for the case where T is diagonal), $\|T\|$ is the larger of the two values on the diagonal of T.

The same calculation works for any diagonal operator T on an n-dimensional space: $\|T\|$ is the largest value on the diagonal of T. Those who have taken a linear algebra course (and have a broader foundational understanding of matrix properties) will be interested in the general answer, even when T is not diagonal. For an n-dimensional matrix T, $\|T\|^2$ is the largest eigenvalue of T^*T, where T^* is the (conjugate) transpose of T. The result matches our derivation in the case when T is diagonal.

Question 5.2.2 For the diagonal matrix $T = \begin{bmatrix} t_1 & 0 & 0 \\ 0 & t_2 & 0 \\ 0 & 0 & t_3 \end{bmatrix}$, prove the norm of the operator $T : \vec{v} \to T\vec{v}$ is the maximum value of t_1, t_2, and t_3. ∎

5.2 Bounded Linear Operators on General Hilbert Spaces

One of the most famous theorems that describes the behavior of matrix multiplication is the Jordan form theorem. It shows how to make a matrix T similar to a matrix that is either diagonal (where the entries off the main diagonal are all zero) or close to diagonal. A matrix T is similar to a matrix J when there exists an invertible matrix P such that $PJP^{-1} = T$. The famous Jordan canonical form describes the matrix in a "close to diagonal" structure, presented here in the Jordan form theorem without proof. Its only assumption is the matrix T has m linearly independent eigenvectors, which requires a preliminary explanation for any reader who has not taken a course on linear algebra. A nonzero vector $\vec{v}$ is an eigenvector for a matrix T if there exists λ so $T\vec{v} = \lambda\vec{v}$. Then λ is an eigenvalue. An eigenvector is special for T because multiplication by the matrix T on the eigenvector has the simplest of all actions: it is equivalent to multiplying the eigenvector by the number λ. A collection of eigenvectors is linearly independent when one cannot be written as a linear combination $\sum c_j \vec{v}_j$ of the others.

Theorem 5.2.1 The Jordan Form Theorem: *If an $n \times n$ matrix T has m linearly independent eigenvectors, then there exists an $n \times n$ invertible matrix P such that $PJP^{-1} = T$, where J is a block diagonal matrix of the form $J = \begin{bmatrix} J_1 & & \\ & \ddots & \\ & & J_m \end{bmatrix}$. (Any blank part of the matrix is assumed to be filled with entries that are all zero.) J_i corresponds to the ith linearly independent eigenvector and is a block matrix of the form $J_i = \begin{bmatrix} \lambda_i & 1 & & \\ & \ddots & 1 \\ & & \lambda_i \end{bmatrix}$, where λ_i is the ith eigenvalue (and the blank components are assumed to be filled with zeros).*

An illustration of the Jordan form theorem is in order. A 4×4 example follows.

Example 5.2.2 Suppose $T = \begin{bmatrix} -17 & 12 & 9 & 10 \\ 0 & 2 & 0 & 0 \\ -25 & 18 & 13 & 14 \\ 0 & 1 & 0 & 2 \end{bmatrix}$. We determine the Jordan form matrix J, observing T has two distinct eigenvalues $\lambda = 2$ and $\lambda = -2$ each satisfying $T\vec{v} = \lambda \vec{v}$. The eigenvector $\vec{v} = \begin{bmatrix} 1 \\ 0 \\ 1 \\ 1 \end{bmatrix}$ corresponds to $\lambda = 2$, since $T\vec{v} = \begin{bmatrix} -17 & 12 & 9 & 10 \\ 0 & 2 & 0 & 0 \\ -25 & 18 & 13 & 14 \\ 0 & 1 & 0 & 2 \end{bmatrix} \begin{bmatrix} 1 \\ 0 \\ 1 \\ 1 \end{bmatrix} = \begin{bmatrix} 2 \\ 0 \\ 2 \\ 2 \end{bmatrix} = 2 \cdot \begin{bmatrix} 1 \\ 0 \\ 1 \\ 1 \end{bmatrix} = \lambda \vec{v}$. The eigenvector $\vec{v} = \begin{bmatrix} 3 \\ 0 \\ 5 \\ 0 \end{bmatrix}$ corresponds to $\lambda = -2$, since $T\vec{v} = \begin{bmatrix} -17 & 12 & 9 & 10 \\ 0 & 2 & 0 & 0 \\ -25 & 18 & 13 & 14 \\ 0 & 1 & 0 & 2 \end{bmatrix} \begin{bmatrix} 3 \\ 0 \\ 5 \\ 0 \end{bmatrix} = \begin{bmatrix} -6 \\ 0 \\ -10 \\ 0 \end{bmatrix} = -2 \cdot \begin{bmatrix} 3 \\ 0 \\ 5 \\ 0 \end{bmatrix} = \lambda \vec{v}$. They turn out to form a linearly independent system.[4]

Hence[5] the Jordan form theorem guarantees the existence of matrices J and P such that $T = PJP^{-1}$. The theorem describes the construction of J from our knowledge of the eigenvalues as $J = \begin{bmatrix} 2 & 1 & 0 & 0 \\ 0 & 2 & 0 & 0 \\ 0 & 0 & -2 & 1 \\ 0 & 0 & 0 & -2 \end{bmatrix}$. You can see how close J is to a diagonal matrix (for T, it is as

---

[4] We present the material in this section in a way that does not depend on the ability to find the eigenvalues and eigenvectors for a given matrix. That derivation is a nontrivial but fairly straightforward process, and it is learned in a linear algebra course. It relies on the calculation of a polynomial associated with T, the characteristic polynomial, found by taking the determinant of a matrix associated with T. The zeros of the characteristic polynomial are the eigenvalues of T, and the multiplicity of a zero corresponds to the sum of the sizes of the matching eigenvalue's Jordan blocks J_i.

[5] In Example 5.2.2, the eigenvalues 2 and -2 each have multiplicity 2 (the characteristic polynomial of T is $(\lambda - 2)^2(\lambda + 2)^2$), which is why (as there is only one linearly independent eigenvector corresponding to each eigenvalue) their corresponding Jordan blocks are 2×2.

close via similarity as you can get), and it consists of two Jordan blocks, namely $J_1 = \begin{bmatrix} 2 & 1 \\ 0 & 2 \end{bmatrix}$ and $J_2 = \begin{bmatrix} -2 & 1 \\ 0 & -2 \end{bmatrix}$. As the theorem guarantees, all other entries in J are 0. Furthermore, it turns out $P = \begin{bmatrix} 1 & 2 & 3 & 4 \\ 0 & 1 & 0 & 0 \\ 1 & 3 & 5 & 7 \\ 1 & 0 & 0 & 0 \end{bmatrix}$ has inverse $P^{-1} = \begin{bmatrix} 0 & 0 & 0 & 1 \\ 0 & 1 & 0 & 0 \\ 7 & -2 & -4 & -3 \\ -5 & 1 & 3 & 2 \end{bmatrix}$, and the product PJP^{-1} produces T. (The construction of P can be determined from knowledge of the eigenvectors and eigenvalues of T.[6]) ∎

Question 5.2.3 Suppose $T = \begin{bmatrix} 4 & -1 & 0 \\ .4 & 3.4 & .2 \\ .2 & -5.8 & 1.6 \end{bmatrix}$. It has only one eigenvalue $\lambda = 3$ (which has multiplicity three as a zero of the characteristic polynomial $p(x) = -(x-3)^3$).

1. Show $\vec{v} = \begin{bmatrix} 1 \\ 1 \\ -4 \end{bmatrix}$ is an eigenvector for T, in that it solves the eigenvector equation $T\vec{v} = \lambda\vec{v}$.
2. Given $\vec{v}$ is the only eigenvector for T, what is the Jordan form matrix J?
3. Using $P = \begin{bmatrix} 1 & 2 & 2 \\ 1 & 1 & 0 \\ -4 & -1 & 1 \end{bmatrix}$, which has inverse $P^{-1} = \frac{1}{5}\begin{bmatrix} 1 & -4 & -2 \\ -1 & 9 & 2 \\ 3 & -7 & -1 \end{bmatrix}$, prove T satisfies $T = PJP^{-1}$ for the matrix J you constructed in part 2. ∎

There are many reasons why we might want to determine the Jordan form for a given matrix T. For one, algebraic manipulations on J are usually simpler than those on T. (It is almost always easier to handle a matrix with many entries that are zero, especially when they are not on the diagonal.) For example, T^2 can be easily handled in terms of J, since $T^2 = T \cdot T = PJP^{-1}PJP^{-1} = PJ^2P^{-1}$. In the same way, $T^{100} = PJ^{100}P^{-1}$, and the same type of calculation works for any power of T.

Many matrices T have a Jordan form matrix J that is diagonal (where each of the blocks J_i are 1×1, and J has only zeros in the off-diagonal entries). Such matrices T are then *diagonalizable*. For example, any real symmetric matrix is diagonalizable. The ability to write T as $T = PJP^{-1}$, where J is diagonal, is the *spectral theorem* for T. Diagonalizable matrices are especially easy to manipulate. For example, an equation such as $T^{100} = PJ^{100}P^{-1}$ involves J^{100}. But J^{100} is also diagonal, and its diagonal entries are the diagonal entries of J (the eigenvalues of T) each raised to the 100th power. Not all $n \times n$ matrices are diagonalizable—the Jordan form theorem says a diagonalizable matrix must have n linearly independent eigenvectors. But those that are can be completely characterized by their eigenvalues and eigenvectors. For example, a diagonalizable matrix T has $T = PJP^{-1}$ where the eigenvalues of T are listed down the diagonal column of J, and the eigenvectors form the columns of the matrix P. This is the spectral theorem in linear algebra.

Section 5.4 will describe a way to diagonalize types of linear operators on an infinite-dimensional Hilbert space. In this way, the spectral theorem can be generalized to deal with more complicated operators than just those on a finite-dimensional space (see [53]). Modern operator theory research continues to investigate types of operators in terms of their spectral theory properties.

---

[6] In this example the first two columns of P can be thought of as being coordinated with the first 2×2 block J_1, and the last two columns with J_2. The first column of P is chosen as the eigenvector corresponding to $\lambda_1 = 2$. In the same way, the third column (which is the first column in the second set of columns) is chosen as the eigenvector corresponding to $\lambda_2 = -2$. Calling the corresponding eigenvector $\vec{e}$, the corresponding eigenvalue λ, and the other column $\vec{v}$, the other two columns satisfy $T\vec{v} = \vec{e} + \lambda\vec{v}$. Straightforward techniques learned in a course on linear algebra can be applied to solve for $\vec{v}$. In the literature, the other columns of P are sometimes called *generalized eigenvectors*.

5.2.2 Operators on Infinite-Dimensional Spaces

This chapter has already studied operators on various L^2 spaces. There are many interesting operators on other spaces, often defined in terms of the Hilbert space structure, even when the space is infinite-dimensional. An example is a rank-one operator—Section 5.1 introduced such operators for an L^2 space, but the point is the definition can prescribe a rank-one operator for any Hilbert space.

Example 5.2.3 Given a Hilbert space $\mathcal{H}$ and an element $w \in \mathcal{H}$, we use the inner product on $\mathcal{H}$ to define a rank-one operator

$$T_w(v) = \langle v, w \rangle \cdot w, \quad v \in \mathcal{H}.$$

In this way, the operator T_w maps an element of $\mathcal{H}$ to a constant multiple of w.

1. When $\mathcal{H} = L^2(\mu)$ for a Borel measure μ, we have seen how such a rank-one operator is determined. In this case, w is a function in the L^2-space, and the inner product is defined according to the integral. Hence for $v(x) \in L^2(\mu)$,

$$T_w(v) = \langle v, w \rangle \cdot w = \left(\int v \cdot w \, d\mu \right) \cdot w(x).$$

2. When $\mathcal{H} = \ell^2$, the element w would be a vector (of infinite length) $\vec{w} = \begin{bmatrix} w_1 \\ w_2 \\ \vdots \end{bmatrix}$ in ℓ^2. The ℓ^2 inner product determines the form of the rank-one operator $T_{\vec{w}}$ as it acts on $\vec{v} = \begin{bmatrix} v_1 \\ v_2 \\ \vdots \end{bmatrix} \in \ell^2$. The operator acts as $T_{\vec{w}}(\vec{v}) = \langle \vec{v}, \vec{w} \rangle \cdot \vec{w} = \left(\sum_{i=1}^{\infty} v_i \cdot w_i \right) \begin{bmatrix} w_1 \\ w_2 \\ \vdots \end{bmatrix}.$

3. Another Hilbert space is a Sobolev space.[7] Given a real interval (a, b), the Sobolev Hilbert space $\mathcal{H} = W^{1,2}$ is the collection of (real) functions f in $L^2(a, b)$ whose first derivative Df exists. Labeling the space as $W^{1,2}$ reflects the fact its formulation involves both the functions' first derivative[8] (hence the index 1) and L^2. The inner product is

$$\langle f, g \rangle_{W^{1,2}} = \int_a^b f(x) \cdot g(x) \, dx + \int_a^b Df(x) \cdot Dg(x) \, dx.$$

For $w \in W^{1,2}$, a rank-one operator $T_w(f)$ on $f \in W^{1,2}$ is therefore of the form

$$T_w(f) = \left(\int_a^b f(x) \cdot w(x) \, dx + \int_a^b Df(x) \cdot Dw(x) \, dx \right) \cdot w(x).$$

---

[7] Named after the Russian mathematician Sergei Sobolev, the spaces can be defined generally in terms of varying norm structures (for example, the Sobolev space $W^{1,1}$ is constructed with norm $\|f\|_{W^{1,1}} = \|f\|_{L^1} + \|Df\|_{L^1}$). Though our presentation here is formulated in terms of the full derivative, Sobolev's motivation was his study of partial differential equations, and Sobolev spaces are most often formulated in terms of partial derivatives of functions f of many variables. Not all Sobolev spaces are Hilbert spaces. Instead, the vector space elements and norms can be defined so as to always form Banach spaces.

[8] The derivatives are defined in the *weak sense*, where Df is the function v such that $\int_a^b f \cdot d\phi/dx \, dx = -\int_a^b \phi \cdot v \, dx$ for every infinitely differentiable function ϕ defined on (a, b). The definition of weak derivative is motivated by the integration by parts formula.

4. To this point we have considered only functional Hilbert spaces of real functions. Hilbert spaces of complex functions arise in theory and in applications. For a complex value $z = x + iy$, where x and y are real, a complex function $f(z)$ is of the form $f(z) = u(x, y) + iv(x, y)$, where u and v are real-valued functions. $u(x, y)$ is the real part of f and $v(x, y)$ is the imaginary part of f. The Hardy space $\mathcal{H} = H^2(\Pi^+)$ is the collection of complex functions f defined on the upper half-plane Π^+, analytic in Π^+, and of finite norm, where $\sup_{y>0} \int_{-\infty}^{\infty} |f(x+iy)|^2 \, dx < \infty$. The three requirements deserve additional explanation. First, the upper half-plane consists of those complex numbers whose imaginary parts are positive; in short, $\Pi^+ = \{x + iy : y > 0\}$. Second, a complex function $f(z)$ is analytic over Π^+ when the derivative $f'(z) = \lim_{h \to 0}[f(z+h) - f(z)]/h$ exists[9] at every point $z \in \Pi^+$. Third, the H^2-norm $\|f\|$ of a function $f \in H^2(\Pi^+)$ is defined according to the square of the norm as $\|f\|^2 = \sup_{y>0} \int_{-\infty}^{\infty} |f(x+iy)|^2 \, dx$.

The inner product of two functions $f, g \in H^2(\Pi^+)$ is then

$$\langle f, g \rangle_{H^2(\Pi^+)} = \sup_{y>0} \int_{-\infty}^{\infty} f(x+iy) \cdot \overline{g(x+iy)} \, dx.\text{[10]}$$

For $w \in H^2(\Pi^+)$, a rank-one operator $T_w(f)$ on $f \in H^2(\Pi^+)$ is therefore of the form

$$T_w(f) = \left[\sup_{y>0} \int_{-\infty}^{\infty} f(x+iy) \cdot \overline{w(x+iy)} \, dx\right] \cdot w(z).$$

5. The Hardy space is also formulated for a collection of complex-valued functions defined on another standard region. They are defined on the unit disk $\mathbb{D}$ in the complex plane, rather than for the upper half-plane Π^+. The disk $\mathbb{D}$ is the set of points $z = x + iy$ within one unit of the origin, so $\mathbb{D} = \{z = x + iy : |z| = \sqrt{x^2 + y^2} < 1\}$. A polar representation of the points is often useful, where the disk is viewed as

$$\mathbb{D} = \{z = re^{it} = r(\cos t + i \sin t) : 0 \leq r < 1\}.$$

The Hardy space $\mathcal{H} = H^2(\mathbb{D})$ is then the collection of complex functions f defined and analytic on the disk $\mathbb{D}$ with finite $H^2(\mathbb{D})$-norm, determined from

$$\|f\|^2_{H^2(\mathbb{D})} = \sup_{0 \leq r < 1} \frac{1}{2\pi} \int_0^{2\pi} |f(re^{it})|^2 \, dt.$$

A function $f \in H^2(\mathbb{D})$ is analytic, and so it has a Maclaurin series expansion valid in $\mathbb{D}$ of the form $f(z) = \sum_{n=0}^{\infty} a_n z^n$. The series representation produces the *boundary function* $f(e^{it})$, which exists almost everywhere, for example on $0 \leq t < 2\pi$. The boundary function can

---

[9] The limit involves the complex value h approaching zero, which has a precise meaning. Since the complex plane is two-dimensional, h is allowed to approach zero along any path. Any standard introductory complex analysis textbook (see [42]) will provide a description of limits in the complex plane.

[10] The value $\overline{g(x+iy)}$ refers to the complex conjugate of g. If $g = u + iv$, then $\overline{g} = u - iv$. The integration is taken just as for real functions, except it involves the real and the imaginary part of the integrand.

5.2 Bounded Linear Operators on General Hilbert Spaces

also be defined as a limit $f(e^{it}) = \lim_{z \to e^{it}} f(z)$.[11] In this sense, then,

$$f(e^{it}) = \sum_{n=0}^{\infty} a_n e^{int}.$$

The inner product of two functions $f, g \in H^2(\mathbb{D})$ can be calculated with the boundary functions as

$$\langle f, g \rangle_{H^2(\mathbb{D})} = \frac{1}{2\pi} \int_0^{2\pi} f(e^{it}) \overline{g(e^{it})} \, dt.$$

The inner product can also be expressed in terms of the boundary functions' series representations: if $f(e^{it}) = \sum_{n=0}^{\infty} a_n e^{int}$ and $g(e^{it}) = \sum_{n=0}^{\infty} b_n e^{int}$, then $\langle f, g \rangle_{H^2(\mathbb{D})} = \sum_{n=0}^{\infty} a_n \overline{b_n}$.

For $w \in H^2(\mathbb{D})$ with boundary function $w(e^{it}) = \sum_{n=0}^{\infty} c_n e^{int}$, a rank-one operator $T_w(f)$ on $f \in H^2(\mathbb{D})$ can therefore be expressed in two forms:

$$T_w(f) = \frac{1}{2\pi} \int_0^{2\pi} f(e^{it}) \overline{w(e^{it})} \, dt \cdot w(z) = \left[\sum_{n=0}^{\infty} a_n \overline{c_n} \right] \cdot w(z).$$

The interplay between a function f in $H^2(\mathbb{D})$ with its boundary function $f(e^{it})$ is so strong that $H^2(\mathbb{D})$ can be characterized as the space of boundary functions. We write it as $H^2(\mathbb{T})$, where $\mathbb{T}$ is the unit circle (the boundary of $\mathbb{D}$) in the complex plane. That is, $H^2(\mathbb{T}) = \{f(e^{it}) : f \text{ is the boundary function of } f(z) \in H^2(\mathbb{D})\}$. ∎

The abstract introduction of the Hilbert spaces as infinite-dimensional can be made more concrete. Each has a countably infinite orthonormal basis as per Definition 3.3.1. For example,

$$\left\{ \frac{\pi^{-1/2}}{x+i} \left(\frac{x-i}{x+i} \right)^n \right\}_{n=-\infty}^{\infty}$$

forms a basis for $L^2(\mathbb{R})$ (an alternate choice to the Hermite functions). In the same way,

$$\left\{ \frac{\pi^{-1/2}}{z+i} \left(\frac{z-i}{z+i} \right)^n \right\}_{n=0}^{\infty}$$

forms a basis for $H^2(\Pi^+)$.[12] The standard basis for $H^2(\mathbb{T})$ is $\{e^{int}\}_{n=0}^{\infty}$.

Many operators on a (separable) Hilbert space $\mathcal{H}$ are compact. The next definition explains what is meant by this term. For example, rank-one operators are compact.

Definition 5.2.1 *For Hilbert spaces $\mathcal{H}$ and $\mathcal{G}$, a bounded linear operator $T : \mathcal{H} \to \mathcal{G}$ is compact when it can be written as*

$$T(f) = \sum_{n=0}^{\infty} \lambda_n \langle g_n, f \rangle \phi_n$$

---

[11] The limit must be taken nontangentially, which means the value z cannot approach e^{it} along a path tangent to the unit circle. A nontangential limit guarantees proper construction of the boundary function. For a reference, see [103, p. 18].

[12] See, for example, [103, p. 112].

for $f \in \mathcal{H}$, where the vectors $g_0, g_1, g_2, \ldots$ and $\phi_0, \phi_1, \phi_2, \ldots$ are (not necessarily complete) orthonormal sets. Here $\lambda_0, \lambda_1, \lambda_2, \ldots$ forms a sequence of complex values. When $g_i = \phi_i$, they are the eigenvalues of T.

Example 5.2.4 If $k(x, y) = \sin(x + y)$, then the Hilbert-Schmidt operator of Example 5.1.2 is seen to be compact on $L^2(0, \pi)$. Namely, we can write $T(f)(x) = \int_0^\pi \sin(x + y) f(y) \, dy$ as $T(f)(x) = \sum_{n=0}^{\infty} \lambda_n \langle g_n, f \rangle \phi_n$ for $f \in L^2(0, \pi)$. To see how[13], apply the identity $\sin(x + y) = \sin x \cdot \cos y + \cos x \cdot \sin y$ and use the linearity of the integral:

$$T(f)(x) = \int_0^\pi \sin(x+y) f(y) \, dy = \int_0^\pi \sin x \cdot \cos y \cdot f(y) \, dy + \int_0^\pi \cos x \cdot \sin y \cdot f(y) \, dy.$$

Since the L^2-inner product is $\langle g, f \rangle = \int_0^\pi g(y) \cdot f(y) \, dy$, we get

$$T(f)(x) = \langle \cos y, f \rangle \cdot \sin x + \langle \sin y, f \rangle \cdot \cos x.$$

The operator T is therefore compact, since $\cos y$ and $\sin y$ are orthogonal in $L^2(0, \pi)$ (see Section 2.4 for a parallel development on $L^2(-\pi, \pi)$). Since $\sqrt{2/\pi} \cos y$ and $\sqrt{2/\pi} \sin y$ are orthonormal in $L^2(0, \pi)$, we see $T(f) = \sum_{n=0}^{\infty} \lambda_n \langle g_n, f \rangle \phi_n$ with $g_0(y) = \sqrt{2/\pi} \cos y$, $g_1(y) = \sqrt{2/\pi} \sin y$, and $g_n = 0$ for $n = 2, 3, \ldots$. Here, $\phi_0(x) = \sqrt{2/\pi} \sin x$, $\phi_1(x) = \sqrt{2/\pi} \cos x$, and $\phi_n = 0$ for $n = 2, 3, \ldots$. The normalization implies $\lambda_0 = \lambda_1 = \pi/2$. Because the infinite series that expresses T has only two nonzero terms, we say the compact operator T has rank two.

This example illustrates a more general fact: namely, a Hilbert-Schmidt operator T as defined in Example 5.1.2 is compact (see [46, pp. 86–87]). Erhard Schmidt determined this in 1907 (see [109].) ∎

We finish this section with a discussion of invariant subspaces. For a Hilbert space operator $T : \mathcal{H} \to \mathcal{H}$, a subspace A of $\mathcal{H}$ is said to be *invariant* under T when $T(A) \subseteq A$. That is, T maps any vector in A to a vector in A. As for any subspace, A is assumed to be closed (in the sense described in Theorem 3.3.1). A host of interesting examples exist of operators on infinite-dimensional Hilbert spaces that have invariant subspaces. A classic case is for a type of integral operator, which the next example presents.

Example 5.2.5 Another type of integral operator is the Volterra operator, named after the Italian mathematician Vito Volterra. An example of a Volterra operator on $L^2(0, 1)$ is

$$T(f)(x) = \int_0^x f(y) \cdot k(x, y) \, dy,$$

where the kernel function k must satisfy $\int_0^1 \int_0^1 |k(x, y)|^2 \, dx \, dy < \infty$ to guarantee T is bounded. Then $T : L^2(0, 1) \to L^2(0, 1)$.

The operator T has many invariant subspaces in $L^2(0, 1)$. For example, for $c \in (0, 1)$, let $A = \{f : f(x) = 0 \text{ for almost all } x \in (0, c)\}$. For $f \in A$,

$$T(f)(x) = \int_0^x f(y) \cdot k(x, y) \, dy = \int_0^x 0 \cdot k(x, y) \, dy = 0$$

if $x \in (0, c)$, which says $T(f)(x) \in A$. So $T(A) \subseteq A$, and A is invariant under T. ∎

---

[13] A similar example appears, for example, in [46, p. 68].

Question 5.2.4 The compact operator T in Example 5.2.4 maps $L^2(0, \pi)$ to itself. Examine the space of functions $A = \{f(x) : f(x) = c_1 \cos x + c_2 \sin x$, where $c_1, c_2 \in \mathbb{R}\}$. Show an element of A is in $L^2(0, \pi)$, and so, since A is a finite-dimensional (it is generated by $\cos x$ and $\sin x$), it is a closed subspace of $L^2(0, \pi)$. Why is A an invariant subspace for T? ∎

A question that remains unanswered after almost a century of effort by thousands of mathematicians, including those at the highest level of talent in operator theory, is whether every bounded linear operator T on a (separable) Hilbert space has a (nontrivial) invariant subspace. The invariant subspace problem is the most famous open problem in operator theory. The answer is known to be the affirmative "Yes, every bounded linear operator T does have an invariant subspace" for large categories of operators. The notes at the end of this chapter describe some of them. For example, every compact operator on a Hilbert space has a nontrivial invariant subspace. The positive results generate more evidence for an affirmative answer to the question in general, and—currently—most mathematicians confidently believe in it. And yet the general answer still remains unsettled.

Solutions to Questions

5.2.1 $T\vec{v} = \begin{bmatrix} 1 & 2 & 3 \\ 4 & 1 & 2 \\ 5 & 4 & 1 \end{bmatrix} \cdot \begin{bmatrix} -1 \\ 6 \\ -2 \end{bmatrix} = \begin{bmatrix} 1(-1)+2(6)+3(-2) \\ 4(-1)+1(6)+2(-2) \\ 5(-1)+4(6)+1(-2) \end{bmatrix} = \begin{bmatrix} 5 \\ -2 \\ 17 \end{bmatrix}.$

5.2.2 Examine $\|T(\vec{v})\| \leq K \cdot \|\vec{v}\|$. First,

$$\|T(\vec{v})\| = \left\| \begin{bmatrix} t_1 & 0 & 0 \\ 0 & t_2 & 0 \\ 0 & 0 & t_3 \end{bmatrix} \begin{bmatrix} v_1 \\ v_2 \\ v_3 \end{bmatrix} \right\| = \left\| \begin{bmatrix} t_1 v_1 \\ t_2 v_2 \\ t_3 v_3 \end{bmatrix} \right\| = \sqrt{(t_1 v_1)^2 + (t_2 v_2)^2 + (t_3 v_3)^2}.$$

Second, the right-hand side is $K\sqrt{v_1^2 + v_2^2 + v_3^2}$. Equating terms and squaring, $t_1^2 v_1^2 + t_2^2 v_2^2 + t_3^2 v_3^2 = K^2(v_1^2 + v_2^2 + v_3^2)$. Without loss of generality, we assume $\|\vec{v}\|$ is a constant, say equal to 1. Hence $v_3^2 = 1 - v_1^2 - v_2^2$ and the equation simplifies to $v_1^2(t_1^2 - t_3^2) + v_2^2(t_2^2 - t_3^2) + t_3^2 = K^2$. There are three cases to consider, depending on which of the diagonal entries t_1, t_2, and t_3 is the largest. Without loss of generality, we assume $t_1 \geq t_2$ and $t_1 \geq t_3$. Then K is maximized when $v_1 = 1$, in which case $K = t_1$. The other cases are similar; in any case we see $\|T\|$ is the maximum of the diagonal values of T.

5.2.3 $T\vec{v} = \begin{bmatrix} 4 & -1 & 0 \\ .4 & 3.4 & .2 \\ .2 & -5.8 & 1.6 \end{bmatrix} \cdot \begin{bmatrix} 1 \\ 1 \\ -4 \end{bmatrix} = \begin{bmatrix} 3 \\ 3 \\ -12 \end{bmatrix} = 3\vec{v}$. Because the only eigenvector is $\vec{v}$, the Jordan form matrix is $J = \begin{bmatrix} 3 & 1 & 0 \\ 0 & 3 & 1 \\ 0 & 0 & 3 \end{bmatrix}$. Multiplying, $\begin{bmatrix} 1 & 2 & 2 \\ 1 & 1 & 0 \\ -4 & -1 & 1 \end{bmatrix} \cdot \begin{bmatrix} 3 & 1 & 0 \\ 0 & 3 & 1 \\ 0 & 0 & 3 \end{bmatrix} \cdot \frac{1}{5} \cdot \begin{bmatrix} 1 & -4 & -2 \\ -1 & 9 & 2 \\ 3 & -7 & -1 \end{bmatrix} =$
$\frac{1}{5} \cdot \begin{bmatrix} 3 & 7 & 8 \\ 3 & 4 & 1 \\ -12 & -7 & 2 \end{bmatrix} \cdot \begin{bmatrix} 1 & -4 & -2 \\ -1 & 9 & 2 \\ 3 & -7 & -1 \end{bmatrix} = T.$

5.2.4 Elements of A are Lebesgue measurable because they are continuous. They are in $L^2(0, \pi)$ because $f \in A$ has $\|f\|^2 = \int_0^\pi (c_1 \cos x + c_2 \sin x)^2 \, dx \leq \int_0^\pi (|c_1| + |c_2|)^2 \, dx = \pi(|c_1| + |c_2|)^2$, which is finite. Hence A is a closed subspace of $L^2(0, \pi)$. A is invariant for T because (see remarks in Example 5.2.4) we can understand T acting on any $f \in L^2(0, \pi)$ (hence certainly for $f \in A$) as in the form $T(f)(x) = \langle \sin y, f \rangle \cdot \cos x + \langle \cos y, f \rangle \cdot \sin x$, which is of the form $T(f)(x) = c_1 \cos x + c_2 \sin x \in A$.

Reading Questions for Section 5.2

1. Describe the Hilbert space inner product on the vector space $\mathbb{R}^2$ and use it to determine the inner product of $\vec{v} = \begin{bmatrix} 1 \\ 2 \end{bmatrix}$ and $\vec{v} = \begin{bmatrix} -3 \\ -4 \end{bmatrix}$.

2. What is the form of a linear operator on a finite-dimensional Hilbert space?

3. What is the product $T\vec{v} = \begin{bmatrix} 5 & -3 \\ 6 & -1 \end{bmatrix} \begin{bmatrix} 1 \\ 2 \end{bmatrix}$?

4. Give four examples of a Hilbert space of infinite dimension.

5. For four different Hilbert spaces, describe the inner product of two elements.

6. What is a rank-one operator on a Hilbert space? Give two examples of a rank-one operator on different Hilbert spaces.

7. What is a compact operator on a Hilbert space?

8. Give an example of a compact operator on an L^2 space.

Exercises for Section 5.2

Exercises 1–10 deal with matrix multiplication on column vectors. Determine the product $T\vec{v}$.

1. $T = \begin{bmatrix} 2 & 1 \\ -3 & 3 \end{bmatrix}$ and $\vec{v} = \begin{bmatrix} 1 \\ 5 \end{bmatrix}$

2. $T = \begin{bmatrix} 0 & 0 \\ 0 & 3 \end{bmatrix}$ and $\vec{v} = \begin{bmatrix} a \\ b \end{bmatrix}$, where $a, b \in \mathbb{R}$

3. $T = \begin{bmatrix} 1 & 0 & 0 \\ 0 & 3 & 0 \\ 0 & 0 & -3 \end{bmatrix}$ and $\vec{v} = \begin{bmatrix} -7 \\ -5 \\ 2 \end{bmatrix}$

4. $T = \begin{bmatrix} 1 & 0 & 5 \\ 2 & 3 & -6 \\ -7 & -2 & -3 \end{bmatrix}$ and $\vec{v} = \begin{bmatrix} a \\ b \\ c \end{bmatrix}$, $a, b, c \in \mathbb{R}$

5. $T = \begin{bmatrix} 1 & 0 & 0 \\ -2 & 1 & 0 \\ 4 & 1 & -3 \end{bmatrix}$ and $\vec{v} = \begin{bmatrix} -1 \\ 1 \\ 2 \end{bmatrix}$

6. $T = \begin{bmatrix} 3 & 1 & 0 \\ 0 & 3 & 0 \\ 0 & 0 & -3 \end{bmatrix}$ and $\vec{v} = \begin{bmatrix} 1 \\ 1 \\ 2 \end{bmatrix}$

7. $T = \begin{bmatrix} 2 & 1 & 0 & 0 \\ 1 & 0 & 0 & 5 \\ 2 & 0 & 3 & 0 \\ 0 & 0 & -3 & 3 \end{bmatrix}$ and $\vec{v} = \begin{bmatrix} 1 \\ 5 \\ 2 \\ 4 \end{bmatrix}$

8. $T = \begin{bmatrix} 2 & 1 & 0 & 0 \\ 0 & 2 & 0 & 0 \\ 0 & 0 & 3 & 1 \\ 0 & 0 & 0 & 3 \end{bmatrix}$ and $\vec{v} = \begin{bmatrix} 1 \\ 5 \\ 2 \\ 4 \end{bmatrix}$

9. $T = \begin{bmatrix} 2 & 1 & 0 & 0 \\ 0 & 2 & 0 & 0 \\ 0 & 0 & 1 & 1 \\ 0 & 0 & 0 & 1 \end{bmatrix}$ and $\vec{v} = \begin{bmatrix} a \\ b \\ c \\ d \end{bmatrix}$, where $a, b, c, d \in \mathbb{R}$

10. $T = \begin{bmatrix} 2 & 1 & 1 & 1 \\ 1 & 2 & 3 & 4 \\ 2 & 0 & 3 & 0 \\ 1 & 1 & -3 & 3 \end{bmatrix}$ and $\vec{v} = \begin{bmatrix} 1 \\ 1 \\ 1 \\ 1 \end{bmatrix}$

Exercises 11–16 deal with matrix multiplication. In each problem, calculate the matrix product.

11. $\begin{bmatrix} 5 & 7 \\ 6 & 8 \end{bmatrix} \cdot \begin{bmatrix} 1 & 2 \\ 3 & 4 \end{bmatrix}$

12. $\begin{bmatrix} -1 & 2 \\ 3 & -4 \end{bmatrix} \cdot \begin{bmatrix} 5 & 1 \\ 0 & 5 \end{bmatrix} \cdot \begin{bmatrix} 2 & 1 \\ 1.5 & .5 \end{bmatrix}$

13. $\begin{bmatrix} 1 & 2 & 0 \\ 0 & 3 & 4 \\ 1 & 4 & 0 \end{bmatrix} \cdot \begin{bmatrix} 5 & 0 & 7 \\ 6 & 0 & 8 \\ 0 & 0 & 1 \end{bmatrix}$

14. $\begin{bmatrix} 1 & 0 & 0 \\ 0 & 4 & 1 \\ 0 & 0 & 4 \end{bmatrix} \cdot \begin{bmatrix} 3 & 0 & 0 \\ 0 & 6 & 1 \\ 0 & 0 & 6 \end{bmatrix} \cdot \begin{bmatrix} 1 & 0 & 0 \\ 0 & 1/4 & -1/16 \\ 0 & 0 & 1/4 \end{bmatrix}$

5.2 Bounded Linear Operators on General Hilbert Spaces

15. $\begin{bmatrix} 1 & 0 & 0 & 0 \\ -3 & 4 & 0 & 0 \\ 2 & 5 & -6 & 0 \\ -1 & 1 & -1 & 1 \end{bmatrix} \cdot \begin{bmatrix} -5 & 7 & 1 & 2 \\ 0 & 1 & 6 & 8 \\ -2 & -3 & -1 & 0 \\ 1/2 & 2 & -3 & 1 \end{bmatrix}$

16. $\begin{bmatrix} 0 & 1 & 0 & 1 \\ 0 & 0 & 0 & 0 \\ 1 & 0 & 1 & 0 \\ 0 & 0 & 0 & 0 \end{bmatrix} \cdot \begin{bmatrix} 1 & 2 & 3 & 4 \\ 2 & 3 & 4 & 1 \\ 3 & 4 & 1 & 2 \\ 4 & 1 & 2 & 3 \end{bmatrix}$

Exercises 17–20 deal with the operator norm for operators on two-dimensional real Hilbert spaces.

17. Find the operator norm $\|T\|$ for the operator $T : \mathbb{R}^2 \to \mathbb{R}^2$ defined as $T\vec{v} = \begin{bmatrix} 5 & 0 \\ 0 & 4 \end{bmatrix} \cdot \begin{bmatrix} a \\ b \end{bmatrix}$, where $a, b \in \mathbb{R}$.

18. Examine the operator $T : \mathbb{R}^2 \to \mathbb{R}^2$ with matrix representation $T = \begin{bmatrix} a & b \\ c & d \end{bmatrix}$. Using the quadratic equation, show the operator norm $\|T\|$ satisfies

$$\|T\|^2 = \frac{N + \sqrt{N^2 - 4|D|^2}}{2},$$

where $N = |a|^2 + |b|^2 + |c|^2 + |d|^2$ and $D = ad - bc$. You may wish to use the fact (see Exercise 5.4.14) that $\|T^*T\| = \|T\|^2$, where T^* in this case is multiplication by $\begin{bmatrix} a & c \\ b & d \end{bmatrix}$.

19. Use the result of the previous exercise to prove, for $T : \mathbb{R}^2 \to \mathbb{R}^2$, $\|T\| \leq 1$ if and only if $N \leq 1 + |D|^2$.

20. Use the result of the previous exercise to determine if $\|T\| \leq 1$, where $T\vec{v} = \begin{bmatrix} 1 & 0 \\ .5 & .5 \end{bmatrix} \cdot \begin{bmatrix} e \\ f \end{bmatrix}$ for $e, f \in \mathbb{R}$.

Exercises 21–26 deal with the Jordan form theorem.

21. Suppose $T = \begin{bmatrix} 4 & 0 & 0 \\ 1 & 4 & 0 \\ 5 & -3 & 1 \end{bmatrix}$. It has two eigenvalues $\lambda = 4$ (which has multiplicity two—in this case, it appears twice on the diagonal and T is lower-triangular) and $\lambda = 1$. Find two eigenvectors for T, one for each eigenvalue.

22. What is the Jordan form matrix J?

23. Show the vector $\vec{v} = \begin{bmatrix} 1 \\ 1 \\ 1 \end{bmatrix}$ satisfies $T\vec{v} = \vec{e} + \lambda\vec{v}$, where λ is the eigenvalue $\lambda = 4$ and $\vec{e} = \begin{bmatrix} 0 \\ 1 \\ -1 \end{bmatrix}$ is an eigenvector for $\lambda = 4$. The vector $\vec{v}$ is a generalized eigenvector and is the second column of P, where the other two columns of P are formed from the other two eigenvectors.

24. Use your results in the last exercise to determine the matrix P such that $T = PJP^{-1}$ for the matrix J you constructed in Exercise 22.

25. Use the matrix $P^{-1} = \begin{bmatrix} 4 & 0 & 0 \\ 1 & 4 & 0 \\ 5 & -3 & 1 \end{bmatrix}$ to write T in Exercise 21 as $T = PJP^{-1}$. Also, show $P \cdot P^{-1} = \begin{bmatrix} 1 & 0 & 0 \\ 0 & 1 & 0 \\ 0 & 0 & 1 \end{bmatrix}$ for the matrix P you obtained in the previous exercise.

26. Write T as $T = PJP^{-1}$ when $T = \begin{bmatrix} 1 & 0 & 0 \\ -.5 & 2.5 & .5 \\ -.5 & -.5 & 1.5 \end{bmatrix}$. Use eigenvalues 2 and 1, and use $P^{-1} = \begin{bmatrix} .5 & .5 & -.5 \\ -.5 & .5 & .5 \\ 1 & 0 & 0 \end{bmatrix}$.

Exercises 27–32 show advantages of the Jordan form and diagonalization.

27. The operator $T\vec{v}$ on $\mathbb{R}^2$, where $T = \begin{bmatrix} 1 & 1 \\ -1 & 3 \end{bmatrix}$, satisfies

$$T = PJP^{-1} = \begin{bmatrix} 1 & 0 \\ 1 & 1 \end{bmatrix} \cdot \begin{bmatrix} 2 & 1 \\ 0 & 2 \end{bmatrix} \cdot \begin{bmatrix} 1 & 0 \\ -1 & 1 \end{bmatrix}.$$

Find J^2, J^3, J^5, and a formula for J^n, where $n \in \mathbb{N}$.

28. With T as in the previous exercise and from direct calculations using matrix multiplication, calculate T^2, T^3, and T^5.

29. Comparing your answers in the two previous exercises, which is easier to find by direct calculations—T^5 or J^5?

30. Using $T^n = PJ^n P^{-1}$ for $n \in \mathbb{N}$, find a formula for T^n with T as in Exercise 27.

31. What is T^{10} for T defined in Exercise 27?

32. What is $T^{10,000}$ for T defined in Exercise 27?

Exercises 33–42 are related to rank-one operators on a Hilbert space $\mathcal{H}$. Each exercise uses the notation for a rank-one operator in Example 5.2.3; namely, for $w \in \mathcal{H}$, $T_w(v) = \langle v, w \rangle \cdot w$ for $v \in \mathcal{H}$.

33. When $\mathcal{H} = L^2(-\pi, \pi)$ and $w(x) = x$, find a formula for $T_w(v)$ for $v(x) \in L^2(-\pi, \pi)$.

34. Using T_w as in the previous exercise, what is $T_w(x^2)$? $T_w(\sin x)$? $T_w(x^n)$ for $n \in \mathbb{N}$? $T_w(p(x))$ for a polynomial $p(x) = \sum_{k=0}^{n} a_k x^k$?

35. When $\mathcal{H} = \ell^2$ and $\vec{w} = \begin{bmatrix} 0 \\ 1 \\ 1 \\ 0 \\ 0 \\ \vdots \end{bmatrix}$, find a formula for $T_{\vec{w}}(\vec{v})$ for $\vec{v} = \begin{bmatrix} v_0 \\ v_1 \\ v_2 \\ \vdots \end{bmatrix} \in \ell^2$.

36. Using $T_{\vec{w}}$ as in the previous exercise, what is $T_{\vec{w}}\left(\begin{bmatrix} 0 \\ 1 \\ 0 \\ 0 \\ \vdots \end{bmatrix}\right)$? $T_{\vec{w}}\left(\begin{bmatrix} 1 \\ 1/2 \\ 1/4 \\ 1/8 \\ \vdots \end{bmatrix}\right)$?

37. When $\mathcal{H} = \ell^2$ and $\vec{w} = \begin{bmatrix} 1 \\ 1/2 \\ 1/4 \\ 1/8 \\ \vdots \end{bmatrix}$, find a formula for $T_{\vec{w}}(\vec{v})$ for $\vec{v} = \begin{bmatrix} v_0 \\ v_1 \\ v_2 \\ \vdots \end{bmatrix} \in \ell^2$.

38. When $\mathcal{H} = \ell^2$ and, for a given fixed constant c with $|c| < 1$, if $\vec{w} = \begin{bmatrix} 1 \\ c \\ c^2 \\ c^3 \\ \vdots \end{bmatrix}$, find a formula for $T_{\vec{w}}(\vec{v})$ for $\vec{v} = \begin{bmatrix} v_0 \\ v_1 \\ v_2 \\ \vdots \end{bmatrix} \in \ell^2$.

39. When $\mathcal{H}$ is the Sobolev space $W^{1,2}$ on the interval $(-1, 1)$ and $w(x) = x$, find a formula for $T_w(f)$ for $f(x) \in W^{1,2}$.

40. Using T_w as in the previous exercise, what is $T_w(x^2)$? $T_w(x^n)$ for $n \in \mathbb{N}$?

41. When $\mathcal{H} = H^2(\mathbb{D})$ and $w(z) = z$, find a formula for $T_w(f)$ for $f(z) \in H^2(\mathbb{D})$. Express your result as an integral times w and as an infinite series times w, assuming $f(e^{it}) = \sum_{n=0}^{\infty} a_n e^{int}$.

42. Using T_w as in the previous exercise, what is $T_w(z^2)$? $T_w(3z + 5z^3)$?

Exercises 43–46 study compact operators.

43. Let T be a compact bounded linear operator on a Hilbert space $\mathcal{H}$ defined as

$$T(f) = \sum_{n=0}^{\infty} \lambda_n \langle \phi_n, f \rangle \phi_n$$

for $f \in \mathcal{H}$, where λ_n is in $\mathbb{C}$ and $\{\phi_n\}$ is an orthonormal set in $\mathcal{H}$. Show $T(\phi_n) = \lambda_n \phi_n$ for $n = 0, 1, 2, \ldots$. In other words, show ϕ_n is an eigenvector for T with eigenvalue λ_n.

44. Determine an infinite number of invariant subspaces for the operator T in the previous exercise.

45. Prove $T : \ell^2 \to \ell^2$ defined as $T(\vec{v}) = \begin{bmatrix} \lambda_0 & 0 & 0 & 0 & \cdots \\ 0 & \lambda_1 & 0 & 0 & \cdots \\ 0 & 0 & \lambda_2 & 0 & \cdots \\ & & \cdots & & \end{bmatrix} \cdot \begin{bmatrix} v_0 \\ v_1 \\ v_2 \\ \vdots \end{bmatrix}$ for any $\vec{v} = \begin{bmatrix} v_0 \\ v_1 \\ v_2 \\ \vdots \end{bmatrix} \in \ell^2$ is compact.

46. Determine an infinite number of invariant subspaces for the operator T in the previous exercise.

Exercises 47–50 study Volterra operators.

47. Let T be the Volterra operator on $L^2(0, 1)$ defined as $T(f)(x) = \int_0^x f(y)\, dy$. Prove T is bounded.

48. In the previous exercise,

$$k(x, y) = \begin{cases} 1 & \text{if } y \leq x \\ 0 & \text{otherwise.} \end{cases}$$

Show $\int_0^1 \int_0^1 |k(x, y)|^2\, dx\, dy = M < \infty$ by finding M. Conclude T, as defined in the previous exercise, is a Hilbert-Schmidt operator and explain why this conclusion is true.

49. Prove T, as defined in Exercise 47, is compact.

50. Define the operator $T : L^2(0, \pi) \to L^2(0, \pi)$ by $T(f)(x) = \int_0^{\pi} \cos(x + y) f(y)\, dy$ for $f \in L^2(0, \pi)$. Prove T is compact.

5.3 The Unilateral Shift Operator

The last section defined several operators on infinite-dimensional Hilbert spaces. They included the Hardy space $H^2(\mathbb{D})$—the Hilbert space of complex-valued analytic functions f on the unit disk $\mathbb{D}$ that have finite $H^2(\mathbb{D})$-norm. This section describes the unilateral shift operator on $H^2(\mathbb{D})$. A famous theorem of the brilliant Swedish mathematician Arne Beurling[14] describes the shift operator, which is more simply characterized as multiplication by z (and often labeled M_z).

Beurling's Theorem, which completely describes the invariant subspaces of M_z on $H^2(\mathbb{D})$, has motivated mathematicians to examine operators analogous to M_z on other Hilbert spaces. Because of the analogy, and because their study generalizes and extends the results of Beurling's Theorem, the operators are also considered unilateral shift operators. This section characterizes a general shift operator, and it describes facts about an infinite dimensional Hilbert space that can be determined by a shift operator's properties. We start with the shift on $H^2(\mathbb{D})$.

5.3.1 The Shift Operator on $H^2(\mathbb{D})$

As described in Example 5.2.3, the Hardy space $\mathcal{H} = H^2(\mathbb{D})$ is the collection of complex functions f analytic on the unit disk $\mathbb{D}$ with

$$\|f\|_{H^2(\mathbb{D})}^2 \equiv \sup_{0 \le r < 1} \frac{1}{2\pi} \int_0^{2\pi} |f(re^{it})|^2 \, dt < \infty.$$

Elements $f(z)$ of $H^2(\mathbb{D})$ can be identified with their boundary functions $f(e^{it}) = \sum_{n=0}^{\infty} a_n e^{int}$ on the unit circle $\mathbb{T}$, and then $\|f\|_{H^2(\mathbb{T})}^2 = \sum_{n=0}^{\infty} |a_n|^2$.

The shift operator M_z on $H^2(\mathbb{D})$ multiplies a function $f(z)$ by z.

Definition 5.3.1 *The shift operator M_z on $H^2(\mathbb{D})$ acts according to*

$$M_z(f)(z) = z \cdot f(z).$$

The boundary function for $M_z(f)(z)$ is $e^{it} f(e^{it})$. In this way, the shift operator M_z induces a shift operator S on $H^2(\mathbb{T})$ defined as $S(f)(e^{it}) = e^{it} f(e^{it})$.

Theorem 5.3.1 *The shift operator is a bounded linear operator with operator norm 1.*

Proof. Linearity follows from the distributive property. For $f, g \in H^2(\mathbb{D})$ and complex constants a and b,

$$M_z(af + bg)(z) = z(af(z) + bg(z)) = azf(z) + bzg(z) = aM_z(f)(z) + bM_z(g)(z).$$

---

[14] Among many other lifetime accomplishments, Beurling deciphered in 1940 the "T52" Nazi communication code, which at that time was the German war machine's tool to communicate secretly between command posts and front-line units. Beurling's decryption allowed Sweden to decode the messages. In late 1940, it decoded a message that described the German army's plan to invade the Soviet Union on June 22, 1941. Allied intelligence alerted the Soviets and urged them to defend against the invasion. Their warnings were correct. On that date, the German Luftwaffe began bombing Soviet-occupied cities in Poland, the start of a massive Soviet front movement. The Nazis were eventually repelled after horrendous numbers of Soviet army deaths.

5.3 The Unilateral Shift Operator

To show $\|M_z\| = 1$, we use the boundary function for $M_z(f)(z)$, when the Taylor series for f is $f(z) = \sum_{n=0}^{\infty} a_n z^n$. From the definition of the operator norm,

$$\|M_z\| = \sup\{\|M_z(f)(z)\|_{H^2(\mathbb{D})}\} = \sup\left\{\left\|e^{it}\sum_{n=0}^{\infty} a_n e^{int}\right\|_{H^2(\mathbb{T})}\right\} = \sup\left\{\left\|\sum_{n=0}^{\infty} a_n e^{i(n+1)t}\right\|_{H^2(\mathbb{T})}\right\}$$

$$= \sup\left\{\left(\sum_{n=0}^{\infty} |a_n|^2\right)^{1/2}\right\},$$

where the supremum is taken over $f \in H^2(\mathbb{D})$ such that $\|f\| = 1$. Since $\|f\| = 1$ implies $\sum_{n=0}^{\infty} |a_n|^2 = 1$, we get $\|M_z\| = \sup\{(\sum_{n=0}^{\infty} |a_n|^2)^{1/2}\} = 1$. ∎

The use of the boundary function $f(e^{it})$ and its Taylor series representation $\sum_{n=0}^{\infty} a_n e^{int}$ also gives us a vectorial way to view the action of the shift operator. Write the Taylor series coefficients of f as a column vector whose transpose is $[a_0 \ a_1 \ a_2 \ldots]$. Do the same for the output vector $M_{e^{it}}(f)(e^{it}) = \sum_{n=0}^{\infty} a_n e^{i(n+1)t}$, which produces, as $M_{e^{it}}(f)$ has no constant coefficient, $[0 \ a_0 \ a_1 \ldots]$. You can see the unilateral shift operator shifts the entries of the column vector for f down one entry notch, while inserting 0 into the first entry. This description of the elements of $H^2(\mathbb{T})$ in terms of their coefficient column vector representation describes the isometric map from $H^2(\mathbb{T})$ to ℓ^2, showing the equivalence of the two spaces. It therefore concretely defines the action of the unilateral shift operator on ℓ^2, which we will explore later in this section.

The last section introduced the concept of an operator's invariant subspaces. We can ask about the shift operator's invariant subspaces: What are the (closed) invariant subspaces for the operator M_z on $H^2(\mathbb{D})$? Beurling's theorem, which we will present momentarily, answers the question. It uses a factorization $f = b \cdot g$ of a function $f \in H^2(\mathbb{D})$, known as the "inner-outer factorization." The function b is an inner function and g an outer function.

Definition 5.3.2 Inner and Outer Functions in $H^2(\mathbb{D})$
An inner function $b(z) \in H^2(\mathbb{D})$ is one that satisfies $|b(e^{it})| = 1$ almost everywhere for $0 \le t < 2\pi$.

An outer function $g(z) \in H^2(\mathbb{D})$ is the solution to the problem of finding the function that maximizes $|g(0)|$ for a function having prescribed values $|g(e^{it})|$ around the unit circle $\mathbb{T}$. Equivalently, an outer function is of the form

$$g(z) = c e^{\frac{1}{2\pi}\int_0^{2\pi} \frac{e^{it}+z}{e^{it}-z} \log|g(e^{it})|\,dt},$$

where $|c| = 1$ and $\int_0^{2\pi} \log|g(e^{it})|\,dt < \infty$.

An inner-outer factorization of a function $f(z) \in H^2(\mathbb{D})$ is of the form $f(z) = b(z) \cdot g(z)$ with $b(z)$ an inner function and $g(z)$ an outer function.

Examples of inner and outer functions on $\mathbb{D}$ give an idea of what this factorization entails. Clearly $b(z) = z$ is an inner function, since $b(e^{it}) = e^{it} = \cos t + i \sin t$ has $|b(e^{it})|^2 = (\cos t)^2 + (\sin t)^2 = 1$. In fact, $b(z) = z$ is a *Blaschke function*, or a *Blaschke product*. A Blaschke function is either a constant λ with $|\lambda| = 1$ or of the form

$$B(z) = \lambda z^m \prod_n \frac{|a_n|}{a_n} \frac{a_n - z}{1 - \overline{a_n} z},$$

where $|\lambda| = 1$, m is a nonnegative integer, $a_n \in \mathbb{D}$ is nonzero, and the product can be over finite or countably infinite values for n. We also require $\sum_{n=1}^{\infty}(1 - |a_n|) < \infty$, which ensures convergence on $\mathbb{D}$ of a Blaschke function's infinite product. The norm-one factor $\frac{|a_n|}{a_n}$ is included to handle similar convergence issues for infinite products.

Blaschke functions form one type of inner function, and there exist inner functions of only one other fundamental type: *singular inner*. This type is characterized by the example $s(z) = e^{\frac{z+1}{z-1}}$. Here, s is clearly nonzero on $\mathbb{D}$ (and hence cannot be multiplicatively decomposed into anything involving a Blaschke function format). A direct calculation (see Exercise 12) shows $|s(e^{it})| = 1$ for $t \in (0, 2\pi)$; the (nontangential) limit as $z \to 1 = e^{i0}$ is more intricate and turns out to be 0. The general description of a singular inner function is

$$S(z) = \lambda e^{-\int_0^{2\pi} \frac{e^{it}+z}{e^{it}-z} d\mu}, z \in \mathbb{D},$$

where $|\lambda| = 1$ and μ is any nonnegative and singular measure. Explicitly, $\mu = \mu_{pm} + \mu_{sc}$, where μ_{pm} is the (possibly infinite) sum of nonnegative point mass measures at values p in the interval $[0, 2\pi)$ and μ_{sc} is a nonnegative continuous singular measure as described in Section 4.2. Using a simple point mass measure at $t = 0$ (in other words, $\mu = \mu_{(p=0)}$) evaluates the integral by replacing t with 0 in the integrand, which produces the example previously mentioned: $s(z) = e^{\frac{z+1}{z-1}}$.

Outer functions g are never zero (since they are exponential). Furthermore, g and $1/g$ are analytic in $\mathbb{D}$. The delightful fact is every (nontrivial) function $f \in H^2(\mathbb{D})$ has an inner-outer factorization.

Theorem 5.3.2 *Every function $f \in H^2(\mathbb{D})$ not identically zero has an inner-outer factorization $f(z) = b(z) \cdot g(z)$, unique up to a constant of norm one.*

Proof. Construct g as $g(z) = e^{\frac{1}{2\pi} \int_0^{2\pi} \frac{e^{it}+z}{e^{it}-z} \log |f(e^{it})| dt}$, where $f(e^{it})$ is the boundary function for $f(z)$. Defining $b(z)$ as $b(z) = f(z)/g(z)$ results in $|b(e^{it})| = 1$ almost everywhere for $t \in [0, 2\pi)$.[15] Given a possibly different inner-outer factorization for f, say $f = b_1 \cdot g_1$, then $|g(e^{it})/g_1(e^{it})| = |b_1(e^{it})/b(e^{it})| = 1$ almost everywhere on $\mathbb{T}$. Hence g/g_1 is an inner function. Similarly, $|g_1(e^{it})/g(e^{it})| = 1$ almost everywhere on $\mathbb{T}$, and g_1/g is inner. Any inner function b satisfies $|b(z)| \leq 1$ for $z \in \mathbb{D}$,[16] and so we have both $|g(z)/g_1(z)| \leq 1$ and $|g_1(z)/g(z)| \leq 1$

---

[15] The particulars of this fact require additional knowledge of complex function theory, which describes the relationship between the structure of the outer function and an object known as the Poisson kernel (see [103, pp. 55–57]). The boundary function of g has the property $|g(e^{it})| = |f(e^{it})|$ a.e. Hence $|b(e^{it})| = 1$ a.e.

[16] This follows immediately from a powerful complex analysis fact, the maximum modulus principle (see [29, pp. 128–130]).

5.3 The Unilateral Shift Operator

on $\mathbb{D}$. The only conclusion possible is $|g(z)| = |g_1(z)|$, and so $g(z) = c \cdot g_1(z)$, where c is a constant with $|c| = 1$. ∎

The inner-outer factorization can be further refined—the inner function part factors. An inner function $b(z) \in H^2(\mathbb{D})$ factors into $b = B \cdot S$, where $B(z)$ is a Blaschke function and $S(z)$ is a singular inner function in $H^2(\mathbb{D})$. The factorization separates the factors of b formed from its zeros (they then form the Blaschke products of this refined factorization, and are collected into the Blaschke function piece $B(z)$). The remaining nonzero piece is the singular inner function $S(z)$. In this way, a function $f(z) \in H^2(\mathbb{D})$ factors as $f(z) = B(z) \cdot S(z) \cdot g(z)$, with B a Blaschke function, S a singular inner function, and g outer.

Example 5.3.1 Let $f(z) = z - i/2$ and note $f \in H^2(\mathbb{D})$. We write $B(z) = \frac{z-i/2}{1+iz/2}$ and $g(z) = 1 + iz/2$ so that $f(z) = B(z) \cdot g(z)$. With $a = i/2$, we realize $B(z) = \frac{z-a}{1-\bar{a}z}$ is a Blaschke factor. Both $g(z)$ and $1/g(z)$ are analytic on $\mathbb{D}$: $g'(z) = i/2$ exists everywhere and the derivative of $(1 + iz/2)^{-1}$, which is $(-i/2)(1 + iz/2)^{-2}$, exists when $z \neq 2i$. Since $2i$ is outside the unit disk, $1/g$ is analytic on $\mathbb{D}$. Hence g is outer.

As constructed in Theorem 5.3.2, we may write $g(z) = 1 + iz/2 = c\, e^{\frac{1}{4\pi} \int_0^{2\pi} \frac{e^{it}+z}{e^{it}-z} \ln(5/4 - \sin t)\, dt}$, where c is some constant with $|c| = 1$, since

$$|1 + (i/2)e^{it}|^2 = (1/2)^2|(2 + i(\cos t + i \sin t))|^2 = (1/4)[(2 - \sin t)^2 + \cos^2 t] = 5/4 - \sin t.$$

In this way, $z - i/2 = B(z) \cdot g(z)$ with B a Blaschke factor and g outer. ∎

Question 5.3.1 Let $f(z) = \frac{z-i/2}{1-z/2}$. Write $f(z) = B(z) \cdot g(z)$ with B a single Blaschke factor and g outer. Then write g in its exponential form described in Definition 5.3.2, where $g(z) = c\, e^{\frac{1}{2\pi} \int_0^{2\pi} \frac{e^{it}+z}{e^{it}-z} \log|f(e^{it})|\, dt}$ with c some norm-one constant. ∎

Arne Beurling was the first person to recognize this factorization has implications about the unilateral shift operator M_z on $H^2(\mathbb{D})$. Namely, it allows us to describe precisely the closed invariant subspaces for M_z.

Theorem 5.3.3 Beurling's Theorem: *The (nontrivial) invariant subspaces of M_z on $H^2(\mathbb{D})$ are characterized as $bH^2 = \{b(z) \cdot f(z) : f \in H^2(\mathbb{D})\}$, where b is an inner function in $H^2(\mathbb{D})$.*

Proof. It is easy to see a subspace of the form bH^2 is invariant under M_z, since $f \in H^2(\mathbb{D})$ has $M_z(b \cdot f)(z) = zb(z) \cdot f(z) = b(z) \cdot [zf(z)] \in bH^2$. ($zf(z)$ is in $H^2(\mathbb{D})$ because $M_z : H^2(\mathbb{D}) \to H^2(\mathbb{D})$.)

The proof of the converse—that any invariant subspace M is of the form $M = bH^2$ with b inner—is more difficult. We outline it in four steps.[17] First, applying M_z repeatedly to elements in an invariant subspace M leads to the conclusion that we may, without loss of generality, assume M contains a function $a(z)$ with nonzero constant coefficient a_0. (See Exercise 17.) We have $a_0 \neq 0$.

---

[17] The proof presented here is now standard (see [45, pp. 82–83]).

Second, write the function $1 = m(z) + n(z)$, where $m \in M$ and $n \in M^\perp$, so that $\langle f(z), n(z) \rangle = 0$ for $f \in M$. The function m is the projection of 1 onto M. Since $z^k m(z) = (M_z)^k(m)(z) \in M$, we have

$$\frac{1}{2\pi} \int_0^{2\pi} e^{ikt}\{m(e^{it}) - |m(e^{it})|^2\} dt = \frac{1}{2\pi} \int_0^{2\pi} e^{ikt} m(e^{it}) \cdot \overline{1 - m(e^{it})} dt$$

$$= \frac{1}{2\pi} \int_0^{2\pi} e^{ikt} m(e^{it}) \cdot \overline{n(e^{it})} \, dt$$

$$= \langle z^k m(z), n(z) \rangle = 0, \quad k = 0, 1, 2, \ldots.$$

Third, $z^k m(z)$ is zero at $z = 0$ for $k = 1, 2, \ldots$. That means the Fourier coefficients of $|m|^2$ are zero for $k = 1, 2, \ldots$, since the equation in step 2 implies $\int_0^{2\pi} e^{ikt} |m(e^{it})|^2 \, dt = \int_0^{2\pi} e^{ikt} m(e^{it}) \, dt = 0$.[18] Hence $|m(z)|^2$ is a constant c. It turns out $c = 0$ would violate the existence of the function a in step 1. Hence $c \neq 0$. Note $|m(z)|^2 = c = \frac{1}{2\pi} \int_0^{2\pi} |m(e^{it})|^2 \, dt = \|m(z)\|^2_{H^2(\mathbb{D})}$. We then define the function $b(z) = m(z)/\|m(z)\|_{H^2(\mathbb{D})}$, which is an inner function because its construction normalizes m to attain norm one on $\mathbb{T}$. Step 4 will show $M = bH^2$.

Fourth, $bH^2 \subseteq M$ because $b \in M$ and so $z^k b(z) \in M$ for $k = 0, 1, 2, \ldots$. Now show $M \subseteq bH^2$. If there exists a function $r(z) \in M$ such that $\langle r, f \rangle = 0$ for all $f \in bH^2$ (so $r \in (bH^2)^\perp$), then $0 = 2\pi \langle r(z), b(z) z^k \|m(z)\| \rangle = \int_0^{2\pi} r(e^{it}) e^{-ikt} \overline{m(e^{it})} \, dt$. This follows from the fact that $b(z) z^k \|m(z)\| \in bH^2$. We know $z^k r(z) \in M$. Since $1 - m(z) = n(z) \in M^\perp$,

$$0 = \langle z^k r(z), 1 - m(z) \rangle = \frac{1}{2\pi} \int_0^{2\pi} e^{ikt} r(e^{it}) - e^{ikt} r(e^{it}) \overline{m(e^{it})} \, dt.$$

Then $\frac{1}{2\pi} \int_0^{2\pi} e^{ikt} r(e^{it}) \overline{m(e^{it})} \, dt = 0$, $k = 1, 2, \ldots$, since $z^k r(z)$ equals zero at $z = 0$ (so $\frac{1}{2\pi} \int_0^{2\pi} e^{ikt} r(e^{it}) \, dt = 0$). We have therefore shown all the Fourier coefficients of the L^1 function $r \cdot \overline{m}$ are zero, and so $r(z) \cdot \overline{m(z)} = 0$. But $|m(z)| > 0$, and so $r(z) = 0$. Since we have shown the zero function is the only element in M also in $(bH^2)^\perp$, it must be the case that $M = bH^2$. ∎

Beurling's theorem illustrates the interplay between the study of operators and the study of functions. The operator M_z is described by its invariant subspaces, and the invariant subspaces are described by the inner-outer factorizations of functions. An inner function $b(z)$ corresponds to an invariant subspace bH^2, and an invariant subspace has the structure given by some inner function. The pairing of inner functions and invariant subspaces sets up a one-to-one correspondence between them, providing a complete characterization.

5.3.2 The Unilateral Shift Operator on a General Hilbert Space

The Hardy space $H^2(\mathbb{D})$ is special because its unilateral shift operator M_z has well understood invariant subspaces. Do other Hilbert spaces have an analogous operator, and can we develop a Beurling-type theorem for them? The answer is "yes." The rest of this section introduces a few fundamental ideas in the general study of the unilateral shift operator. It

---

[18] When an analytic function $h(z)$ has $h(0) = 0$, the integral $\int_0^{2\pi} h(e^{it}) \, dt$ is always zero. This remarkable fact is one version of the mean value theorem.

5.3 The Unilateral Shift Operator

begins by stating, for a general Hilbert space $\mathcal{H}$, which operators are analogous to M_z on $H^2(\mathbb{D})$.

Definition 5.3.3 *(a) A bounded linear operator S on a Hilbert space $\mathcal{H}$ is an isometry if $\langle Sf, Sg \rangle_{\mathcal{H}} = \langle f, g \rangle_{\mathcal{H}}$ for all $f, g \in \mathcal{H}$.*

(b) The adjoint of a bounded linear operator S on a Hilbert space $\mathcal{H}$ is the operator S^* satisfying $\langle Sf, g \rangle_{\mathcal{H}} = \langle f, S^*g \rangle_{\mathcal{H}}$ for all $f, g \in \mathcal{H}$.

(c) A bounded linear operator S on a Hilbert space $\mathcal{H}$ is a unilateral shift operator[19] *if S is an isometry and $\lim_{n \to \infty} \|(S^*)^n f\|_{\mathcal{H}} = 0$ for all $f \in \mathcal{H}$.*

If we take $\mathcal{H} = H^2(\mathbb{D})$ and $S = M_z$, it is not difficult to see that M_z is a unilateral shift operator in the sense of Definition 5.3.3. Theorem 5.3.1 has shown M_z is an isometry. The adjoint of $S = M_z$ is described as

$$S^* : f(z) \to \frac{f(z) - f(0)}{z} \text{ for any } f(z) \in H^2(\mathbb{D}).$$

When $f, g \in H^2(\mathbb{D})$ have $f(z) = \sum_{k=0}^{\infty} a_k z^k$ and $g(z) = \sum_{k=0}^{\infty} b_k z^k$, then

$$\langle zf(z), g(z) \rangle_{H^2(\mathbb{D})} = \left\langle \sum_{k=0}^{\infty} a_k z^{k+1}, \sum_{k=0}^{\infty} b_k z^k \right\rangle$$
$$= a_0 b_1 + a_1 b_2 + \cdots$$
$$= \left\langle \sum_{k=0}^{\infty} a_k z^k, \sum_{k=1}^{\infty} b_k z^{k-1} \right\rangle = \langle f(z), [g(z) - g(0)]/z \rangle.$$

Finally, $\|(S^*)^n f\|^2_{H^2(\mathbb{D})} = \|\sum_{k=n}^{\infty} a_k z^{k-n}\|^2 = |a_n|^2 + |a_{n+1}|^2 + \cdots$. The tail to the convergent series for $\|f\|^2$ must approach zero as $n \to \infty$. By Definition 5.3.3, $S = M_z$ is therefore a unilateral shift operator on $H^2(\mathbb{D})$.

Two surprising and powerful theorems can be stated about unilateral shift operators. They give the most general results possible on specific topics for arbitrary Hilbert spaces. The first is the *universal model*, which we refer to as the Rota-de Branges-Rovnyak model, named after Gian-Carlo Rota, who, along with the two American mathematicians Louis de Branges and James Rovnyak, first fully developed the idea.[20] The universal model identifies the use of the adjoint S^* of a unilateral shift operator S on a Hilbert space $\mathcal{G}$ as a way to represent any bounded linear operator on a Hilbert space $\mathcal{H}$. Here's how it works: for an invariant subspace M of S^*, examine every operator of the form $S^*|_M$, by which we mean S^* restricted to (applied only to) elements in the subspace M of $\mathcal{G}$. That class of operators $S^*|_M$ can represent every bounded linear operator on $\mathcal{H}$. The formal statement of this remarkable fact is the next theorem. In it, the operator T is normalized so that $\|T\| \leq 1$. There is one caveat required in order for the theorem

---

[19] This development is presented by Marvin Rosenblum and James Rovnyak in [101].

[20] Italian-born G. C. Rota, who was educated and lived in America and taught at MIT, was the first to give the universal model for Hilbert space operators (cf. [104]). Paul Halmos published the insight that the shift operator could represent bounded linear operators, and saw the merit in the strategic approach to use the representation to solve the invariant subspace problem (cf. [52]).

to be true. It involves the multiplicity of the shift operator S acting on $\mathcal{G}$. The multiplicity of S is defined as the dimension of the subspace $\mathcal{K} = \{f : f \in \mathcal{G} \text{ and } S^*f = 0\}$. (The subspace $\mathcal{K}$ of $\mathcal{G}$ is the kernel of S^*.) The universal model works when S produces $\mathcal{K}$ with dimension at least as large as the dimension of the closure of $(I - T^*T)\mathcal{H}$. (This is the closure described in Section 3.3 of the set of objects $f - T^*Tf$ for $f \in \mathcal{H}$.)

Theorem 5.3.4 The Rota-de Branges-Rovnyak Universal Model: *Suppose T is a bounded linear operator on a Hilbert space $\mathcal{H}$ with $\|T\| \leq 1$ and $\|T^n f\| \to 0$ as $n \to \infty$ for all $f \in \mathcal{H}$. Suppose also we can find a unilateral shift operator S on a Hilbert space $\mathcal{G}$ with multiplicity at least the size of the dimension of the closure of $(I - T^*T)\mathcal{H}$. Then there is an invariant subspace M of S^* such that T is unitarily equivalent to $S^*|_M$.*

Proof. The proof is straightforward but advanced. See [101, p. 5] for the full result. The key insight is that we can define an operator W mapping $\mathcal{H}$ into $\mathcal{G}$ according to

$$Wf = \sum_{j=0}^{\infty} S^j K (I - T^*T)^{1/2} T^j f,$$

where K can be chosen as an isometry that maps the closure of $(I - T^*T)^{1/2} \mathcal{H}$ into $\mathcal{K}$. Then W is an isometry, and $Tf = W^{-1} S^* W f$ for $f \in \mathcal{H}$. Setting $M = W\mathcal{H}$, we obtain M as an invariant subspace of S^*, and the result follows. ∎

The second surprising and powerful theorem is a generalization of Beurling's theorem. The first such generalization was produced by the American Peter David Lax, who extended the result to vector-valued structures (see [79]). A more complete generalization[21] involves a partial isometry, which is an isometry on a subspace of the Hilbert space and which maps everything to 0 off of the subspace.

Definition 5.3.4 *An operator T mapping a Hilbert space $\mathcal{H}$ to a Hilbert space $\mathcal{K}$ is a partial isometry if there is a subspace N of $\mathcal{H}$ such that $T|_N$ is an isometry and $\ker T \equiv \{f : f \in \mathcal{H} \text{ and } Tf = 0\} = N^\perp$.*

Partial isometries work like inner functions in Beurling's theorem. Every invariant subspace of a unilateral shift operator S on a general Hilbert space $\mathcal{H}$ is of the form $T\mathcal{H} \equiv \{Tf : f \in \mathcal{H}\}$, where T is a partial isometry that commutes with S. (Because of its analogy—playing the same role as does an inner function in Beurling's

Figure 5.1. Peter David Lax in the 1970s.

---

[21] The level of generality presented here was developed through the efforts of other mathematicians, including Marvin Rosenblum and James Rovnyak (see [102]), and Joseph Ball and William Helton (see [7]).

5.3 The Unilateral Shift Operator

theorem—T is known as "S-inner.") The proof is involved and is omitted here; for a reference, see [101, p. 98].

Theorem 5.3.5 Generalized Beurling-Lax Theorem: *Let S be a unilateral shift operator on a Hilbert space $\mathcal{H}$. A subspace M of $\mathcal{H}$ is invariant under S if and only if $M = T\mathcal{H}$, where T is a partial isometry (on $\mathcal{H}$) that commutes with S; in other words, $TSf = STf$ for $f \in \mathcal{H}$.*

We finish the section with an example of how the generalization applies.

Example 5.3.2 Consider the Hilbert space $\mathcal{H}$ that is a direct sum of two $H^2(\mathbb{D})$ spaces, so

$$\mathcal{H} = \left\{ \begin{bmatrix} f_1(z) \\ f_2(z) \end{bmatrix} : f_1, f_2 \in H^2(\mathbb{D}) \right\}.$$

In this case, $\|f\|_{\mathcal{H}}^2 = \|f_1\|_{H^2(\mathbb{D})}^2 + \|f_2\|_{H^2(\mathbb{D})}^2$ with inner product

$$\left\langle \begin{bmatrix} f_1(z) \\ f_2(z) \end{bmatrix}, \begin{bmatrix} g_1(z) \\ g_2(z) \end{bmatrix} \right\rangle_{\mathcal{H}} = \langle f_1(z), g_1(z) \rangle_{H^2(\mathbb{D})} + \langle f_2(z), g_2(z) \rangle_{H^2(\mathbb{D})}.$$

Then for $f = \begin{bmatrix} f_1(z) \\ f_2(z) \end{bmatrix} \in \mathcal{H}$, the operator $S : \mathcal{H} \to \mathcal{H}$ defined as $Sf = \begin{bmatrix} zf_1(z) \\ zf_2(z) \end{bmatrix}$ is a shift operator on $\mathcal{H}$. According to the generalized Beurling-Lax theorem, any S-invariant subspace is generated by a partial isometry that commutes with S. Here are three examples, where each partial isometry is determined on an arbitrary element $f = \begin{bmatrix} f_1(z) \\ f_2(z) \end{bmatrix} \in \mathcal{H}$:

1. Let $b(z)$ be an inner function in $H^2(\mathbb{D})$. Then $Tf = \begin{bmatrix} b(z)f_1(z) \\ 0 \end{bmatrix}$ is a partial isometry. The subspace M of vectors $\begin{bmatrix} b(z)f_1(z) \\ 0 \end{bmatrix}$ is invariant for S: for $f \in M$, $Sf = \begin{bmatrix} zb(z)f_1(z) \\ 0 \end{bmatrix} \in M$ since $b(z)zf_1(z) \in bH^2$. We have $\ker T = \left\{ \begin{bmatrix} 0 \\ g(z) \end{bmatrix} \right\}$, where $g \in H^2(\mathbb{D})$, since T maps an element in the set to the zero vector. For $N = (\ker T)^\perp$, $T|_N$ acts as an isometry: for $f = \begin{bmatrix} f_1(z) \\ 0 \end{bmatrix} \in N$,

$$\|Tf\|_{\mathcal{H}}^2 = \|b \cdot f_1\|_{H^2(\mathbb{D})}^2 = \|b\|_{H^2(\mathbb{D})}^2 \|f_1\|_{H^2(\mathbb{D})}^2 = \|f_1\|_{H^2(\mathbb{D})}^2 = \|f\|_{\mathcal{H}}^2.$$

   T commutes with S because, for $f \in \mathcal{H}$,

$$STf = S\begin{bmatrix} b(z)f_1(z) \\ 0 \end{bmatrix} = \begin{bmatrix} zb(z)f_1(z) \\ 0 \end{bmatrix} = \begin{bmatrix} b(z)zf_1(z) \\ 0 \end{bmatrix} = T\begin{bmatrix} zf_1(z) \\ zf_2(z) \end{bmatrix} = TSf.$$

   The generalized Beurling-Lax theorem described these relationships correctly.

2. Let $b(z)$ be an inner function in $H^2(\mathbb{D})$. Then $Tf = \begin{bmatrix} 0 \\ b(z)f_2(z) \end{bmatrix}$ is a partial isometry with M the set of vectors of the form $\begin{bmatrix} 0 \\ b(z)f_2(z) \end{bmatrix}$. Issues of M's invariance and properties of T with S parallel those developed in part 1.

3. Let $b_1(z)$ and $b_2(z)$ be inner functions in $H^2(\mathbb{D})$. Then $Tf = \begin{bmatrix} b_1(z)f_1(z) \\ b_2(z)f_2(z) \end{bmatrix}$ is a partial isometry with M the set of vectors in $T\mathcal{H}$. Issues of M's invariance are similar to those in part 1. ∎

Question 5.3.2 Define the operator $Sf = \begin{bmatrix} z^2 f_1(z) \\ zf_2(z) \end{bmatrix}$ on $\mathcal{H}$, where $f(z) = \begin{bmatrix} f_1(z) \\ f_2(z) \end{bmatrix} \in \mathcal{H} = H^2(\mathbb{D}) \times H^2(\mathbb{D})$. Show S is a unilateral shift operator in the sense of Definition 5.3.3. ∎

Solutions to Questions

5.3.1 $f(z) = \frac{z-i/2}{1-z/2} = B(z)g(z)$, where $B(z) = \frac{z-i/2}{1+iz/2}$ and $g(z) = \frac{1+iz/2}{1-z/2}$. $B(z)$ is a Blaschke factor with $a = i/2$ and g is outer (as it has no zeros in $\mathbb{D}$ and, by the quotient rule, both it and $1/g(z)$ are analytic on $\mathbb{D}$). We find

$$g(z) = \frac{1+iz/2}{1-z/2} = ce^{\frac{1}{4\pi}\int_0^{2\pi} \frac{e^{it}+z}{e^{it}-z} \ln\left(\frac{5-4\sin t}{5-4\cos t}\right) dt},$$

where c is some constant with $|c| = 1$, since

$$\left|\frac{1+ie^{it}/2}{1-e^{it}/2}\right|^2 = \left|\frac{2+i(\cos t + i \sin t)}{2 - \cos t - i \sin t}\right|^2 = \frac{(2-\sin t)^2 + \cos^2 t}{(2-\cos t)^2 + \sin^2 t} = \frac{5-4\sin t}{5-4\cos t}.$$

5.3.2 For a $f, g \in \mathcal{H}$ (with coordinate functions labeled as in the stem of the question),

$$\langle Sf, Sg \rangle_{\mathcal{H}} = \langle z^2 f_1, z^2 g_1 \rangle_{H^2(\mathbb{D})} + \langle zf_2, zg_2 \rangle_{H^2(\mathbb{D})} = \langle f_1, g_1 \rangle_{H^2(\mathbb{D})} + \langle f_2, g_2 \rangle_{H^2(\mathbb{D})} = \langle f, g \rangle_{\mathcal{H}}.$$

Hence S is an isometry. For such f with $f_1 = \sum_{k=0}^{\infty} a_k z^k$, $\|(S^*)f\|_{\mathcal{H}} = \|\frac{f_1(z)-f_1(0)}{z} - a_1\|_{H^2(\mathbb{D})} + \|\frac{f_2(z)-f_2(0)}{z}\|_{H^2(\mathbb{D})}$. The terms go to zero as S^* is applied repeatedly (see the comments after Definition 5.3.3), and so $\lim_{n\to\infty} \|(S^*)^n f\|_{\mathcal{H}} = 0$.

Reading Questions for Section 5.3

1. What is the shift operator M_z defined on $H^2(\mathbb{D})$?

2. For a Hilbert space $\mathcal{H}$, what is a unilateral shift operator S on $\mathcal{H}$?

3. What are the two types of inner functions on $H^2(\mathbb{D})$? Give one example of each type.

4. What is an outer function on $H^2(\mathbb{D})$?

5. What is Beurling's theorem?

6. What does the Rota-de Branges-Rovnyak universal model say about a bounded linear operator on a Hilbert space?

7. What is a partial isometry, and how does a partial isometry on a Hilbert space $\mathcal{H}$ play a role in the generalized Beurling-Lax theorem?

Exercises for Section 5.3

In Exercises 1–8, explain why, according to Definition 5.3.3, the operator S is a unilateral shift operator on its given underlying Hilbert space.

1. $S(f)(e^{it}) = T_{e^{it}}(f)(e^{it}) = e^{it} f(e^{it})$, which is multiplication by e^{it} on any $f \in H^2(\mathbb{T})$.

2. $S\vec{v} = \begin{bmatrix} v_0 \\ v_1 \\ v_2 \\ \vdots \end{bmatrix} \to \begin{bmatrix} 0 \\ v_0 \\ v_1 \\ \vdots \end{bmatrix}$ on ℓ^2.

3. $S(f)(x) = f(x) - \int_0^x e^{-(1/2)(x-t)} f(t)\, dt$ on $L^2(0, \infty)$. (A basis for $L^2(0,\infty)$ is $\{e^{-t/2} L_n(t)\}_0^{\infty}$.)

4. $S(f)(z) = \frac{z-i}{z+i} f(z)$ on $H^2(\Pi^+)$. (As mentioned in Section 5.2, $\left\{\frac{\pi^{-1/2}}{z+i} \left(\frac{z-i}{z+i}\right)^n\right\}_{n=0}^{\infty}$ forms a basis for $H^2(\Pi^+)$.)

5.3 The Unilateral Shift Operator

5. $S(f)(x) = \frac{x-i/2}{x+i/2} f(x)$ on $H^2(\mathbb{R})$, which can be defined as the Hilbert space with orthonormal basis $\left\{ \frac{i(2\pi)^{-1/2}}{x+i/2} \left(\frac{x-i/2}{x+i/2} \right)^n \right\}_{n=0}^{\infty}$.

6. $Sf = \begin{bmatrix} zf_1(z) \\ z^2 f_2(z) \end{bmatrix}$ on $\mathcal{H}$, where $f(z) = \begin{bmatrix} f_1(z) \\ f_2(z) \end{bmatrix} \in \mathcal{H} = H^2(\mathbb{D}) \times H^2(\mathbb{D})$ is as in Question 5.3.2.

7. $S = M_{z^2}$ on $H^2(\mathbb{D})$, defined according to $S(f)(z) = z^2 \cdot f(z)$. (This shift operator has multiplicity two.)

8. $Sf = \begin{bmatrix} z^2 f_1(z) \\ z^2 f_2(z) \end{bmatrix}$ on $\mathcal{H}$, where $f(z) = \begin{bmatrix} f_1(z) \\ f_2(z) \end{bmatrix} \in \mathcal{H} = H^2(\mathbb{D}) \times H^2(\mathbb{D})$ is as in Question 5.3.2.

Exercises 9–12 combine to show the function $s(z) = e^{\frac{z+1}{z-1}}$ is inner—it satisfies $|s(e^{it})| = 1$ for $t \in (0, 2\pi)$.

9. For a complex number $u = a + ib$, show $|u|^2 = a^2 + b^2 = u \cdot \bar{u}$.

10. Use the previous exercise to prove, for a complex number $w = c + id$, $|e^w| = e^c$. (If we write $c = Re[w]$, which stands for the "real part of w," $|e^w| = e^{Re[w]}$.)

11. Prove $Re\left[\frac{z+1}{z-1}\right] = \frac{|z|^2-1}{|z|^2-2x+1}$, where $z = x + iy$. (*Hint*: multiply numerator and denominator by $\bar{z} - 1$ and simplify.)

12. Use the previous two exercises and the fact that $|z| \to 1$ as $z \to e^{it}$ to prove $|s(e^{it})| = 1$ for $t \in (0, 2\pi)$. What happens for your argument at $t = 0$?

Exercises 13–17 combine to prove a fact needed in the proof of Beurling's theorem: given a (nontrivial) invariant subspace $M \neq \{0\}$ of the operator M_z on $H^2(\mathbb{D})$, we may assume M contains a function $a(z)$ with nonzero constant coefficient a_0.

13. Suppose $M \neq \{0\}$ is an invariant subspace for M_z on $H^2(\mathbb{D})$. Why must there be a nonzero element $f \in M$ with $f(z) = a_k z^k + a_{k+1} z^{k+1} + \ldots$ for some k in $\{0, 1, 2, \ldots\}$?

14. For f as identified as in the previous exercise, the axiom of choice says we can choose an $f \in M$ with the least such k. Why can we write $M = z^k \mathcal{M}$ for some subspace $\mathcal{M} \subset H^2(\mathbb{D})$?

15. Why is the subspace $\mathcal{M}$, formed as in the previous exercise, invariant for M_z on $H^2(\mathbb{D})$?

16. The subspace $\mathcal{M}$, as in the previous exercise, has a function $a(z)$ with nonzero constant coefficient a_0. Why?

17. If we can prove $\mathcal{M}$, as in the previous exercise, is of the form $\mathcal{M} = b(z)H^2$ with b inner, then why have we also proved $M = b(z)H^2$ for some b inner? (In this way, we realize we may as well assume $\mathcal{M} = M$, and thus that M contains a function $a(z)$ with nonzero constant coefficient a_0.)

Exercises 18–22 deal with the multiplication operator $M_f : H^2(\mathbb{D}) \to H^2(\mathbb{D})$, where $f \in H^2(\mathbb{D})$, so M_f is defined by $M_f(g)(z) = f(z) \cdot g(z)$ for $g \in H^2(\mathbb{D})$. The investigation uses the concept of the *essential supremum* of the function f, which can be thought of as the supremum of $|f|$ but taken when ignoring f on sets of Lebesgue measure zero. Technically, it is defined (on the boundary function $f(e^{it})$) as

$$\text{ess sup}\, |f| = \inf\{a \in \mathbb{R} : m(\{t : |f(e^{it})| > a\}) = 0\}.$$

18. Let $M_f : H^2(\mathbb{D}) \to H^2(\mathbb{D})$ as described above. For $\varepsilon > 0$, say why the Lebesgue measure of the set $S = \{t : t \in [0, 2\pi) \text{ and } |f(t)| \geq (\text{ess sup } f) - \varepsilon\}$ is greater than zero.

19. For χ_S the characteristic function of S, show
$$\|M_f(\chi_S)(z)\| \geq m(S)(\text{ess sup } f - \varepsilon).$$
(Hint: $\|M_f(\chi_S)(z)\| = \int f(t)^2 \chi_S(t)^2 \, dt = \int_S f(t)^2 \, dt$.)

20. Use the previous exercise to prove the operator norm of M_f satisfies, for $\varepsilon > 0$,
$$\|M_f\| = \sup_{g \in H^2} \frac{\|M_f(g)\|}{\|g\|} \geq \text{ess sup } f - \varepsilon.$$

21. Use the previous exercise to conclude $\|M_f\| \geq \text{ess sup } f$.

22. Now prove $\|M_f\| \leq \text{ess sup } f$ to show $\|M_f\| = \text{ess sup } f$.

In Exercises 23–26, factor f into an inner-outer factorization $f(z) = b(z) \cdot g(z)$ for $z \in \mathbb{D}$.

23. $f(z) = z - 1/2$

24. $f(z) = (z - i/2)(z - 1/2)$

25. $f(z) = \frac{2z-i}{z-2}$

26. $f(z) = .4z^3 - .2z^2$

In Exercises 27–30, say why the subspace M is invariant for the indicated shift operator S.

27. $M = e^{it} H^2(\mathbb{T})$ for the shift operator $S(f)(e^{it}) = e^{it} f(e^{it})$, where $f \in H^2(\mathbb{T})$.

28. $M = \left\{ \begin{bmatrix} b(z)g \\ 0 \end{bmatrix} : g \in H^2(\mathbb{D}) \right\}$, where b is an inner function in $H^2(\mathbb{D})$, for the shift operator S defined in Exercise 8.

29. Let b be an inner function on $H^2(\mathbb{D})$. $M = b(z) H^2(\mathbb{D})$ for $S = M_{z^2}$ on $H^2(\mathbb{D})$, defined by $S(f)(z) = z^2 \cdot f(z)$.

30. Let b_1 and b_2 be inner functions on $H^2(\mathbb{D})$. $M = \left\{ \begin{bmatrix} b_1(z) f_1(z) \\ b_2(z) f_2(z) \end{bmatrix} : f_1, f_2 \in H^2(\mathbb{D}) \right\}$ on $\mathcal{H}$, where $f(z) = \begin{bmatrix} f_1(z) \\ f_2(z) \end{bmatrix}$ is an element of $\mathcal{H} = H^2(\mathbb{D}) \times H^2(\mathbb{D})$ and S is as defined in Exercise 6.

5.4 Application: A Spectral Theorem Example

This section describes the spectral theorem without proof, and presents an example of how it is applied to understand large categories of linear operators. The spectral theorem is sometimes described as the diagonalization of an operator because in a linear algebra course it is a diagonalization of a linear transformation, as a matrix, on a finite-dimensional vector space. This section focuses on self-adjoint operators.

An operator T on a Hilbert space $\mathcal{H}$ is *self-adjoint* when $\langle Tf, g \rangle = \langle f, Tg \rangle$ for $f, g \in \mathcal{H}$. Thus $T^* = T$, and so T is its own adjoint. We consider only self-adjoint operators of multiplicity one. An operator T on $\mathcal{H}$ has multiplicity one when it acts on a vector $x \in \mathcal{H}$ in a way that generates the whole space. That is, the closed span of $x, Tx, T^2x, T^3x, \ldots$ is the whole space $\mathcal{H}$. It turns out any self-adjoint operator T with multiplicity one produces a measure μ on $\mathbb{R}$ and

5.4 Application: A Spectral Theorem Example

an equivalence of Hilbert-space structures (using an isometry $U : \mathcal{H} \to L^2(\mu)$) such that T acts as multiplication by x on $f(x) \in L^2(\mu)$. The function f can be complex valued, even though μ acts on a real variable x. (A simple example is the complex valued function $f(x) = (3 + 2i)x$ for $-1 < x < 1$, when the measure $\mu(x)$ is supported on $(-1, 1)$.) The isometry establishes the two Hilbert spaces as isometrically isomorphic to one another, and then the operators T and multiplication by x are isometrically isomorphic. Specifically, the measure μ and isometry U relate to the self-adjoint operator T by

$$T[U^{-1}f(x)] = U^{-1}[xf(x)] \text{ for } f \in L^2(\mu).$$

From Section 5.1 and Definition 5.3.3, an isometry must preserve all inner products between the Hilbert spaces. So a concrete spectral theory for T occurs when the measure μ (the *spectral measure of T*) is determined along with the isometry U that provides the equivalence between operators. In the infinite-dimensional setting, multiplication by x acts much as multiplication by a diagonal matrix in finite dimensions. We state this as

Theorem 5.4.1 The Spectral Theorem: *If T is a self-adjoint bounded linear operator on a Hilbert space $\mathcal{H}$ with multiplicity one, then there exists a Borel measure μ on $\mathbb{R}$ and a Hilbert space isometry $U : \mathcal{H} \to L^2(\mu)$ such that $UTU^{-1}f(x) = xf(x)$ for $f \in L^2(\mu)$.*

In Section 5.1, we studied the operator T_w on ℓ^2, described in terms of a matrix $T = [a_{ij}]$ of Toeplitz form, so the matrix for T_w was constant along its diagonals. The elements $a_0, a_1, a_2, \ldots$ of the matrix show up in a complex-valued series expansion of $w(e^{it})$; the correspondence is $w(e^{it}) = \sum_{n=-\infty}^{\infty} a_n e^{int}$, and $a_{-n} = \overline{a_n}$ when T_w is self-adjoint.

When T acts on a finite-dimensional vector space, the spectral theorem is a statement that a self-adjoint T can be diagonalized—its Jordan form matrix is diagonal. In that case, the conjugate-transpose of T is T itself and the measure μ is a point mass measure that occurs at each eigenvalue for T. Actually, many more matrices can be diagonalized, not just self-adjoint ones. The next question explores such a situation.

Question 5.4.1 Let multiplication by $T = \begin{bmatrix} 21 & -12 \\ 30 & -17 \end{bmatrix}$ act on the space $\mathcal{H}$ of two-dimensional real vectors.

1. Diagonalize T, finding its eigenvalues and eigenvectors, and then determine P and D so $PTP^{-1} = D$ with D diagonal.
2. Use the results of part 1 and a definition of μ as a point mass measure on the eigenvalues to express multiplication by T (on the standard two-dimensional Hilbert space of vectors $\begin{bmatrix} x_1 \\ x_2 \end{bmatrix}$) as multiplication by x, where $UTU^{-1}f(x) = xf(x)$ for $f \in L^2(\mu)$. ∎

In this section, we describe a concrete spectral theory for the Toeplitz operator T_w when w is a continuous real-valued function on $\mathbb{R}$. The results, which can be generalized in many ways (for example, to certain discontinuous functions), are due to Marvin Rosenblum (see [101, pp. 63–73]). We present them, without proof, in the following theorem.

Theorem 5.4.2 *Suppose T_w is a Toeplitz operator on ℓ^2 as described in Section 5.1 and where $w(e^{it})$ is a (nonconstant) continuous real-valued function for e^{it} in the unit circle $\mathbb{T}$ that is bounded below and where $|w|$ is integrable over $\mathbb{T}$. For real x assume the set $E_x = \{t : w(e^{it}) \geq x\}$ is almost everywhere an arc of $\mathbb{T}$. Then T_w is self-adjoint, and so the spectral theorem applies.*[22] *In addition:*

1. *The spectral measure μ is absolutely continuous, so $\mu(x) = m(x)\,dx$ for some (almost everywhere) continuous function $m(x)$ on $\mathbb{R}$. Hence, a description of the spectral measure μ is determined from a description of its weight function $m(x)$.*
2. *The support of the spectral measure μ (the portion of the real line where $m(t) \neq 0$) is the spectrum of T and is denoted $sp(T)$. It is equal to the interval $[c, d]$, where $c = \min w(e^{it})$ and $d = \max w(e^{it})$.*
3. *Write the set $E_x = \{e^{it} : w(e^{it}) \geq x\} = \{e^{it} : a(x) \leq t \leq b(x)\}$ for some real functions $a(x)$ and $b(x)$. Then*

$$m(x) = \pi^{-1} \sin\left(\frac{b(x) - a(x)}{2}\right) \text{ on } [c, d].$$

As an example, we look at T_w when $w(e^{it}) = \cos t$, which was introduced briefly in Section 5.1. From Theorem 5.4.1, clearly $c = \min w(e^{it}) = -1$ and $d = \max w(e^{it}) = 1$, and so $sp(T_{\cos t}) = [-1, 1]$. What is the measure μ? To find it, we find $m(x)$, which is determined from the sets E_x. In this case,

$$E_x = \{e^{it} : w(e^{it}) \geq x\} = \{e^{it} : \cos t \geq x\} = \{e^{it} : -\arccos x \leq t \leq \arccos x\} \text{ for } t \in [-\pi, \pi].$$

And so, when $w(e^{it}) = \cos t$, in the above theorem we have $a(x) = -\arccos x$ and $b(x) = \arccos x$.

Using the result in part 3 of Theorem 5.4.2, when $w(e^{it}) = \cos t$, we have:

$$\mu(x) = m(x)\,dx = \pi^{-1} \sin\left(\frac{\arccos x - (-\arccos x)}{2}\right) dx = \pi^{-1} \sin(\arccos x)\,dx$$

$$= \pi^{-1}\sqrt{1 - x^2}\,dx.$$

So $T_{\cos t}$ acts as multiplication by x on the Hilbert space $L^2(-1, 1)$ equipped with the absolutely continuous measure $\pi^{-1}\sqrt{1 - x^2}\,dx$.

Here are some additional results for any such self-adjoint Toeplitz operators T_w (also from [101, pp. 63–73]). We start by factoring $|w(e^{it}) - x|$ as a function of two variables $|w(e^{it}) - x| = |g(e^{it}, x)|^2$, where the second variable $x \in \mathbb{R}$ and for which $\frac{1}{2\pi}\int_0^{2\pi} |\log|w(e^{it}) - x||\,dt < \infty$. Such a factorization exists for almost all real x.[23] In addition, for any such x and in terms of the first variable, the function $g(e^{it}, x)$ is the boundary function for an outer function $g(z, x)$ on the disk $\mathbb{D}$ (so $z \in \mathbb{D}$). At the origin we have $g(0, x) > 0$. A formula for g follows from Definition 5.3.2.

$$g(z, x) = c\, e^{\frac{1}{4\pi}\int_0^{2\pi} \frac{e^{it}+z}{e^{it}-z} \log|w(e^{it})-x|\,dt},$$

---

[22] The hypotheses guarantee T_w has multiplicity one, which is not true for every Toeplitz operator (see [100]). For spectral theory in the general case, see [101].

[23] See [101, p. 64–65] for the complete proof.

5.4 Application: A Spectral Theorem Example

with c a norm-one constant. Out of the outer function g comes the isometry U, which you recall will map ℓ^2 (on which T_w is acting) to $L^2(\mu)$ so T_w acts as multiplication by x. For T_w, U is determined by its action on the kernel function $\vec{k}_z$ for ℓ^2 (see Question 3.2.3), which can be formed for $z \in \mathbb{D}$. Here, $\vec{k}_z = \begin{bmatrix} 1 \\ \bar{z} \\ \bar{z}^2 \\ \vdots \end{bmatrix}$. We can calculate the isometry's action on these elements, obtaining[24] the output $U(\vec{k}_z)$ as a real-valued function in $L^2(\mu)$. Namely, for almost every $x \in \mathbb{R}$ and any $z \in \mathbb{D}$,

$$U(\vec{k}_z)(x) = [\overline{g(z,x)}\sqrt{1 - \bar{z}e^{ia(x)}}\sqrt{1 - \bar{z}e^{ib(x)}}]^{-1}.$$

We can also provide the inverse mapping, which isometrically (preserving inner products) takes $L^2(\mu)$ onto ℓ^2. The inverse mapping is defined for $f \in L^2(\rho)$, and it is given by

$$U^{-1}(f)(z) = \int_c^d f(x)\overline{U(\vec{k}_z)(x)}\, d\mu(x),$$

but put into a vector format (where the coefficients in the power series expansion of $U^{-1}(f)(z) = \sum_{n=0}^{\infty} a_n z^n$ are formed into the vector $\begin{bmatrix} a_0 \\ a_1 \\ a_2 \\ \vdots \end{bmatrix}$). We summarize the results in

> **Theorem 5.4.3** *Suppose T_w and $w(e^{it})$ are as in Theorem 5.4.2. Then the isometry U, defined on kernels $\vec{k}_z \in \ell^2$ and for almost all real x according to*
>
> $$U(\vec{k}_z)(x) = [\overline{g(z,x)}\sqrt{1 - \bar{z}e^{ia(x)}}\sqrt{1 - \bar{z}e^{ib(x)}}]^{-1},$$
>
> *sets up T_w as unitarily equivalent (via the Hilbert space isomorphism U) to multiplication by x on $L^2(\mu)$, where μ is constructed as in Theorem 5.4.2. That is, $UTU^{-1}f(x) = xf(x)$ for $f \in L^2(\mu)$. Here, $|w(e^{it}) - x| = |g(e^{it}, x)|^2$, where g is outer and $x \in \mathbb{R}$ satisfies $\frac{1}{2\pi}\int_0^{2\pi} |\log|w(e^{it}) - x||\, dt < \infty$. Furthermore, $U^{-1}(f)(z)$ is the vector in ℓ^2 formed from the Taylor series coefficients of $h(z) = \int_c^d f(x)\overline{U(\vec{k}_z)(x)}\, d\mu(x)$.*

Example 5.4.1 This example applies Theorem 5.4.3 and thus finishes the complete concrete spectral theory for the Toeplitz operator $T_{\cos t}$ (see [100]). When $w(e^{it}) = \cos t$, we first find the outer function g:

$$|w(e^{it}) - x| = |\cos t - x| = \frac{1}{2}|e^{it} + e^{-it} - 2x| = \frac{1}{2}|e^{it} + e^{-it} - 2x| \cdot |e^{it}|$$

$$= \frac{1}{2}|1 - 2xe^{it} + e^{2it}| = \left(\frac{1}{\sqrt{2}}|1 - 2xe^{it} + (e^{it})^2|^{1/2}\right)^2.$$

Then $g(e^{it}, x) = \frac{1}{\sqrt{2}}(1 - 2xe^{it} + (e^{it})^2)^{1/2}$, which says $g(z, x) = \frac{1}{\sqrt{2}}(1 - 2xz + z^2)^{1/2}$ is the outer function out of which U is constructed.

Since $\sqrt{1 - \bar{z}e^{-i\arccos(x)}}\sqrt{1 - \bar{z}e^{i\arccos(x)}} = \left(1 + \bar{z}^2 - \bar{z}[e^{-i\arccos(x)} + e^{i\arccos(x)}]\right)^{1/2}$, we get

$$U(\vec{k}_z)(x) = [\overline{g(z,x)}\sqrt{1 - \bar{z}e^{ia(x)}}\sqrt{1 - \bar{z}e^{ib(x)}}]^{-1} \tag{5.1}$$

$$= \sqrt{2}(1 - 2x\bar{z} + \bar{z}^2)^{-1/2}(1 + \bar{z}^2 - \bar{z}[e^{-i\arccos(x)} + e^{i\arccos(x)}])^{-1/2}. \tag{5.2}$$

---

[24] See [101, p. 71].

The result simplifies considerably. Note

$$e^{i\arccos x} + e^{-i\arccos x} = \cos(\arccos x) + i\sin(\arccos x) + \cos(-\arccos x)$$
$$+ i\sin(-\arccos x) = 2x.$$

Therefore

$$(1 + \bar{z}^2 - \bar{z}[e^{-i\arccos(x)} + e^{i\arccos(x)}])^{-1/2} = (1 - 2x\bar{z} + \bar{z}^2)^{-1/2}.$$

Substituting into (5.2), the remaining spectral theory details follow. The isometry $U : \ell^2 \to L^2(\mu)$ satisfies

$$U(\vec{k}_z)(x) = \sqrt{2}(1 - 2x\bar{z} + \bar{z}^2)^{-1}.$$

In addition, $U^{-1}(f)(z)$ is the vector in ℓ^2 formed from the series coefficients of

$$\frac{\sqrt{2}}{\pi} \int_{-1}^{1} f(x)(1 - 2xz + z^2)^{-1} \sqrt{1 - x^2}\, dx.$$

Now it turns out

$$(1 - 2xz + z^2)^{-1} = \sum_{n=0}^{\infty} z^n U_n(x),$$

where $\{U_n(x)\}_{n=0}^{\infty}$ are Chebyshev polynomials of the second kind defined recursively as $U_0(x) = 1$, $U_1(x) = 2x$, $U_{n+1}(x) = 2xU_n(x) - U_{n-1}(x)$.[25] Just as the Legendre polynomials form a basis for $L^2(-1, 1)$ (as discussed in Question 3.3.3), the Chebyshev polynomials form a basis for $L^2(\mu)$, where $\mu = \pi^{-1}\sqrt{1 - x^2}\, dx$ is the spectral measure for $T_{\cos x}$ on $(-1, 1)$. More specifically, $\{\sqrt{2}U_n(x)\}_{n=0}^{\infty}$ forms a complete orthonormal basis for $L^2(\mu)$. Therefore, if we write $f(x) \in L^2(\mu)$ in its series expansion with these basis elements as $f(x) = \sum_{n=0}^{\infty} a_n \sqrt{2} U_n(x)$, we get

$$\frac{\sqrt{2}}{\pi} \int_{-1}^{1} f(x)(1 - 2xz + z^2)^{-1} \sqrt{1 - x^2}\, dx = \langle f(x), \sqrt{2}(1 - 2xz + z^2)^{-1}\rangle_{L^2(\mu)}$$

$$= \left\langle \sum_{n=0}^{\infty} a_n \sqrt{2} U_n(x), \sum_{n=0}^{\infty} z^n \sqrt{2} U_n(x) \right\rangle_{L^2(\mu)}$$

$$= \sum_{n=0}^{\infty} a_n z^n.$$

Therefore, as an ℓ^2 vector, $U^{-1}(f)(z) = \begin{bmatrix} a_0 \\ a_1 \\ a_2 \\ \vdots \end{bmatrix}$. T_w behaves as multiplication by x on $L^2(\mu)$. ∎

Question 5.4.2 Show the spectral measure μ for the Toeplitz operator $T_{\sin t}$ is the same as for $T_{\cos t}$ as in Example 5.4.1. In other words, show $\mu = \frac{1}{\pi}\sqrt{1 - x^2}\, dx$ with $x \in [-1, 1]$. In addition, show why $U(\vec{k}_z)(x) = \sqrt{2}(1 - 2ix\bar{z} - \bar{z}^2)^{-1}$. (See [99, p. 595].) ∎

---

[25] The polynomials are named after the 19th-century Russian mathematician Pafnuty Chebyshev, who determined that they are solutions to the differential equations $(1 - x^2)\, y'' - 3x\, y' + n(n + 2)\, y = 0$. They can also be defined by the trigonometric formula $U_n(\cos \theta) = \frac{\sin((n+1)\theta)}{\sin \theta}$.

5.4 Application: A Spectral Theorem Example

The operator $T_{\cos t}$ that Example 5.4.1 considered is historically quite well known. As part of his 1908 Ph.D. dissertation [122], the famous mathematician Hermann Weyl diagonalized, by determining the complete spectral theory solution, an operator sum composed of $T_{\cos t}$ and a rank-one operator, namely (for $\vec{v} \in \ell^2$) the operator $B\vec{v} = T_{\cos t}\vec{v} + \langle \vec{v}, \begin{bmatrix} c \\ 0 \\ 0 \\ \vdots \end{bmatrix} \rangle \begin{bmatrix} c \\ 0 \\ 0 \\ \vdots \end{bmatrix}$, where c is a nonnegative constant. For some values of c the spectral theorem's measure for B remains absolutely continuous, but only up to a point. When $c > 1/2$, the spectral measure for Weyl's operator B is (see [63]) a sum of an absolutely continuous measure plus a point mass measure at $p = (4c^2 + 1)/4c$. In fact, absolutely continuous (typically considered the simplest format) spectral measures can change quite a bit, even for simple rank-one perturbations—the case where a rank-one operator is added to the original operator. Simple perturbations can drastically change the structure of a spectral measure to whatever complexity a spectral measure can possibly have, even starting with absolutely continuous spectral measures. The drastic changes can be illustrated with the Toeplitz operators whose concrete spectral theory we have just determined. For example, given a singular continuous measure μ, you can find an element $\vec{u}$ of ℓ^2 so (for $\vec{x} \in \ell^2$) the operator $B\vec{x} = T_w \vec{x} + \langle \vec{x}, \vec{u} \rangle \vec{u}$ produces μ as its spectral measure (see [63, pp. 264–266]). In short, you change T_w's absolutely continuous spectral measure to B's singular continuous measure in just one rank-one step.

This chapter has considered just a few of the many issues involved with the theory of bounded linear operators. It has restricted its attention only to operators on Hilbert spaces. The rich and more general setting of Banach spaces expands the ideas considerably. Also, an infinite-dimensional Hilbert space has many unsolved questions about specific types of operators and the general theory of operators. We have only touched on the connections between the properties of complex functions and the study of classes of operators (such as the connections between inner-outer factorizations of functions in H^2 and the study of the invariant subspaces of the shift operator). A full course in function theory or operator theory, typically offered at the graduate level, will provide a more complete development of the ideas touched on in this chapter, as well as a host of interesting applications. We hope this section and this book have provided you with the motivation and resources to carry on eagerly into these advanced studies.

Solutions to Questions

5.4.1 (1) An eigenvalue e for T satisfies $T\vec{v} = e\vec{v}$ for some eigenvector $\vec{v}$. Hence $(T - eI)\vec{v} = 0$, where I is the identity matrix (and so $T - eI = \begin{bmatrix} 21-e & -12 \\ 30 & -17-e \end{bmatrix}$). A standard result from linear algebra says this happens only when the determinant of $T - eI$ equals zero; in other words, when $(21 - e)(-17 - e) - 30(-12) = 0$. Solving, the eigenvalues are 1 and 3. Using $e = 1$ to find its eigenvector $\vec{v} = \begin{bmatrix} x_1 \\ x_2 \end{bmatrix}$, we must have $\begin{bmatrix} 21 & -12 \\ 30 & -17 \end{bmatrix}\begin{bmatrix} x_1 \\ x_2 \end{bmatrix} = T\vec{v} = e\vec{v} = \begin{bmatrix} e \cdot x_1 \\ e \cdot x_2 \end{bmatrix} = \begin{bmatrix} x_1 \\ x_2 \end{bmatrix}$, and so $21x_1 - 12x_2 = x_1$ and $30x_1 - 17x_2 = x_2$. We get $x_1 = 3x_2/5$, and thus may, for example, choose $\vec{v} = \begin{bmatrix} 3 \\ 5 \end{bmatrix}$. Similarly, we find $\begin{bmatrix} 2 \\ 3 \end{bmatrix}$ is an eigenvector for the eigenvalue 3. As explained in Example 5.2.2, the matrix P must be constructed with eigenvector columns; hence $P = \begin{bmatrix} 3 & 2 \\ 5 & 3 \end{bmatrix}$. Inverting (e.g., using a calculator such as the TI-83) $P^{-1} = \begin{bmatrix} -3 & 2 \\ 5 & -3 \end{bmatrix}$, and we see $P^{-1}TP = \begin{bmatrix} -3 & 2 \\ 5 & -3 \end{bmatrix}\begin{bmatrix} 21 & -12 \\ 30 & -17 \end{bmatrix}\begin{bmatrix} 3 & 2 \\ 5 & 3 \end{bmatrix} = \begin{bmatrix} 1 & 0 \\ 0 & 3 \end{bmatrix} = D$. (2) The point mass μ is defined as the sum of point mass measures on the eigenvalue points 1 and 3 and with point masses 1 and 3, respectively, for the (real) vector space $L^2(\mu)$. As described in Question 4.3.1, we may understand $L^2(\mu)$ as two-dimensional, whose elements are (real) functions with domains including $x = 1$ and $x = 3$. An orthogonal basis can be chosen as $e_1(x) = \mathcal{X}_{\{1\}}(x)$ and $e_2(x) = \mathcal{X}_{\{3\}}(x)$. Using the standard basis elements $\vec{v}_1 = \begin{bmatrix} 1 \\ 0 \end{bmatrix}$ and $\vec{v}_2 = \begin{bmatrix} 0 \\ 1 \end{bmatrix}$ for $\mathcal{H}$, we define the Hilbert space isomorphism $W : L^2(\mu) \to \mathcal{H}$ as $W(e_1(x)) = \vec{v}_1$ and $W(e_2(x)) = \vec{v}_2$. Under this isometric isomorphism (and since W is linear), any $L^2(\mu)$ function $f(x) = c_1 e_1(x) + c_2 e_2(x)$ corresponds to the vector $\vec{w} = W(f) = c_1 \vec{v}_1 + c_2 \vec{v}_2$. Then the function $s_1(x) = W^{-1} P^{-1} T P W e_1(x)$ has $s_1(x) = W^{-1} D W e_1(x) = W^{-1} D(\vec{v}_1) = W^{-1} D \begin{bmatrix} 1 \\ 0 \end{bmatrix} = W^{-1} \begin{bmatrix} 1 \\ 0 \end{bmatrix} = 1 e_1(x) + 0 e_2(x) = \mathcal{X}_{\{1\}}(x)$. We find $s_1(1) = 1$ and $s_1(3) = 0$. Comparing with $x e_1(x) = x \mathcal{X}_{\{1\}}(x)$, we have $x e_1(x)|_{x=1} = 1$ and $x e_1(x)|_{x=3} = 0$. Hence they are equal μ-almost everywhere. In the same way, we find $s_2(x) = W^{-1} P^{-1} T P W e_2(x) = W^{-1} D(\vec{v}_2) = 0 e_1(x) + 3 e_2(x) = 3 \mathcal{X}_{\{3\}}(x)$, and so $s_2(1) = 0$ and $s_2(3) = 3$, which shows $s_2(x) = x e_2(x)$ μ-almost everywhere. Because the operator $W^{-1} P^{-1} T P W$ agrees with multiplication by x on the basis elements, it must be the case that $U^{-1} T U f(x) = x f(x)$ for any $f \in L^2(\mu)$, where $U = PW$.

5.4.2 Apply Theorem 5.4.2 with $w(e^{it}) = \sin t$. Then $sp(T) = [\min\{\sin t\}, \max\{\sin t\}] = [-1, 1]$. Also, $E_x = \{e^{it} : \sin t \geq x\} = \{e^{it} : \arcsin x \leq t \leq \pi - \arcsin x\}$ for $t \in [-\pi/2, 3\pi/2]$. (The graph of sine and the fact that $\arcsin x$ has range values restricted to $[-\pi/2, \pi/2]$ illustrate well the need to express E_x this way.) Hence $\mu(x) = m(x)\,dx = \pi^{-1} \sin(\frac{\pi - \arcsin x - (\arcsin x)}{2})\,dx = \pi^{-1} \sin(\pi/2 - \arcsin x)\,dx = \pi^{-1} \sin(\arccos x)\,dx = \pi^{-1} \sqrt{1 - x^2}$ for $x \in [-1, 1]$. $|w(e^{it}) - x| = |\sin t - x| = |\frac{1}{2i}(e^{it} - e^{-it}) - x| = \frac{1}{2}|e^{it} - e^{-it} - 2ix| \cdot |-e^{it}| = \frac{1}{2}|1 + 2ix e^{it} - e^{2it}| = (\frac{1}{\sqrt{2}}|1 + 2ix e^{it} - (e^{it})^2|^{1/2})^2$, so $g(e^{it}, x) = \frac{1}{\sqrt{2}}(1 + 2ix e^{it} - (e^{it})^2)^{1/2}$. Hence $g(z, x) = \frac{1}{\sqrt{2}}(1 + 2ixz - z^2)^{1/2}$. From Theorem 5.4.3, $U(\vec{k}_z)(x) = \sqrt{2}(1 - 2ix\bar{z} - \bar{z}^2)^{-1/2}(1 + \bar{z}^2 e^{i\pi} - \bar{z}[e^{i \arcsin(x)} + e^{i(\pi - \arcsin(x))}])^{-1/2}$. Since $e^{i\pi} = -1$ and $e^{i \arcsin x} + e^{i(\pi - \arcsin x)} = \cos(\arcsin x) + i\sin(\arcsin x) - \cos(\arcsin x) + i\sin(\arcsin x) = 2ix$, we get $U(\vec{k}_z)(x) = (1 - 2ix\bar{z} - \bar{z}^2)^{-1}$.

Reading Questions for Section 5.4

1. What does the spectral theorem say about a self-adjoint bounded linear operator on a given Hilbert space $\mathcal{H}$?

2. What type of measure is the spectral measure μ of a self-adjoint Toeplitz operator T_w, where $w(e^{it})$ is a continuous real-valued function for e^{it} in the unit circle?

3. What is the support of the spectral measure μ for a self-adjoint Toeplitz operator T_w, where $w(e^{it})$ is a continuous real-valued function for e^{it} in the unit circle?

4. What is a formula for the spectral measure's weight function $m(x)$, corresponding to a self-adjoint Toeplitz operator T_w, where $w(e^{it})$ is a continuous real-valued function for e^{it} in the unit circle?

5. For $T_{\cos t}$, what is the spectral measure μ?

Exercises for Section 5.4

Exercises 1–9 discuss the adjoint of an operator T.

1. What is the adjoint T^* of the rank-one operator $T : \mathbb{R}^2 \to \mathbb{R}^2$ defined, for $\vec{w} = \begin{bmatrix} a \\ b \end{bmatrix} \in \mathbb{R}^2$, by $T_{\vec{w}}(\vec{v}) = \langle \vec{v}, \vec{w} \rangle \vec{w}$? (*Hint*: examine, for $\vec{u}, \vec{v} \in \mathbb{R}^2$, the equation $\langle T_{\vec{w}}(\vec{v}), \vec{u} \rangle = \langle \vec{v}, T^*(\vec{u}) \rangle$).

2. Show the operator T on $\mathbb{R}^2$ is self-adjoint when $T(\begin{bmatrix} a \\ b \end{bmatrix}) = \begin{bmatrix} c & d \\ d & e \end{bmatrix} \cdot \begin{bmatrix} a \\ b \end{bmatrix}$ is matrix multiplication using the symmetric matrix $\begin{bmatrix} c & d \\ d & e \end{bmatrix}$, where $c, d, e \in \mathbb{R}$.

5.4 Application: A Spectral Theorem Example

3. Let T be the operator on $\mathbb{R}^2$ defined by $T([{}^{v_1}_{v_2}]) = [{}^a_c\,{}^b_d] \cdot [{}^{v_1}_{v_2}]$ where $a, b, c, d \in \mathbb{R}$. When is $T^* = T$? When is $T^* = T^{-1}$?

4. Find the adjoint of the operator T on $\mathbb{R}^3$ when $T([{}^a_b_c]) = \begin{bmatrix} d & e & f \\ g & h & i \\ j & k & l \end{bmatrix} \cdot [{}^a_b_c]$ is matrix multiplication by a matrix with real entries.

5. When is the operator in the previous problem self-adjoint?

6. What is the adjoint T^* of the operator $T : H^2(\mathbb{D}) \to H^2(\mathbb{D})$ defined as multiplication by $f \in H^2(\mathbb{D})$? In other words, find T^* when $T(g)(z) = f(z) \cdot g(z)$ for $g \in H^2(\mathbb{D})$.

7. Define the reproducing kernel function $k_w(z) \in H^2(\mathbb{D})$ by $k_z(w) = (1 - \bar{z}w)^{-1}$, where $w, z \in \mathbb{D}$. It has the property (see [32, p. 17]) $\langle f(w), k_z(w) \rangle_{H^2(\mathbb{D})} = f(z)$. Use this to show, for an operator $C : H^2(\mathbb{D}) \to H^2(\mathbb{D})$, $C^*(f)(z) = \langle f(w), C(k_z)(w) \rangle$.

8. For $\varphi(z) \in H^2(\mathbb{D})$, define the composition operator $C_\varphi : H^2(\mathbb{D}) \to H^2(\mathbb{D})$ by $C_\varphi(f)(z) = f(\varphi(z))$. Use the result of the previous exercise to prove
$$C_\varphi^*(f)(z) = \frac{1}{2\pi} \int_0^{2\pi} \frac{f(e^{it})}{1 - \overline{\varphi(e^{it})}z}\, dt.$$
(This formula was first proved by Carl Cowen and Eva Gallardo-Guttiérez. See [33].)

9. Find $C_\varphi^*(k_w)(z)$, where $k_w(z)$ is the reproducing kernel defined in Exercise 7, when $C_\varphi(f)(z) = f(\varphi(z))$ for $f \in H^2(\mathbb{D})$ and φ is the fractional linear transformation $\varphi(z) = (az + b)(cz + d)^{-1}$, where $ad - bc = 1$. Compare your result to $T_g C_\psi T_h^*(k_w)(z)$, where T_g is multiplication by the function $g(z) = (-\bar{b}z + \bar{d})^{-1}$, $\psi(z) = (\bar{a}z - \bar{c})(-\bar{b}z + \bar{d})^{-1}$, and T_h^* is multiplication by $\overline{h(z)} = cz + d$. Conclude $C_\varphi^* = T_g C_\psi T_h^*$. (This result is also due to Cowen; see [31] and [32, p. 322].)

In Exercises 10–15, use the definition of adjoint in terms of the inner product to prove the facts about adjoints of bounded linear operators A and B on a Hilbert space.

10. $(A + B)^* = A^* + B^*$.

11. $(AB)^* = B^* A^*$.

12. $A^{**} = A$.

13. $\|A^*A\| = \|A\|^2$.

14. $\|A^*\| = \|A\|$.

15. A^*A is the identity I when A is an isometry.

Exercises 16–19 refer to the result in Question 5.4.1 and continue the study of the Toeplitz operator $T_{\sin t}$.

16. For a function $f \in L^2(\mu)$, write an integral expression for the function $\sum_{n=0}^{\infty} a_n z^n$ having Taylor series coefficients equal to the vector components of the ℓ^2 element $U^{-1}(f)(z)$.

17. Based on your solution to the previous exercise, write $U^{-1}(f)(z)$ as an element in ℓ^2, when $f = \sum_{n=0}^{\infty} a_n \sqrt{2} U_n(x)$ is an element in $L^2(\mu)$.

18. Determine the entries in the Toeplitz matrix (as described after Example 5.1.3 and as the operator is formulated to act on ℓ^2) for the operator $T_{\sin t}$.

19. Determine the entries in the Toeplitz matrix for the operator $T_{\sin t + \cos t}$.

Exercises 20–22 refer to the Toeplitz operator T_w when w is a characteristic function.

20. Examine T_w when $w(e^{it}) = \begin{cases} 1 & \text{if } -\pi/2 < x < \pi/2 \\ 0 & \text{otherwise.} \end{cases}$ Show the spectral measure for T_w is $\frac{1}{\pi} dx$ supported on the interval $[0, 1]$.

21. Let T_w be the Toeplitz operator in the previous exercise. For $f \in L^2(\mu)$, write an integral expression for the function $\sum_{n=0}^{\infty} a_n z^n$ having Taylor series coefficients equal to the vector components of the ℓ^2 element $U^{-1}(f)(z)$.

22. Examine T_w when $w(e^{it}) = \begin{cases} 1 & \text{if } -k < x < k \\ 0 & \text{otherwise,} \end{cases}$ where k is a given constant in $(0, \pi)$. Show the spectral measure for T_w is $\mu = \frac{1}{\pi} \sin k \, dx$ supported on the interval $[0, 1]$. (See [100].)

Notes for Chapter 5

As we learned in Section 5.2, a subspace A is said to be invariant under an operator T when $T(A) \subseteq A$. One of the most famous problems in all of mathematics is about invariant subspaces: does every bounded linear operator T on a (separable) Hilbert space have a (nontrivial) invariant subspace? The answer has long been known to be a resounding "yes" when, for example, the Hilbert space is of finite dimension; namely, invariant subspaces exist for any linear transformation (of, say, a complex-valued vector space) of finite dimension into itself. The result, often studied in a linear algebra course, was first proved by Carl Friedrich Gauss (he used a factorization of polynomials to get the result). The general problem is known throughout the mathematical community as "the invariant subspace problem."

Other partial versions of the invariant subspace problem have been proven. In the early 1900s, David Hilbert found any isometry T (where $\|Tx\| = \|x\|$ for every x in the Hilbert space) has a nontrivial invariant subspace. In the 1930s, John von Neumann showed the result for every compact operator. Von Neumann did not publish his proof, and it was independently discovered by Nachman Aronszajn (an accomplished Polish-American mathematician) much later, in 1954 [4]. A normal operator T is one for which $T^*T = TT^*$ (where T^* is the adjoint operator), and a subnormal operator is defined as the restriction of a normal operator (which always has a nontrivial invariant subspace) to one of its invariant subspaces. In 1978, the American operator theorist Scott Brown proved every subnormal operator has an invariant subspace [15].

Operators on Hilbert spaces of analytic functions having a complex-valued domain have also been studied. The Hardy space H^2 for the complex plane's unit disk is the Hilbert space of functions represented in the disk by power series, where the squares of the moduli of the coefficients form a finite sum (this property is known as "square-summable" [16].) The Hilbert space H^2 is easily and naturally thought of as isometrically isomorphic to ℓ^2. As this chapter has described, Arne Beurling [11] constructed invariant subspaces for large classes of operators

Figure 5.2. David Hilbert, John von Neumann, and Arne Beurling.

on H^2 by linking them to functions—analytic and bounded by one in the unit disk—where the invariant subspace is formed via a factorization of the function. Other Hilbert spaces of analytic complex functions exist. The space $A^2(\mathbb{D})$ uses a norm in terms of the area integral. That is, functions $f \in A^2(\mathbb{D})$ satisfy $(\int_{\mathbb{D}} |f(x+iy)|^2 \, dx \, dy)^{1/2} < \infty$.

Many other results on invariant subspaces of bounded linear operators are known (see [94] for descriptions and details of advances). Yet the proof of the general theorem that all bounded linear operators on a Hilbert space have a nontrivial invariant subspace eludes the mathematical community.

Otto Toeplitz earned his Ph.D. in 1905 and subsequently began working on Toeplitz operators at the University of Göttingen. He had arrived in Göttingen in 1909, when David Hilbert was producing vast amounts of work on integral equations and integral operators. Toeplitz understood operators on infinite-dimensional Hilbert spaces as generalizations of matrix multiplication on finite-dimensional vector spaces, and he used the insight to produce significant research results in this early stage of his distinguished career. By 1928, Toeplitz was awarded the notable chair position at the University of Bonn. He also wrote on the history of mathematics, and published a popular book titled *The Enjoyment of Mathematics*. In 1935, he became

Figure 5.3. A young and fit Otto Toeplitz (left) around 1920 with the Russian mathematician Alexander Ostrowski.

yet another victim of the Nazi rise to power. Hitler's elimination of anyone of Jewish origin from any Nazi-controlled academic post led that year to Toeplitz's dismissal from Bonn. He fled to Jerusalem in 1939, where he was appointed to the Hebrew University but died from tuberculosis less than a year later. His book on the history of calculus, *The Calculus: a Genetic Approach* was published posthumously in 1949 and translated to English in 1963.

This chapter has discussed operators on Hilbert spaces, and it is important to note operators on Banach spaces are just as important and interesting, and the amount of material in that broader framework is astounding. In addition to the invariant subspace problem, many other basic, unsolved (open) questions can be formulated and described. This text has only scratched the surface in its presentation of operator theory. In addition, the study of operators on Hilbert spaces is much richer than presented here. Finally, we remark this chapter's presentation is limited to theoretical results. Operator theory is an important aspect of functional analysis, and it certainly forms a basis for a modern, general approach to a theoretical study of functions. It also shows up in applications. For example, it models many phenomena that deal with probability, and those that appear in modern engineering fields such as image processing, signal processing, and information and systems theory. We have seen how Hilbert spaces form structures to study theoretical physics and mechanics, and operator theory plays a crucial role in those applications as well (see [95]). Hopefully this book has presented a jumping off point for further investigations in either theoretical or applied operator theory and functional analysis.

Solutions to Selected Problems

These explanations are not meant to represent properly a full, detailed presentation of any solution, but only to give a final answer to a calculation or to provide a key idea (a crucial element) in the solution.

Section 1.1

1. False; e.g., the empty set.
3. True; S can be written as a sequence by deleting the elements of $T = \{t_1, t_2, t_3, \ldots\}$ not in S.
5. False; any such union is uncountable.
7. $|S| = 4$. For example, $s_1 =$ Bob, $s_2 =$ Chris, $s_3 =$ Jamie, and $s_4 =$ Rob.
9. $2^A = \{\Phi, \{a\}, \{b\}, \{c\}, \{a,b\}, \{a,c\}, \{b,c\}, \{a,b,c\}\}$. Hence $|2^A| = 8$.
11. Countable, since $P = \{2, 3, 5, 7, 11, 13, \ldots\}$.
13. Countable: if $a \in \mathbb{Q}$, then $S \subseteq \{a, a+1/2, a+1/3, a+1+1/2, a+2+1/2, a+1+1/3, a+2/3, a+1/4, a+1+2/3, a+2+1/3, a+3+1/2, a+4+1/2, \ldots\}$; in fact S is this sequence with any value larger than b removed. A similar argument works if $a \notin \mathbb{Q}$.
15. Countable: write $\mathbb{Q} \times \mathbb{Q} = \bigcup_{q \in \mathbb{Q}} \{(q, r) : r \in \mathbb{Q}\}$ and apply the countable union theorem.
17. Countable: write $\mathbb{Q} \times \mathbb{Q} \times \mathbb{Q} = \bigcup_{q \in \mathbb{Q}} (\bigcup_{r \in \mathbb{Q}} \{(q, r, s) : s \in \mathbb{Q}\})$ and apply the countable union theorem.
19. Countable: write $S = \bigcup_{q_0 \in \mathbb{Q}} \bigcup_{q_1 \in \mathbb{Q}} \cdots \bigcup_{q_{n-1} \in \mathbb{Q}} \{q_0 + q_1 x + \cdots + q_{n-1} x^{n-1} + q_n x^n : q_n \in \mathbb{Q}\}$ and apply the countable union theorem.
21. Uncountable.
23. $|A| = 2$.
25. Countable; the number of such polynomials is countable, and each equation has one solution.
27. Countable; the number of such polynomials is countable.
29. The answer depends on the sequence constructed.
30. Given sets $S_1, S_2, \ldots, S_n$, if $S_1 \cup S_2 \cup \ldots \cup S_n$ is uncountable, then at least one set S_j (with $j = 1, 2, \ldots, n$) is uncountable.
31. The elements of S are natural numbers. By definition, $S \subset 2^{\mathbb{N}}$.
33. It cannot be the case that S is in $2^{\mathbb{N}}$ and also not equal to some S_n (as the assumption would then make S not a subset of $2^{\mathbb{N}}$).

35. $|A| = 4$.
37. A proof of $|C| = \aleph_1$ is similar to Exercises 31–34's proof of $|2^{\mathbb{N}}| = \aleph_1$.

Section 1.2

1. 2
3. 24.4
5. $\sqrt{3} + \sqrt{5}$
7. 0
9. Given $\varepsilon > 0$, the intervals $(-3 - \varepsilon/6, -3 + \varepsilon/6)$, $(7 - \varepsilon/6, 7 + \varepsilon/6)$, and $(31 - \varepsilon/6, 31 + \varepsilon/6)$ form an open cover with bounded total measure equal to ε.
11. Similar to the solution to Exercise 9, but use $\varepsilon/52$ as the intervals' radii.
13. Similar to the solution to Exercise 9, but use $\varepsilon/2^{n+1}$ as the nth interval's radius (centered at the value $2n$).
15. Write $(0, 2] = \{q_1, q_2, q_3, \ldots\}$ and form an open cover with the nth interval centered at q_n and with interval radius $\varepsilon/2^{n+1}$.
17. $\{0\}$, which has measure zero since it is finite.
19. $\{-2\}$, which has measure zero since it is finite.
21. See Question 1.2.3, part 2. Each ternary digit is either 0 or 2.
23. For example, $a = .2[3] = 2/3$, $b = .222[3] = 2/3 + 2/9 + 2/27 = 26/27$, $c = .02020202\ldots[3] = .25$, and $d = .202020\ldots[3] = 16/27$.
25. $65/81$
27. $\sum_{n=1}^{\infty} 2^{n-1}(1/4)^n = 1/2$
29. SVC does not contain any intervals.
31. $(1 - 3r)/(1 - 2r)$
33. (a) 32 (b) 32 (c) 32
35. For example, for an integer n, form the rectangular interval $I_n = \{(x, y) : n < x < n + 2, y_0 - \varepsilon/2^{|n|+3} < y < y_0 + \varepsilon/2^{|n|+3}\}$ and realize that $m(I_n) = \varepsilon/2^{|n|+1}$.

Section 1.3

7. $\int h = -2\pi + 2\sqrt{2} - 2 \approx -5.455$ and $\int t = 135$.
9. Using Definition 1.3.3, $\int h + s = -2\pi + 2\sqrt{2} - 2 - 10(5 - \sqrt{2}) - 30 - 25 - 30 - 10$
11. Using Definition 1.3.3, $\int 3f = 12 \cdot 2 - 18 + 6 - 9 + 3 = 6$
13. $-2\pi + 2\sqrt{2} + 134.52$
15. We may express f in the format of Definition 1.3.2, e.g., as $f(x) = 0 \cdot \mathcal{X}_{[1,2]}$
17. The two sides agree for $x \in \mathbb{R}$. For example, if $x \in (b, c)$, then $x \in (a, c)$ and $x \in (b, c)$ but $x \notin (a, b]$. Hence both sides equal $k_1 + k_2$ for $x \in (b, c)$. The other cases for x are similar.
19. If $a \leq b$, then $a \cdot m(I) \leq b \cdot m(I)$.
21. $g = -2 \cdot \mathcal{X}_{[-1,1)} + 3 \cdot \mathcal{X}_{[1,2)} + 5 \cdot \mathcal{X}_{[2,3)}$
23. 6
25. $\pi(24 - 6\sqrt{2})$
27. $18\pi + 8e - 128\pi^2 + 100\sqrt{10}$
29. If $f = \sum_{j=1}^{m} c_j \cdot \mathcal{X}_{I_j}$ for disjoint I_j, then $|f| = \sum_{j=1}^{m} |c_j| \cdot \mathcal{X}_{I_j}$. Hence $\int |f| = \sum_{j=1}^{m} |c_j| \cdot m(I_j)|$, which (by repeated applications of the triangle inequality) is at least as large as $|\int f| = |(\sum_{j=1}^{m} c_j \cdot m(I_j))|$.

31. Assume $f = \sum_{j=1}^{m} c_j \cdot \mathcal{X}_{I_j}$ with each I_j disjoint. Then $f^+ = \sum_{k=1}^{M} c_k \cdot \mathcal{X}_{I_k}$, where each c_k in this sum appears as any one of the M positive values $c_j > 0$ in the representation for f. (The negative values are left out.) In this way, we have a step-function representation for f^+.
33. Take the difference of the two step-function representations obtained in the solutions to Exercises 31 and 32, showing you obtain the original step-function representation of f.
35. By Exercise 28, $|f|$ and $|g|$ are step functions. The sum of two step functions is always a step function. A similar argument shows $|f + g|$ is a step function.
37. Use the hint.

Section 1.4

1. Given $\varepsilon > 0$, set $N = 11/(3\varepsilon)$.
3. Given $\varepsilon > 0$, (assuming $\varepsilon < 5$) set $N = 5/\varepsilon - 1$.
5. Given $\varepsilon > 0$, set $N = 2/\varepsilon$.
7. Given $\varepsilon > 0$, (assuming $\varepsilon < 1$) set $N = \ln(\varepsilon)/\ln(5/9)$.
9. Given $\varepsilon > 0$, set $\delta = \varepsilon/2$.
11. Given $\varepsilon > 0$, set $\delta = \varepsilon/5$.
13. Given $\varepsilon > 0$, set $\delta = \min\{1, \varepsilon/19\}$.
15. Given $\varepsilon > 0$, set $\delta = \min\{1, 2\varepsilon\}$.
17. Use Definition 1.4.3 with $\delta = \varepsilon/8$.
19. As $x \to x_0$ for $x_0 \in \mathbb{R}$, use Definition 1.4.3 with $\delta = \varepsilon/8$.
21. Given $\varepsilon > 0$, set $\delta = \min\{1, \varepsilon/5\}$.
23. Given $\varepsilon > 0$, set $\delta = \min\{1, 2\varepsilon\}$.
25. $x_0 = 0$ is not in the domain of f.
27. Evaluating each function f_n at $x = 2$ produces the sequence of function outputs $27/2$, $27/2$, $45/8$, $45/8$, $378/64$, $378/64$, $765/128$, ..., which converges to 6. Hence $f(2) = 6$.
29. Evaluating g_n at $x = 3/2$ produces the sequence of function outputs 0, 1, $9/4$, $9/4$, $9/4$, $9/4$, ..., which converges to $9/4$. Hence $f(3/2) = 9/4$.
31. Use the hint.

Section 1.5

1. Define ϕ_n, where $n = 1, 2, \ldots$, on 2^n equal-sized subintervals of $[1, 2]$ as $\phi_n(x) = (k-1)/2^n + 1$ for x in the kth subinterval, where $k = 1, 2, \ldots, 2^n$. Then $\int f = \lim_{n \to \infty} \int \phi_n = \lim_{n \to \infty} \sum_{k=1}^{2^n} [\frac{k-1}{2^n} + 1] \cdot m(I_n) = 3/2$.
3. Define ϕ_n, where $n = 1, 2, \ldots$, on 2^n equal-sized subintervals of $[1, 2]$ as $\phi_n(x) = (k-1)/2^n$ for x in the kth subinterval, where $k = 1, 2, \ldots, 2^n$. Then $\int h = \lim_{n \to \infty} \phi_n = \lim_{n \to \infty} \int \sum_{k=1}^{2^n} [\frac{k-1}{2^n}] \cdot (1/2^n) = 1/2$.
5. Define ϕ_n, where $n = 1, 2, \ldots$, on 2^n equal-sized subintervals of $[0, 1]$ as $\phi_n(x) = (k-1)/2^n + 2$ for x in the kth subinterval, where $k = 1, 2, \ldots, 2^n$. Then $\int_0^1 x + 2 \, dx = \lim_{n \to \infty} \int \phi_n = \lim_{n \to \infty} \sum_{k=1}^{2^n} [\frac{k-1}{2^n} + 2] \cdot (1/2^n) = 5/2$.
7. Define ϕ_n, where $n = 1, 2, \ldots$, on 2^n equal-sized subintervals of $[-1, 5]$ as $\phi_n(x) = 3 \cdot [6(k-1)/2^n - 1]$ for x in the kth subinterval, where $k = 1, 2, \ldots, 2^n$. Then $\int_{-1}^{5} 3x \, dx = \lim_{n \to \infty} \int \phi_n = \lim_{n \to \infty} \sum_{k=1}^{2^n} 3[\frac{6(k-1)}{2^n} - 1] \cdot (6/2^n) = 36$.

9. Define ϕ_n, where $n = 1, 2, \ldots$, on 2^n equal-sized subintervals of $[0, 2]$ as $\phi_n(x) = (2(k - 1)/2^n)^2$ for x in the kth subinterval, where $k = 1, 2, \ldots, 2^n$. Then $\int_0^2 x^2 = \lim_{n\to\infty} \int \phi_n = \lim_{n\to\infty} \sum_{k=1}^{2^n}(2(k-1)/2^n)^2 \cdot (2/2^n) = \lim_{n\to\infty} \frac{(2^n-1)2^n(2\cdot 2^n+1)}{6\cdot 2^{3n-3}} = 8/3$.

11. Define ϕ_n, where $n = 1, 2, \ldots$, on 2^n equal-sized subintervals of $[1, 4]$ as $\phi_n(x) = [3(k - 1)/2^n + 1]^2 + 1$ for x in the kth subinterval, where $k = 1, 2, \ldots, 2^n$. Then $\int_1^4 x^2 + 1 \, dx = \lim_{n\to\infty} \int \phi_n = \lim_{n\to\infty} \sum_{k=1}^{2^n}[(\frac{3(k-1)}{2^n} + 1)^2 + 1] \cdot (3/2^n) = 24$.

13. Define ϕ_n, where $n = 1, 2, \ldots$, on 2^n equal-sized subintervals of $[0, x]$ as $\phi_n(t) = 2x(k-1)/2^n + 1$ for t in the kth subinterval, where $k = 1, 2, \ldots, 2^n$. Then $\int_0^x 2t + 1 \, dt = \lim_{n\to\infty} \int \phi_n = \lim_{n\to\infty} \sum_{k=1}^{2^n}[2x(k-1)/2^n + 1] \cdot (x/2^n) = \lim_{n\to\infty} \left(\frac{2x^2}{2^{2n}}\right)\left(\frac{2^n(2^n-1)}{2}\right) + x = x^2 + x$.

15. Define ϕ_n, where $n = 1, 2, \ldots$, on 2^n equal-sized subintervals of $[0, 1]$ as $\phi_n(x) = [2(k-1)/2^n]^3$ for x in the kth subinterval, where $k = 1, 2, \ldots, 2^n$. Use the hint to calculate $\int_0^1 2x^3 \, dx = \lim_{n\to\infty} \int \phi_n = \lim_{n\to\infty} \sum_{k=1}^{2^n}[2(\frac{k-1}{2^n})^3] \cdot (1/2^n) = 1/2$.

17. Define ϕ_n, where $n = 1, 2, \ldots$, on 2^n equal-sized subintervals (with right endpoints included) of $[1, 5]$ as $\phi_n(x) = -3[(4k/2^n) + 1]$ for x in the kth subinterval, where $k = 1, 2, \ldots, 2^n$. Then $\int_1^5 -3x \, dx = \lim_{n\to\infty} \int \phi_n = \lim_{n\to\infty} \sum_{k=1}^{2^n} -3((4k/2^n) + 1) \cdot (4/2^n) = \lim_{n\to\infty} -\frac{48(2^{2n}+2^n)}{2\cdot 2^{2n}} - 12 = -36$.

19. Define ϕ_n, where $n = 1, 2, \ldots$, on 2^n equal-sized subintervals (with right endpoints included) of $[-1, -3]$ as $\phi_n(x) = (2k/2^n - 1)^2$ for x in the kth subinterval, where $k = 1, 2, \ldots, 2^n$. Then $\int_{-3}^{-1} x^2 \, dx = \lim_{n\to\infty} \int \phi_n = \lim_{n\to\infty} \sum_{k=1}^{2^n}(2k/2^n - 1)^2 \cdot (2/2^n) = 26/3$.

21. Define ϕ_n, where $n = 1, 2, \ldots$, on 2^n equal-sized subintervals (with right endpoints included for subintervals over negative x values and left endpoints for subintervals over positive values) of $[-1, 1]$ as $\phi_n(x) = (1 - 2k)/2^n$ for x in the kth subinterval when $k = 1, 2, 3 \ldots, 2^{n-1}$ and $\phi_n(x) = -1 + (2k - 1)/2^n$ for x in the kth subinterval when $k = 2^{n-1} + 1, 2^{n-1} + 2, \ldots, 2^n$. Then $\int_{-1}^1 |x| \, dx = \lim_{n\to\infty} \sum_{k=1}^{2^{n-1}}(\frac{1-2k}{2^n}) \cdot (2/2^n) + \sum_{k=2^{n-1}+1}^{2^n}(-1 + \frac{2k-1}{2^n}) \cdot (2/2^n) = 1$.

23. Similar to (21) but using x^2; $\int_{-1}^1 x^2 \, dx = 2/3$.

25. Similar to (21) and (23) but easiest if using $3 \cdot 2^n$ subintervals; $\int_{-1}^2 4 - x^2 \, dx = 9$.

27. Define, for $n = 1, 2, \ldots$, the nondecreasing sequence of step functions $\phi_n(x) = j/(j+1)$ whenever $x \in (\frac{1}{j+1}, \frac{1}{j}]$, where $j = 1, 2, 3, \ldots, n$ (and 0 otherwise). $\lim_{n\to\infty} -\phi_n(x) = -f(x)$ almost everywhere, and by Definition 1.5.2, $\int -f = -\lim_{n\to\infty} \int \phi_n = -\sum_{j=1}^\infty j/(j+1) \cdot 1/(j(j+1)) = -\sum_{j=1}^\infty 1/(j+1)^2$, which is a convergent p-series with $p = 2$.

29. Since $\mathcal{X}_\mathbb{I} = 1$ almost everywhere, the integral is 1.

31. The proof follows the argument in the hint.

33. Write $f = g - h$ for $g, h \in L^0$ and obtain step-function sequences ϕ_n and ψ_n for g and h, respectively. Then the result of the previous exercise can be applied to the integrals of $\phi_n(x) - \psi_n(x)$ and $\phi_n(x + k) - \psi_n(x + k)$. Taking limits and using Definition 1.5.2 proves the result.

35. Similar to (33), except integrate $\phi_n(x) - \psi_n(x)$ and $|k| \cdot (\phi_n(k \cdot x) - \psi_n(k \cdot x))$.

Section 2.1

1. For example, $\mathbb{R}$.
3. For example, $(-\infty, 5)$.
5. For example, $\{6, 6\frac{1}{2}, 6\frac{2}{3}, 6\frac{3}{4}, \ldots\}$.
7. $\inf B = 0 \in B$, and $\sup B = 19 \notin B$.
9. $\inf D = 0 \in D$, and $\sup D = 3 \notin D$.
11. $\inf F = 3 \notin F$, and $\sup F = 8 \in F$.
13. $\sup H = 620/121 \notin H$, and $\inf H = 577/121 \in H$.
15. Define ϕ_n and ψ_n on 2^n equal-sized subintervals of $[4, 10]$ as $\phi_n(x) = (4 + 6(k-1)/2^n)^2 + 3 \cdot (4 + 6(k-1)/2^n) + 5$ and $\psi_n(x) = (4 + 6k/2^n)^2 + 3 \cdot (4 + 6k/2^n) + 5$ for x in the kth subinterval $[4 + 6(k-1)/2^n, 4 + 6k/2^n)$, where $k = 1, 2, \ldots, 2^n$.
17. Define ϕ_n and ψ_n on 2^{n+1} equal-sized subintervals of $[-1, 1]$ as $\phi_n(x) = -3(-1 + k/2^n)^2 + 7$ and $\psi_n(x) = -3(-1 + (k-1)/2^n)^2 + 7$ if $x < 0$ (if $x \in [-1 + (k-1)/2^n, -1 + k/2^n)$, where $k = 1, 2, \ldots, 2^n$), and $\phi_n(x) = -3((k-1)/2^n)^2 + 7$ and $\psi_n(x) = -3(k/2^n)^2 + 7$ if $x \geq 0$ (if $x \in [(k-1)/2^n, k/2^n)$, where $k = 1, 2, \ldots, 2^n$).
19. Define ϕ_n and ψ_n on 2^n equal-sized subintervals of $[-\pi/2, \pi/2]$ as $\phi_n(x) = \sin(-\pi/2 + \pi(k-1)/2^n)$ and $\psi_n(x) = \sin(-\pi/2 + \pi k/2^n)$ for x in the kth subinterval, where $k = 1, 2, \ldots, 2^n$.
21. Define $\psi_n = 2k/2^n$ if $x \in [2(k-1)/2^n, 2k/2^n)$, where $k = 1, 2, \ldots, 2^n$, so $R\text{-}\int_0^2 x \, dx = \lim_{n \to \infty} \sum_{k=1}^{2^n} 2k/2^n \cdot (2/2^n) = \lim_{n \to \infty} (4/2^{2n}) \frac{2^n(2^n+1)}{2} = 2$.
23. Define $\psi_n = 2(k/2^n)^2$ if $x \in [(k-1)/2^n, k/2^n)$, where $k = 1, 2, \ldots, 2^n$, so $R\text{-}\int_0^1 2x^2 \, dx = \lim_{n \to \infty} \sum_{k=1}^{2^n} 2(k/2^n)^2 \cdot (1/2^n) = \lim_{n \to \infty} (2/2^{3n}) \frac{2^n(2^n+1)(2 \cdot 2^n+1)}{6} = 2/3$.
25. Define $\psi_n = 3(1 + k/2^n)^2$ if $x \in [1 + (k-1)/2^n, 1 + k/2^n)$, where $k = 1, 2, \ldots, 2^n$, so $R\text{-}\int_1^2 3x^2 \, dx = \lim_{n \to \infty} \sum_{k=1}^{2^n} 3(1 + k/2^n)^2 \cdot (1/2^n) = \lim_{n \to \infty} 3[\frac{2^n(2^n+1)(2 \cdot 2^n+1)}{6 \cdot 2^{3n}} + 2\frac{2^n(2^n+1)}{2 \cdot 2^{2n}} + 1] = 7$.
27. Define $\psi_n = (-1 + 4k/2^n)^3 - 4$ if $x \in [-1 + 4(k-1)/2^n, -1 + 4k/2^n)$, where $k = 1, 2, \ldots, 2^n$, so $R\text{-}\int_1^2 x^3 - 4 \, dx = \lim_{n \to \infty} \sum_{k=1}^{2^n} [(-1 + 4k/2^n)^3 - 4] \cdot (4/2^n) = \lim_{n \to \infty} 4 [64(\frac{2^n(2^n+1)}{2 \cdot 2^{2n}})^2 - 48\frac{2^n(2^n+1)(2 \cdot 2^n+1)}{6 \cdot 2^{3n}} + 12\frac{2^n(2^n+1)}{2 \cdot 2^{2n}}] - 20 = 4$.
29. $f(x) = 4\sin(1/x)$ is continuous and bounded on $(0, 2\pi)$, so the integral exists.
31. The points of discontinuity are $x = 2, 3$. Since $m(\{2, 3\}) = 0$, the integral exists.
33. The set of points of discontinuity $\{\frac{1}{2^n} : n \in \mathbb{N}\}$ is countable and thus has measure zero. Hence the integral exists.
35. (a) Choose an irrational $a \in [0, 1]$. Use the fact that, given $\varepsilon > 0$, there exists $\delta > 0$ such that for all $|x - a| < \delta$, x is either irrational or $x = m/n$ in lowest terms with $n > 1/\varepsilon$. For such x, $|f(x) - f(a)| = 1/n < 1/(1/\varepsilon) = \varepsilon$. The function's set of discontinuities is therefore $\mathbb{Q}$.
    (b) f is integrable.
    (c) $R\text{-}\int_0^1 f(x) \, dx = \int_0^1 f(x) \, dx = \int_0^1 0 = 0$.
37. Suppose i is as in the alternate characterization: given $\varepsilon > 0$, there exists a number $y \in S$ such that $y < i + \varepsilon$. Then a value greater than i must be greater than some y in S. Hence a value greater than i cannot be a lower bound of S. Thus i must be the greatest lower bound in the sense of Definition 2.1.1. Conversely, suppose i is an infimum for S as in Definition 2.1.1. Then i is a lower bound for S and $i > m$ for any other lower bound m. Then, given

Section 2.2

1. $9881/5$
3. $2/3$
5. 0
7. $\pi/4$
9. 1
11. $-3x^5/25 + (1/5)x^5 \ln(x^3)\big|_1^e = 12e^5/25 - 3/25$.
13. $\ln 2$
15. If $b = 0$, the integral is 0. Otherwise, it is $\dfrac{b + e^{a\pi}(-b\cos b\pi + a\sin b\pi)}{a^2 + b^2}$.

    In the solutions to 17–25, the values for n are $n = 1, 2, 3, \ldots$.
17. Use $f_n(x) = 1/\sqrt{3x}\, \mathcal{X}_{[1/n,1]}(x)$.
19. Use $f_n(x) = a \cdot s^a \cdot x^{-(a+1)} \mathcal{X}_{[1,n]}(x)$.
21. Use $f_n(x) = x^{-2} \mathcal{X}_{[2,n]}(x)$.
23. Use $f_n(x) = |\sin x|/x^2 \mathcal{X}_{[\pi,(2n+1)\pi/2]}(x)$.
25. Use $f_n(x) = x/(1+x^2)^2 \mathcal{X}_{[-n,0]}(x)$.
27. Write $\int_0^\infty t^{x-1} e^{-t}\, dt = \int_0^1 t^{x-1} e^{-t}\, dt + \int_1^\infty t^{x-1} e^{-t}\, dt$. (Take note of the case $0 < x < 1$.) For the first integral, use $f_n(x) = t^{x-1} e^{-t} \mathcal{X}_{[1/n,1]}(t)$. For the second, use $f_n(x) = t^{x-1} e^{-t} \mathcal{X}_{[1,n]}(t)$.
29. $\Gamma(1) = \int_0^\infty e^{-t}\, dt = 1$ by Example 2.2.2.
31. $\Gamma(1/2) = \int_0^\infty t^{-1/2} e^{-t}\, dt$. Now use the hint to show this value is $\sqrt{\pi}$.
33. $\mathcal{X}_S$ is a nonnegative function in L^1, and so it satisfies the hypotheses of the theorem in the previous exercise. Hence $\mathcal{X}_S = 0$ a.e. That means $m(S) = 0$, since S is exactly the set where $\mathcal{X}_S \neq 0$.
35. Without loss of generality, assume $f \geq g$. Then $\max\{f, g\} = f$, and $(f+g)/2 + |f-g|/2 = (f+g)/2 + (f-g)/2 = f$. The calculation for the minimum is similar. Because $f, g \in L^1$, we have $f+g$ and $f-g$ are in L^1. Then, by Theorem 2.2.4, the terms in the sums for $\max\{f, g\}$ and $\min\{f, g\}$ are in L^1. Therefore $\max\{f, g\}$ and $\min\{f, g\}$ are in L^1.

Section 2.3

1. $\int_1^\infty \sin x/x^2\, dx$
3. $\int_0^\infty \lim_{n \to \infty} e^{-x} \sin 0\, dx = \int_0^\infty 0\, dx = 0$.
5. $\int_0^1 0\, dx = 0$.
7. $1/2$
9. $1 - 1/e$
11. Monotone convergence (the integrand is positive). Using (also) integration by parts, the integral is 1.
13. LDCT with $g(x) = e^{-x}$. The integral is $1/2$.
15. Monotone convergence (the integrand is positive). For each n, the integral is $\dfrac{1}{n-1}$.
17. By LDCT applied to its integral definition, $\lim_{n \to \infty} F(n) = 0$. Then $0 = C - \lim_{n \to \infty} \arctan n = C - \pi/2$, so $C = \pi/2$.

19. $f_1(.25) = .5$ and $f_2(.25) = 0$. $f_1(.5) = 0$ and $f_2(.5) = 1$. Since the sequence is not monotone, the monotone convergence theorem cannot apply.
21. f_n is continuous and bounded, and vanishes off of $[0, 1]$. $\int f_n = (1/2) \cdot n \cdot \int_0^{1/n} 1 \, dx = 1/2$.
23. $\lim_{n \to \infty} f_n(x) = 1$ when $x \in \mathbb{Q} \cup [0, 1]$. But $\lim_{n \to \infty} f_n(x)$ does not exist when x is irrational in $[0, 1]$.

Section 2.4

1. $e^{it} + e^{-it} = \cos t + i \sin t + \cos(-t) + i \sin(-t) = \cos t + i \sin t + \cos t - i \sin t = 2 \cos t$.
3. $c_0 = a_0 = \frac{1}{2\pi} \int_{-\pi}^{\pi} f(t) \, dt = \frac{1}{2\pi} \int_{-\pi}^{\pi} f(t) e^{-i0t} \, dt$, since $e^0 = 1$.
5. $a_0 + \sum_{n=1}^{\infty}(a_n \cos nx + b_n \sin nx) = c_0 + \sum_{n=1}^{\infty}(a_n \frac{e^{inx}+e^{-inx}}{2} + b_n \frac{e^{inx}-e^{-inx}}{2i}) = c_0 + \frac{1}{2}\sum_{n=1}^{\infty}[(a_n - ib_n)e^{inx} + (a_n + ib_n)e^{-inx}] = \sum_{n=-\infty}^{\infty} c_n e^{inx}$.
7. $f(x) \sim \sum_{n=1}^{\infty}(-1)^{n+1}(\frac{2\pi^2}{n} - \frac{12}{n^3}) \sin nx$.
9. $f(x) \sim \pi/2 + \sum_{k=1}^{\infty} \frac{-4}{\pi(2k-1)^2} \cos(2k-1)x$.
11. $h(x) \sim 1 - \pi/2 + \sum_{n=1}^{\infty} \frac{2(1-(-1)^n)}{\pi n^2} \cos nx$.
13. $s(x) \sim \pi^2/3 + \sum_{n=1}^{\infty} \frac{4(-1)^n}{n^2} \cos nx$.
15. $u(x) \sim 2/\pi - \frac{1}{\pi}\sum_{n=1}^{\infty} \frac{2(1+(-1)^n)}{n^2-1} \cos nx$.
17. $w(x) \sim 2/\pi + \frac{1}{\pi}\sum_{n=2}^{\infty} \frac{(-1)^{n/2+1}2(1+(-1)^n)}{n^2-1} \cos nx$.
19. $c_0 = 0$ and $c_n = i(-1)^n/n$ for $n \neq 0$.
21. $c_0 = \pi/2$, $c_n = 0$ when n is nonzero even and $c_n = \frac{-2}{\pi n^2}$ when n is odd.
23. f is nondecreasing. The values of the series at π and $-\pi$ are $f(-\pi) = f(\pi) = 1$.
25. h is nondecreasing. The value of the series at each of π and $-\pi$ is c.
27. $s(x) = \sin x = a(x) - b(x)$, where $a(x) = \begin{cases} -1 & \text{if } -\pi \leq x \leq \pi/2 \\ \sin x & \text{if } -\pi/2 < x < \pi/2 \\ 1 & \text{if } \pi/2 < x < \pi \end{cases}$ and

$b(x) = \begin{cases} -1-\sin x & \text{if } -\pi \leq x \leq \pi/2 \\ 0 & \text{if } -\pi/2 < x < \pi/2 \\ 1-\sin x & \text{if } \pi/2 < x < \pi. \end{cases}$ The value of the series at π and $-\pi$ is 0.

29. $u(x) = \sec x/4 = a(x) - b(x)$, where $a(x) = \begin{cases} 0 & \text{if } -\pi \leq x \leq 0 \\ \sec x/4 & \text{if } 0 < x < \pi \end{cases}$ and

$b(x) = \begin{cases} -\sec x/4 & \text{if } -\pi \leq x \leq 0 \\ 0 & \text{if } 0 < x < \pi. \end{cases}$ The value of the series at π and $-\pi$ is $\sqrt{2}$.

Section 3.1

1. $\sqrt{\pi/2}$
3. $\|h_n\| = \sqrt{2\pi}$, $n \in \mathbb{Z}$.
5. $1/\sqrt{30}$
7. $2\sqrt{\pi}$
9. $\lim_{n \to \infty} \sqrt{\frac{1}{2n+1}} = 0$.
11. $\lim_{n \to \infty} \sqrt{\frac{1}{3n^2}} = 0$.
13. As stated in Section 2.1, $\int |D(x)| = \int D(x) = 0$. For any function f, $\|f\| = 0$ exactly when $f = 0$ in the metric space, which means, for L^1, that $f = 0$ a.e.
15. $\|x^2\| = \sqrt{\int_{-1}^{1} x^4} = \sqrt{2/5}$. Similarly, $\|\sqrt{2} \cdot x^2\| = \sqrt{4/5}$.

17. $\|1+x\| = \sqrt{7/3} < 1.53 < 1 + .57 < 1 + \sqrt{1/3} = \|1\| + \|x\|$.
19. If $f(x) \geq 0$, then mid$\{-c\mathcal{X}_I, |f(x)|, c\mathcal{X}_I\}$ =mid$\{-c\mathcal{X}_I, f(x), c\mathcal{X}_I\}$ = $\max\{\min\{f(x), c\mathcal{X}_I\}, -c\mathcal{X}_I\} = \min\{f(x), c\mathcal{X}_I\} = |\min\{f(x), c\mathcal{X}_I\}| = |\max\{\min\{f(x), c\mathcal{X}_I\}, -c\mathcal{X}_I\}| = |\text{mid}\{-c\mathcal{X}_I, f(x), c\mathcal{X}_I\}|$. A similar equality works if $f(x) < 0$.

Section 3.2

1. Proof for a general complex function: $\langle f, f \rangle = \int f \cdot \bar{f} = \int |f|^2 = \|f\|^2$.
3. $\langle af + bg, h \rangle = \int (af + bg) \cdot \bar{h} = a \int f \cdot \bar{h} + b \int g \cdot \bar{h} = a\langle f, h \rangle + b\langle g, h \rangle$.
5. $\|f\| = 2$, $\|g\| = 2\sqrt{2}$, $\theta = \arccos(2\sqrt{2}/3) \approx .3398$.
7. $\|f\| = \sqrt{\pi} = \|g\|$, $\theta = \pi/2$. The functions are orthogonal.
9. $\|f\| = 2\pi = \|g\|$, $\theta = \pi/2$.
11. $\|f\| = \sqrt{(5^{2n+1} - 1)/(2n+1)}$, $\theta = \arccos(\frac{(5^{m+n+1}-1)\sqrt{(2n+1)(2m+1)}}{(m+n+1)\sqrt{(5^{2m+1}-1)(5^{2n+1}-1)}})$.
13. $\|f\| = \sqrt{b-a}$, $\|g\| = \sqrt{d-c}$, $\theta = \pi/2$.
15. $\langle 1, \sin x \rangle = \int_{-\pi}^{\pi} \sin x \, dx = -\cos x|_{-\pi}^{\pi} = -(1-1) = 0$. Similarly, $\int_{-\pi}^{\pi} \cos x \, dx = 0$ and $\int_{-\pi}^{\pi} \sin x \cos x \, dx = \frac{1}{2}\sin^2 x|_{-\pi}^{\pi} = 0$.
17. $\|\frac{1}{n}\sin nx - 0\|^2 = \frac{1}{n^2}\int_{-\pi}^{\pi}\sin^2 nx \, dx = \frac{1}{n^2}(\pi - \frac{\sin 2\pi n}{2n})$, which approaches 0 as n goes to infinity.
19. $\sqrt{\sum_{n=0}^{\infty}(1/2^n)^2} = \sqrt{4/3}$.
21. $\langle e_j, e_k \rangle = 0$ when $j \neq k$ because each product of corresponding entries equals 0.
23. $\|\vec{k}_{1/2}\| = 2/\sqrt{3}$, $\|\vec{k}_{1/3}\| = 3/\sqrt{8}$, and $\langle \vec{k}_{1/2}, \vec{k}_{1/3} \rangle = 6/5$. Hence $\theta = \arccos(\frac{\sqrt{24}}{5}) \approx .2014$.
25. $\|g\|^2 = \sum_{n=1}^{\infty} 1/(2n)^2$, which is finite (e.g., by the well-known p-series test with $p = 2$).
27. $\|f\| = \sqrt{\pi} = \|g\|$ and $\|f + g\| = \sqrt{2\pi}$. The Pythagorean theorem follows immediately.
29. $\|f\| = \sqrt{(e^2 - 1)/2}$ and $\|g\| = \sqrt{1/3}$. The law says $\|f + g\|^2 + \|f - g\|^2 = (e^2 - 1) + (2/3)$, which is borne out by $\|f + g\|^2 = e^2/2 + 11/6$ and $\|f - g\|^2 = e^2/2 - 13/6$.

Section 3.3

1. When $n + m$ is odd, $\langle x^n, x^m \rangle = \int_{-1}^{1} x^{n+m} \, dx = 0$. When $n + m$ is even, $\langle x^n, x^m \rangle = \int_{-1}^{1} x^{n+m} \, dx = 2/(n + m + 1) \neq 0$.
3. $e_n(x) = f_n(x) \cdot \sqrt{(2n+1)/2}$ has norm 1.
5. $e_0 = 1/\sqrt{2\pi}$, $e_1 = \sqrt{3/(2\pi^3)} \cdot x$, $e_2 = \sqrt{15/(16\pi^5)}(x^2 + \pi^2/3)$.
7. The reader can check the inner-product properties are satisfied; e.g., the second property follows because $\langle \vec{v}, \vec{v} \rangle = \sum_{n=1}^{4} a_n^2 = 0$ implies each $a_n = 0$, which means $\vec{v} = \vec{0}$. In the same way, the reader can check the properties of the norm are satisfied when $\|\vec{v}\|^2 = \sum_{n=1}^{4} a_n^2$. The vector space is complete because the entries come from the set of real numbers, which is complete.
9. The Gram-Schmidt process returns exactly the same elements as it starts with; i.e., the four orthonormal basis vectors for $\mathcal{K}$ are immediately understood as $e_1 = 1$, $e_2 = x$, $e_3 = x^2$, and $e_4 = x^4$.
11. $H_0(x) = 1$, $H_1(x) = 2x$, $H_2(x) = 4x^2 - 2$, $H_3(x) = 8x^3 - 12x$, $H_4(x) = 16x^4 - 48x^2 + 12$.
13. $\|f_0\| = \pi^{1/4}$, $\|f_1\| = (2\sqrt{\pi})^{1/2}$, $\|f_2\| = (8\sqrt{\pi})^{1/2}$, $\|f_3\| = (48\sqrt{\pi})^{1/2}$, and $\|f_4\| = (8\sqrt{6\pi})^{1/2}$. Set $e_n(x) = f_n(x)/\sqrt{2^n\sqrt{\pi}n!}$. (See Theorem 3.4.2.)

15. Since the vector space M is finite-dimensional over the reals (the polynomial elements have real-valued coefficients), it is closed. The fact that $\langle e_n, e_m \rangle = 0$ when $n \neq m$ for the normalized elements e_0, e_1, and e_2 for the three-dimensional vector space means the elements form an orthonormal basis for M: if $p(x) = a_0 e_0 + a_1 e_1 + a_2 e_2$ is perpendicular to every e_n, then $a_n = 0$, and so $p(x) = 0$.

17. $e_0 = 1 e^{-x^2/2}/\pi^{1/4}$ and $e_1 = 2x e^{-x^2/2}/(2\sqrt{\pi})^{1/2}$ are both of the required form.

19. $M^\perp = \{ f - \langle f, e_0 \rangle e_0 - \langle f, e_1 \rangle e_1 : f \in L^2(\mathbb{R}) \}$, where e_0 and e_1 are as in the solution to Exercise 17.

21. $e_5 = \sqrt{11/128}(63x^5 - 70x^3 + 15x)$.

23. $M^\perp = \{ f - \langle f, e_0 \rangle e_0 - \langle f, e_1 \rangle e_1 : f \in L^2(-1, 1) \}$, where e_0 and e_1 are stated in Exercise 4.

25. $\langle \vec{e}_j, \vec{e}_k \rangle = 0$ when $j \neq k$ because each product of corresponding entries equals 0. $\|\vec{e}_n\| = 1$ because the nth entry produces a 1 when squared in the summation. If a vector $\vec{v} \in \ell^2$ satisfies $\langle \vec{v}, \vec{e}_n \rangle = 0$ for all $n = 1, 2, \ldots$, then each nth entry of $\vec{v}$ would be 0, and so $\vec{v} = \vec{0}$.

27. This fact follows upon verification of $\int_{-\pi}^{\pi} \cos nx \cdot \cos mx \, dx = 0$ and $\int_{-\pi}^{\pi} \sin nx \cdot \sin mx \, dx = 0$ when $m, n \in \mathbb{N}$ has $m \neq n$, and $\int_{-\pi}^{\pi} \cos^2 nx \, dx = \pi = \int_{-\pi}^{\pi} \sin^2 nx \, dx$.

29. Verify the following list of integral values, when $m, n \in \mathbb{N}$ has $m \neq n$: $\int_{-a}^{a} \cos n\pi x/a \cdot \cos m\pi x/a \, dx = 0$, $\int_{-a}^{a} \sin n\pi x/a \cdot \sin m\pi x/a \, dx = 0$, $\int_{-a}^{a} \cos n\pi x/a \cdot \sin m\pi x/a \, dx = 0$, $\int_{-a}^{a} (1/\sqrt{2a}) \sin m\pi x/a \, dx = 0$, $\int_{-a}^{a} (1/\sqrt{2a}) \cos m\pi x/a \, dx = 0$, $\int_{-a}^{a} (2a)^{-1} \, dx = 1$, $\int_{-a}^{a} \cos^2 m\pi x/a \, dx = \pi$, and $\int_{-a}^{a} \sin^2 m\pi x/a \, dx = \pi$.

31. Let $\vec{e}_n \in \ell^2$ be defined as in Exercise 25 for $n = 1, 2, 3, \ldots$.
    (a) For each normalized Laguerre function e_n (see Example 3.3.2), define $T(e_n) = \vec{e}_{n+1}$.
    (b) For each normalized Hermite function e_n (defined as in Exercise 17), define $T(e_n) = \vec{e}_{n+1}$.
    (c) For each normalized Legendre function e_n (defined as in Exercise 4), define $T(e_n) = \vec{e}_{n+1}$.

33. Since $f \in \bar{M}$, it must be the strong limit of a Cauchy sequence $\{ f_n \}$ of functions in M. Hence $\lim_{n \to \infty} \| f - f_n \| = 0$. By the definition of limit, for any $\varepsilon > 0$, there exists a value N such that $\| f - f_N \| < \varepsilon$. The function $\hat{f} = f_N$ has the desired property.

35. $\langle g, \hat{f} \rangle = 0$ because $g \in M^\perp$ and $\hat{f} \in M$. Hence, for $\varepsilon > 0$, there exists $\hat{f} \in M$ such that $|\langle g, f \rangle| = |\langle g, f \rangle - \langle g, \hat{f} \rangle| = |\langle g, f - \hat{f} \rangle| \leq \|g\| \varepsilon$. Since ε is arbitrary, we must have $\langle g, f \rangle = 0$.

Section 3.4

1. $H_4(u) = 16u^4 - 48u^2 + 12$, $H_5(u) = 32u^5 - 160u^3 + 120u$, $H_6(u) = 64u^6 - 480u^4 + 720u^2 - 120$, $H_7(u) = 128u^7 - 1344u^5 + 3360u^3 - 1680u$, $H_8(u) = 256u^8 - 3584u^6 + 13440u^4 - 13440u^2 + 1680$.

3. $g_1''(u) + g_2''(u) - 2u \cdot (g_1'(u) + g_2'(u)) + 2p(g_1(u) + g_2(u)) = g_1''(u) + g_2''(u) - 2ug_1'(u) - 2ug_2'(u) + 2pg_1(u) + 2pg_2(u) = [g_1''(u) - 2ug_1'(u) + 2pg_1(u)] + [g_2''(u) - 2ug_2'(u) + 2pg_2(u)] = 0 + 0 = 0$.

5. A graphing calculator provides the solution.

7. We get $[\pi \sqrt{fm/h}]^2 \psi''(u) + (8\pi^2 m/h^2)(E - k \cdot u^2/2)\psi(u) = 0$, where $u = [\pi \sqrt{fm/h}]x$. Dividing by $[\pi \sqrt{fm/h}]^2$ produces $\psi''(u) + (2p + 1 - u^2)\psi(u) = 0$.

9. $f(0) = 1 = H_0(x)$.

11. When n is even, $f^{(n)}(t) = f(t)[2^n(x-t)^n + c_{n-2}2^{n-2}(x-t)^{n-2} + \cdots + c_0]$, where the constants c_n force the evaluation $f^{(n)}(0) = H_n(x)$. A similar result happens for n odd.
13. If n is even, then $g_1(x)$, defined in Section 3.4.1, satisfies $g_1(-x) = g_1(x)$, and $H_n(x) = 2^n g_1(x)$ when $p = n$.
15. The nth coefficient of a Maclaurin series for an analytic function $f(x)$ is $a_n = f^{(n)}(0)/n!$. Substituting $z = x - t$ and noting $\partial z/\partial t = -1$, the chain rule implies
$H_n(x) = e^{x^2} \frac{\partial^n}{\partial t^n} e^{-(x-t)^2}\big|_{t=0} = e^{x^2} \frac{\partial^n}{\partial z^n} e^{-z^2}\big|_{z=x} \cdot (-1)^n = (-1)^n e^{x^2} \frac{d^n}{dx^n} e^{-x^2}$.
17. From the hint, $H'_{2k}(x) = \frac{d}{dx}(1 - (4k/2!)x^2 + \cdots - (4^k 2k(2k-2) \cdots 2/(4k)!)x^{2k}) = -2(2k)x + \cdots - 2k \cdot 4^k 2k(2k-2) \cdots 2/(4k)! x^{2k-1} = 4kcx - \cdots + c \cdot \frac{2^{[(2k-1)-1]/2}((2k-1)-1)(2k-4)\ldots(2)}{(2k-1)!} x^{2k-1} = 2(2k)H_{2k-1}(x)$, where the constant $c = \frac{4k 2^{2k-1}(2k-1)!}{2^{k-1}((2k-1)-1)(2k-4)\ldots(2)}$ makes the leading coefficient $4k \cdot 2^{2k-1}$.
19. Use the Rodrigues formula. See Exercise 11 in Section 3.3.
21. By Exercise 20, the leading coefficient c_n of H_n is $2c_{n-1} = 2^2 c_{n-2} = \cdots = 2^n c_{n-n} = 2^n c_0 = 2^n \cdot 1 = 2^n$.

Section 4.1

1. A is measurable by Theorem 4.1.2. $m(A) = 4.5$.
3. Define S as the complement of C. Then S is a countable union of intervals, and so is measurable by Theorem 4.1.2. Then C is measurable and has measure zero, since $m(C) = \int \mathcal{X}_C = \int \mathcal{X}_{[0,1]} - \mathcal{X}_S = 1 - \int \mathcal{X}_S = 0$ (as calculated in Question 1.2.3).
5. E is measurable by Theorem 4.1.2. $m(E) = 1$.
7. G is measurable by Theorem 4.1.2. $m(G) = 100,000$.
9. I is measurable because it can be written as a single interval. $m(I) = .5$.
11. $\max\{\mathcal{X}_S(x), \mathcal{X}_T(x)\} = 1$ if $x \in S \cup T$ and equals 0 otherwise. By the definition of a characteristic function, $\mathcal{X}_{S \cup T}(x) = 1$ if $x \in S \cup T$ and equals 0 otherwise. Hence the two functions are the same.
13. No. For example, $[0, 1]$ is measurable, but the Vitali set V of Example 3.1.1 is nonmeasurable, and $[0, 1] = V \cup ([0, 1] \setminus V)$.
15. When A and B are disjoint or share only an endpoint.
17. When $b_1 \le a_1 \le a_2 \le b_2$.
19. It is not possible. For if it was, then $0 > m(A) + m(B) - m(A \cap B) = m(A \cup B)$.
21. Case 1: If $f \ge g$, then $\max\{f, g\} = f = .5(f + g) + .5(f - g) = .5[(f + g) + |f - g|]$. The other case is similar.
23. From the hint, $\text{mid}\{-\phi, f_n + g_n, \phi\} \le \phi$ because ϕ is either the largest of the three terms (and then dominates) or $\phi < f_n + g_n$ (and then $\text{mid}\{-\phi, f_n + g_n, \phi\} \le \phi$. The convergence follows because $\lim_{n \to \infty} f_n = f$ and $\lim_{n \to \infty} g_n = g$. Then LDCT implies $\text{mid}\{-\phi, f + g, \phi\} \le \phi$ is integrable, proving the result.
25. For a nonnegative step function ϕ, $\text{mid}\{-\phi, .5f, \phi\} = .5\text{mid}\{-2\phi, f, 2\phi\}$. Since f is measurable and 2ϕ is a step function, $\text{mid}\{-\phi, .5f, \phi\}$ is integrable, and the result follows.

Section 4.2

1. $1/2$
3. $w(x) = 1$.
5. $1/3$

Solutions to Selected Problems   269

7. $e^{-a} - e^{-b}$
9. $(b^2 - a^2)/2$
11. $\int \mathcal{X}_I \, d\mu = 3$ if $1/2 \in I$ and equals 0 otherwise.
13. $\|f\| = \sqrt{3 \cdot f(1/2)^2}$ and $\|x\| = \sqrt{3}/2$.
15. $\int \mathcal{X}_I d\mu = 2$ if 1 and 2 are both in I, it is 1 if exactly one of 1 or 2 is in I, and it is 0 otherwise.
17. $\sqrt{5}$
19. $\mu([0, 1]) = 1$.
21. $\mu_F([0, 1/3]) = 1/2$, $\mu_F([0, 1/9]) = 1/4$, and $\mu_F([0, 1/27]) = 1/8$.
23. $\int \mathcal{X}_{[0,1/3]} \, d\mu_F = 1/2$.
25. $\int x^2 \, d\mu_F(x) = 1/3$.
27. $\int \mathcal{X}_{[8/9,1]} \, d\mu_F = 1/4$.
29. $\int a\mathcal{X}_{[0,(1/3)^n]} + b\mathcal{X}_{[1-(1/3)^n,1]} \, d\mu_F = a/2^n + b/2^n$.
31. $F(1/3) = \lim_{k \to \infty} \int_0^{1/3} \prod_{n=0}^{k} \frac{3}{2} \cdot f(3^n t) \, dt = \lim_{k \to \infty} 1/2 = 1/2$. Similar for $F(1/2)$.
33. From the origin to $(1/4, (1+t)^2/4)$, then to $(1/2, (1+t)/2)$, then to $(3/4, (1+t)(3-t)/4)$, then to $(1, 1)$.
35. $\nu([0, 1]) = 1$ and $\nu([0, 1/2]) = 3/4$.
37. $3/4$.

Section 4.3

1. Any real set is μ_5-measurable. $\int f \, d\mu_5 = f(5)$ and $\|f\|_{L^2(\mu_5)} = |f(5)|$. $\langle f, g \rangle = f(5) \cdot g(5)$.
3. Any real set is μ-measurable, and any function f with 5, 10, and 15 in its domain is μ-measurable. $\|f\|_{L^2(\mu)} = \sqrt{f(5)^2 + f(10)^2 + f(15)^2}$ and $\langle f, g \rangle = f(5) \cdot g(5) + f(10) \cdot g(10) + f(15) \cdot g(15)$.
5. $\|f\|_{L^2(\mu)} = \sqrt{f(5)^2 + 2f(10)^2 + 3f(15)^2}$ and $\langle f, g \rangle = f(5) \cdot g(5) + 2f(10) \cdot g(10) + 3f(15) \cdot g(15)$.
7. Any function f whose domain contains $\mathbb{N}$ is μ-measurable. $\|f\|_{L^2(\mu)} = \sqrt{\sum_{n=1}^{\infty} f(n)^2/2^n}$ and $\langle f, g \rangle = \sum_{n=1}^{\infty} f(n) \cdot g(n)/2^n$.
9. Any function that is Lebesgue measurable (Definition 3.1.1) and whose domain contains π.
11. $\|f\|_{L^2(\mu)}^2 = \int f(x)^2 \, d\mu = \int f(x)^2 \, k \cdot dx = k \cdot \|f\|^2$.
13. For $n \in \mathbb{N}$, $\|\sin nx\|_{L^2(\mu)}^2 = \int \sin^2 nx \, d\mu = \frac{1}{\pi} \int_{-\pi}^{\pi} \sin^2 nx \, dx = 1$.
15. $\|x\|_{L^2(\mu)}^2 = \int x^2 \, d\mu = \frac{1}{\pi} \int_{-\pi}^{\pi} x^2 \, dx = 2\pi^2/3$. For $n \in \mathbb{N}$, $\langle x, \cos nx \rangle = \frac{1}{\pi} \int_{-\pi}^{\pi} x \cdot \cos nx \, dx = 0$ and $\langle x, \sin nx \rangle = \frac{1}{\pi} \int_{-\pi}^{\pi} x \cdot \sin nx \, dx = (-2 \cos n\pi)/n$.
17. For $n \in \mathbb{N}$, $\|H_n(x)\|_{L^2(\mu)}^2 = \int H_n(x)^2 \, d\mu = \int H_n(x)^2 \cdot e^{-x^2} \, dx = \sqrt{\pi} 2^n n!$.
19. $\|c\|_{L^2(\mu)}^2 = \int c^2 \, d\mu = \int c^2 \mathcal{X}_A(x) \, dx = c^2 m(A)$.

Section 4.4

1. $S = \{0, 1\}$, $\mathbb{F} = \{\emptyset, S, \{0\}, \{1\}\}$, and $\mu(\emptyset) = 0$, $\mu(S) = 1$, $\mu(\{0\}) = 1/2$, and $\mu(\{1\}) = 1/2$.
3. $S = \{1, 2, 3, 4, 5, 6, 7, 8\}$, $\mathbb{F}$ is the collection of the 256 subsets of S, and μ satisfies $\mu(\{x\}) = 1/8$ for $x \in S$, and (for any general $A \subset \mathbb{F}$) $\mu(A) = n/8$, where $|A| = n$.
5. $S = \{0, 1, 2\}$, $\mathbb{F} = \{\emptyset, S, \{0\}, \{1\}, \{2\}, \{0, 1\}, \{1, 2\}, \{0, 2\}\}$, and $\mu(\emptyset) = 0$, $\mu(S) = 1$, $\mu(\{0\}) = 188/221$, $\mu(\{1\}) = 32/221$, $\mu(\{2\}) = 1/221$, $\mu(\{0, 1\}) = 220/221$, $\mu(\{1, 2\}) = 33/221$, and $\mu(\{0, 2\}) = 189/221$.

7. $S = \{1, 2, 3, \ldots, n\}$, $\mathbb{F}$ is the collection of the 2^n subsets of S, and μ is determined by $\mu(\{x\}) = x!/(n!(n-x)!2^n)$ for $x \in S$.
9. $S = [a, b]$, $\mathbb{F}$ consists of all the Lebesgue measurable subsets of S, and, for $A \subset \mathbb{F}$, $\mu(A) = m(A)/(b-a)$.
11. $P[X \leq x] = 0$ if $x < 0$, $P[X \leq x] = 1/2$ if $0 \leq x < 1$, and $P[X \leq x] = 1$ if $x \geq 1$.
13. $P[Y \leq y] = 0$ if $y < a$, $P[Y \leq y] = (y-a)/(b-a)$ if $a \leq y \leq b$, and $P[Y \leq y] = 1$ if $y > b$.
15. $P[K \leq k] = 0$ if $k < 0$, $P[K \leq k] = 188/221$ if $0 \leq k < 1$, $P[K \leq k] = 220/221$ if $1 \leq k < 2$, and $P[K \leq k] = 1$ if $k \geq 2$.
17. $P[X \leq x] = 0$ if $x < 0$, $P[X \leq x] = 625/1369$ if $0 \leq x < 1$, $P[X \leq x] = 1225/1369$ if $1 \leq k < 2$, and $P[X \leq x] = 1$ if $x \geq 2$.
19. $P[W \leq w]$ can be determined by analyzing $w = h + r$ for different combinations of $h = 0$ or 1 and $r = 1, 2, 3, 4, 5,$ or 6. As part of the list required, $P[W \leq w] = 1/4$ if $2 \leq x < 3$, $P[X \leq x] = 5/12$ if $3 \leq k < 4$, and $P[X \leq x] = 11/12$ if $6 \leq x < 7$.

Section 5.1

1. $S(\vec{v}) = \begin{bmatrix} 0 \\ 1 \\ 0 \\ 0 \\ \vdots \end{bmatrix}$.

3. $S(\vec{v}) = \begin{bmatrix} 0 \\ 1 \\ \vdots \\ 1/2^{n-1} \end{bmatrix}$.

5. $\|\vec{v}\|_{\ell^2}^2 = \sum_{n=1}^{\infty} |v_n|^2 = \|S(\vec{v})\|_{\ell^2}^2$.
7. For $f, g \in L^2(0, 1)$ and constants c, d, $T(cf + dg) = \langle cf + dg, r \rangle \cdot r(x) = c\langle f, r \rangle \cdot r(x) + d\langle g, r \rangle \cdot r(x) = cT(f) + dT(g)$.
9. For each n, $\|g_n\|_{L^2(0,1)} = 1$.
11. $|r(x_0)| \cdot \|r\|$
13. $T(f)(x) = \int_0^1 (x + y) \cdot f(y)\, dy$. The term M is $\int_0^1 \int_0^1 |x + y|^2\, dx\, dy = \int_0^1 \int_0^1 (x^2 + 2xy + y^2)\, dx\, dy = \int_0^1 (1/3 + y + y^2)\, dy = 7/6$.
15. $T(f)(x) = \int_0^1 e^{x+y} \cdot f(y)\, dy$. The term M is $\int_0^1 \int_0^1 e^{2x+2y}\, dx\, dy = (e^2 - 1)^2/4$.
17. $T(f)(x) = \int_{-1}^1 x \cdot f(y)\, dy$. The term M is $\int_{-1}^1 \int_{-1}^1 x^2\, dx\, dy = 4/3$.
19. $T(f)(x) = \int_{-\pi}^{\pi} \sin(x + y) \cdot f(y)\, dy$. The term M is $\int_{-\pi}^{\pi} \int_{-\pi}^{\pi} \sin^2(x + y)\, dx\, dy = 2\pi^2$.
21. $T(f)(x) = \int_{-\pi}^{\pi} y \sin(x + y)\, dy = 2\pi \cos x$.
23. $w(t) = \cos(-i \log t) = \frac{1}{2}[e^{i(-i\log((2x-i)/(2x+i)))} + e^{-i(-i\log((2x-i)/(2x+i)))}] = \frac{1}{2}\left[\frac{2x-i}{2x+i} + \frac{2x+i}{2x-i}\right] = \frac{(2x-i)^2+(2x+i)^2}{(2x+i)(2x-i)} = \frac{4x^2-1}{4x^2+1} = W(x)$.
25. $T_k(L_0 e^{-x/2}) = T_k(1 \cdot e^{-x/2}) = \frac{1}{2}(1 - x)e^{-x/2} = \frac{1}{2}L_1(x)e^{-x/2}$.
27. The (matrix multiplication) solution to Exercise 26 shows $T_w(\vec{e}_0) = \frac{1}{2}\vec{e}_1$.
29. $\frac{1}{2}(\vec{e}_0 + \vec{e}_2)$.

Section 5.2

1. $\begin{bmatrix} 7 \\ 12 \end{bmatrix}$
3. $\begin{bmatrix} -7 \\ -15 \\ -6 \end{bmatrix}$
5. $\begin{bmatrix} -1 \\ 3 \\ -9 \end{bmatrix}$

Solutions to Selected Problems

7. $\begin{bmatrix} 7 \\ 21 \\ 8 \\ 6 \end{bmatrix}$

9. $\begin{bmatrix} 2a+b \\ 2b \\ c+d \\ d \end{bmatrix}$

11. $\begin{bmatrix} 26 & 38 \\ 30 & 44 \end{bmatrix}$

13. $\begin{bmatrix} 17 & 0 & 23 \\ 18 & 0 & 28 \\ 29 & 0 & 39 \end{bmatrix}$

15. $\begin{bmatrix} -5 & 7 & 1 & 2 \\ 15 & -17 & 21 & 26 \\ 2 & 37 & 38 & 44 \\ \frac{15}{2} & -1 & 3 & 7 \end{bmatrix}$

17. $\|T\| = 5$.

19. $\frac{N+\sqrt{N^2-4|D|^2}}{2} = \|T\|^2 \leq 1$ when $N + \sqrt{N^2 - 4|D|^2} \leq 2$, which means $N \leq 1 + |D|^2$.

21. For example, $\begin{bmatrix} 0 \\ 1 \\ -1 \end{bmatrix}$ and $\begin{bmatrix} 0 \\ 0 \\ 1 \end{bmatrix}$.

23. Straightforward calculation.

25. Here $J = \begin{bmatrix} 4 & 1 & 0 \\ 0 & 4 & 0 \\ 0 & 0 & 1 \end{bmatrix}$.

27. $J^n = \begin{bmatrix} 2^n & n2^{n-1} \\ 0 & 2^n \end{bmatrix}$.

29. J^5

31. $\begin{bmatrix} -4096 & 5120 \\ -5120 & 6144 \end{bmatrix}$

33. $T_w(v) = x \int_{-\pi}^{\pi} x \cdot v(x)\,dx$.

35. $T_{\vec{w}}(\vec{v}) = (v_1 + v_2) \cdot \vec{w}$.

37. $T_{\vec{w}}(\vec{v}) = \left(\sum_{n=0}^{\infty} v_n/2^n\right) \cdot \vec{w}$.

39. $T_w(f) = x\left(\int_{-1}^{1} x \cdot f(x)\,dx + f(1) - f(0)\right)$.

41. $T_w(f) = \frac{z}{2\pi} \int_0^{2\pi} f(e^{it})e^{-it}\,dt = a_1 z$.

43. $T(\phi_n) = \sum_{k=0}^{\infty} \lambda_k \langle \phi_k, \phi_n \rangle \phi_k = \lambda_n \phi_n$ since every inner product equals zero except when $k = n$, and then it equals 1.

45. $T(\vec{v}) = \sum_{n=0}^{\infty} \lambda_n \langle \vec{w}_n, f \rangle \vec{w}_n$, where $\vec{w}_n$ is the standard orthonormal basis element for ℓ^2 as in Example 3.3.7. Now use Definition 5.2.1.

47. The proof mimics the description in Example 5.1.2.

49. Write k, as in Exercise 48, as $k(x, y) = \left(\sum_{i=1}^{\infty} \sum_{j=1}^{\infty} \langle k(x, y), \phi_{i,j}(x, y) \rangle\right) \cdot \phi_{i,j}(x, y)$, where $\phi_{i,j}(x, y) = e_i(x)e_j(y)$ is the product of the standard orthonormal basis elements e_n for $L^2(0, 1)$ and the inner product is defined as $\langle f(x, y), g(x, y) \rangle = \int_0^1 \int_0^1 f(x, y) \cdot g(x, y)\,dx\,dy$.[1] Now define the operators $T_n(f)(x) = \int_0^1 k_n(x, y)f(y)\,dy$ for $n = 1, 2, \ldots$, where $k_n(x, y) = \left(\sum_{i=1}^{n} \sum_{j=1}^{n} \langle k(x, y), \phi_{i,j}(x, y) \rangle\right) \cdot \phi_{i,j}(x, y)$. Then $T_n(f)$ is an element of $\vee\{e_1, \ldots, e_n\}$, and T is the limit of T_n (in the sense of the norm limit). This shows T is of the correct form and is compact.

Section 5.3

1. $\langle Sf, Sg \rangle = \langle e^{it}f(e^{it}), e^{it}g(e^{it}) \rangle = \sum_{n=0}^{\infty} a_n \overline{b_n} = \langle f, g \rangle$, where the notation is as in Example 5.2.3 part 5, since multiplication by e^{it} does not change the coefficients in the power series. The rest of the proof mimics that for the shift operator on $H^2(\mathbb{D})$ as described after Definition 5.3.3, but using the boundary value e^{it} in the domain variable instead of z.

---

[1] The inner product is technically for the Hilbert space $L^2((0, 1) \times (0, 1))$, of which $k(x, y)$ and $\phi_{i,j}(x, y)$ are members.

3. The result follows if S acts on the basis elements according to $S: e^{-\frac{1}{2}t}L_n(t) \to e^{-\frac{1}{2}t}L_{n+1}(t)$, because then Definition 5.3.3 holds generally by linearity and approximation. But that follows immediately by applying the Laguerre integration identity $\int_0^x L_n(t)\,dt = L_n(x) - L_{n+1}(x)$ to the formula for S.

5. Similar to the solution for Exercise 3, the result follows if S acts on the basis elements (identified in the problem) according to $S(e_n) = e_{n+1}$. That fact is immediately apparent.

7. $\langle Sf, Sg\rangle = \langle z^2 f(z), z^2 g(z)\rangle = \langle 0 + 0z + \sum_{n=2}^\infty a_{n-2} z^n, 0 + 0z + \sum_{n=2}^\infty b_{n-2} z^n\rangle = \sum_{n=0}^\infty a_n \overline{b_n} = \langle f, g\rangle$. Hence S is an isometry. We see $S^*: f(z) \to \sum_{n=2}^\infty a_n z^{n-2}$, and $\lim_{n\to\infty} \|(S^*)^n f\| = 0$ follows in a similar way as for the multiplicity-one shift on $H^2(\mathbb{D})$.

9. Applying the distance formula from u to $0 = 0 + 0i$, $|u|^2 = a^2 + b^2$. Also, $u \cdot \bar{u} = (a + ib)(a - ib) = a^2 + iab - iab - i^2 b^2 = a^2 + b^2$.

11. By the formula in Exercise 9, $\frac{(z+1)(\bar{z}-1)}{(z-1)(\bar{z}-1)} = \frac{|z|^2 - 1 + \bar{z} - z}{|z|^2 - 2x + 1} = \frac{|z|^2 - 1 - 2iy}{|z|^2 - 2x + 1}$. The result follows.

13. There must be a nonzero element of M by assumption, and every element in $H^2(\mathbb{D})$ can be written in terms of a power series as $f(z) = \sum_{n=0}^\infty a_n z^n$. If $f \neq 0$, then at least one a_n is nonzero.

15. For any $f \in M$, $z^k f(z) \in M$, and so $M_z(z^k f(z)) = z^{k+1} f(z) = z^k[z f(z)] \in M$, which implies $M_z(f) = z f(z) \in M$.

17. Because $z^k b(z)$ is inner for any inner function b.

19. Follows from the hint and the application of the result in Exercise 18.

21. Because ε is arbitrary in Exercise 20, the result follows. (If it didn't, then a contradiction to the result in Exercise 20 would follow.)

23. $b(z) = \frac{z - 1/2}{1 - z/2}$ and $g(z) = 1 - z/2$.

25. $b(z) = \frac{z - i/2}{1 + iz/2}$ and $g(z) = 2\frac{1 + iz/2}{z - 2}$.

27. Any function $g \in M$ is of the form $g(e^{it}) = e^{it} f(e^{it})$, where $f \in H^2(\mathbb{T})$, so $S(g)(e^{it}) = e^{it}[e^{it} f(e^{it})]$, which is in M.

29. The result follows from $z^2 b(z) f(z) = b(z) z^2 f(z)$ and the fact that $z^2 f(z) \in H^2(\mathbb{D})$ for any $f \in H^2(\mathbb{D})$.

Section 5.4

1. Suppose $\vec{v} = \begin{bmatrix} c \\ d \end{bmatrix}$ and $\vec{u} = \begin{bmatrix} e \\ f \end{bmatrix}$. Then $\langle T_{\vec{w}}(\vec{v}), \vec{u}\rangle = \langle (ac + bd)\vec{w}, \vec{u}\rangle = (ac + bd)\langle \vec{w}, \vec{u}\rangle = (ac + bd)(ae + bf) = c(ae + bf)a + d(ae + bf)b = \langle \vec{v}, (ae + bf)\vec{w}\rangle = \langle \vec{v}, (\langle \vec{u}, \vec{w}\rangle)\vec{w}\rangle$. Hence $T^*_{\vec{w}}(\vec{u}) = \langle \vec{u}, \vec{w}\rangle \vec{w}$. In short, $T^* = T$.

3. $T^* = T$ when $\begin{bmatrix} a & b \\ c & d \end{bmatrix} = \begin{bmatrix} a & c \\ b & d \end{bmatrix}$. $T^* = T^{-1}$ when $\begin{bmatrix} a & b \\ c & d \end{bmatrix} = \frac{1}{ad - bc}\begin{bmatrix} d & -b \\ -c & a \end{bmatrix}$, which happens when $ad - bc = -1$ and $a = -d$, or when $ad - bc = 1$ and $b = c = 0$.

5. When $\begin{bmatrix} d & e & f \\ g & h & i \\ j & k & l \end{bmatrix} = \begin{bmatrix} d & g & j \\ e & h & k \\ f & i & l \end{bmatrix}$.

7. $\langle g(z), \langle f(w), C(k_z)(w)\rangle\rangle = \langle g(z), \langle C^*(f)(w), k_z(w)\rangle\rangle = \langle g(z), C^*(f)(z)\rangle = \langle C(g(z)), f(z)\rangle$.

9. $C^*_\varphi(k_w)(z) = k_{\varphi(w)}(z)$. And $T_g C_\psi T^*_h(k_w)(z) = \overline{h(w)} T_g C_\psi(k_w)(z)$, which is seen, when substituting formulas, to be $k_{\varphi(w)}(z)$.

11. For f, g in the Hilbert space, $\langle (AB)f, g\rangle = \langle A(Bf), g\rangle = \langle B(f), A^*(g)\rangle = \langle f, B^*(A^*(g))\rangle$ and so $(AB)^* = B^* A^*$.

13. For x in the Hilbert space, $\|Ax\|^2 = \langle Ax, Ax\rangle = \langle A^* Ax, x\rangle \leq \|A^* A\| \|x\|^2$, so $\|A\|^2 \leq \|A^* A\|$. For the other direction, $\|A^* A\| \leq \|A^*\| \|A\| = \|A\|^2$, because $\|A\| = \|A^*\|$.

15. For x and y in the Hilbert space, $\langle A^*Af, g\rangle = \langle Af, Ag\rangle = \langle f, g\rangle$. Since the steps are reversible, we see A is an isometry if and only if $A^*A = I$.

17. Using Question 5.4.2, $\frac{\sqrt{2}}{\pi}\int_{-1}^{1} f(x)(1+2ixz-z^2)^{-1}\sqrt{1-x^2}\,dx = \langle f(x), \sqrt{2}(1+2ixz-z^2)^{-1}\rangle_{L^2(\mu)} = \langle \sum_{n=0}^{\infty} a_n\sqrt{2}U_n(x), \sum_{n=0}^{\infty} z^n\sqrt{2}i^n U_n(x)\rangle_{L^2(\mu)} = \sum_{n=0}^{\infty} i^n a_n z^n$. As an element of ℓ^2, $U^{-1}(f)(z) = \begin{bmatrix} a_0 \\ ia_1 \\ -a_2 \\ \vdots \end{bmatrix}$.

19. To get this matrix, simply replace the entries $\frac{1}{2}$ in the matrix for $T_{\cos t}$ with $\frac{1}{2} - \frac{i}{2}$ above the main diagonal and $\frac{1}{2} + \frac{i}{2}$ below the main diagonal. (See Question 5.1.2.)

21. Since $U(k_a)(x) = e^{\pi b/2}(1+e^{-2\pi b})^{1/2}(1+iz)^{-1/2-ib}(1-iz)^{-1/2-ib}$ (see [99]) and $(1+iz)^{-1/2-ib}(1-iz)^{-1/2-ib} = \sum_{n=0}^{\infty} P_n(x)z^n$, where $b = -\frac{-1}{2\pi}\mathrm{Log}(1/x-1)$, and $P_n(x)$ is a specific Meixner-Pollaczek polynomial of order $1/2$ (see [70]), we get $\int_0^1 f(x)\overline{U(k_a)(x)}\frac{1}{\pi}\,dx = \langle \sum_{n=0}^{\infty} a_n P_n(x), \sum_{n=0}^{\infty} z^n P_n(x)\rangle_{L^2(\mu)} = \sum_{n=0}^{\infty} a_n z^n$. As an element of ℓ^2 expressed in this way in terms of the Meixner-Pollaczek basis, $U^{-1}(f)(z) = \begin{bmatrix} a_0 \\ a_1 \\ a_2 \\ \vdots \end{bmatrix}$

Bibliography

[1] Abbott, Stephen, *Understanding Analysis*, Springer-Verlag, Berlin-New York, 2001. ISBN: 0-387-95060-5

[2] Adams, Malcolm, and Guillemin, Victor, *Measure Theory and Probability*, Birkhäuser, Boston, 1996. ISBN: 978-0817638849

[3] Apostol, Tom M., *Mathematical Analysis*, Addison-Wesley, Reading, Mass., 1957.

[4] Aronszajn, Nachman, and Smith, Kennan T., Invariant subspaces of completely continuous operators, *Ann. of Math.*, 60:345–350, 1954.

[5] Axler, Sheldon, Review of *The Life of Stefan Banach*, *American Mathematical Monthly*, 104:577–579, 1997.

[6] Ball, Joseph A., Review of *Hardy Classes and Operator Theory*, *Bulletin of the AMS*, 16:149–152, January, 1987.

[7] Ball, Joseph A., and Helton, J. William, Factorization results related to shifts in an indefinite metric, *Integral Equations and Operator Theory*, 5:632–658, 1982.

[8] Banach, Stefan, *Theory of Linear Operators*, translated by Jellett, F., Dover, New York, 2009. ISBN: 978-0486469836

[9] Banach, Stefan, and Tarski, Alfred, Sur la décomposition des ensembles de points en parties respectivement congruentes, *Fundamenta Mathematicae* 6: 244à277, 1924. matwbn.icm.edu.pl/ksiazki/fm/fm6/fm6127.pdf.

[10] Bartle, Robert G., *The Elements of Integration and Lebesgue Measure*, John Wiley & Sons, New York, 1966. ISBN: 978-0471042228

[11] Beurling, Arne, On two problems concerning linear transformations in Hilbert space, *Acta Math.*, 81:239à255, 1949.

[12] Billingsley, Patrick, *Probability and Measure*, 3rd ed., John Wiley & Sons, New York, 1995. ISBN: 978-0471007104

[13] Bohm, David, *Quantum Theory*, Prentice-Hall, Upper Saddle River, N.J., 1951.

[14] Boyer, Carl B., and Merzbach, Uta C., *A History of Mathematics*, 2nd ed., John Wiley & Sons, New York, 1991. ISBN: 978-0471543978

[15] Brown, Scott, Invariant subspaces for subnormal operators, *Integral Equations and Operator Theory*, 4:1–9, 1978.

[16] de Branges, Louis, and Rovnyak, James, *Square Summable Power Series*, Holt, Rinehart and Winston, New York, 1966.

[17] Bressoud, David M., *A Radical Approach to Lebesgue's Theory of Integration*, Cambridge University Press, New York, 2008. ISBN: 978-0521711838

[18] Burk, Frank, *Lebesgue Measure and Integration: An Introduction*, John Wiley & Sons, New York, 1998. ISBN: 978-0471179788

[19] Burkill, J. C., Henri Lebesgue: 1875–1941. Obituary, *J. Lond. Math. Soc.*, 4:483–490, 1944.

[20] Burton, David M., *The History of Mathematics*, Wm. C. Brown, Dubuque, Iowa, 1991. ISBN: 0-697-11196-2

[21] Cantor, Georg, *Contributions to the Founding of the Theory of Transfinite Numbers*, Dover, 1915. www.archive.org.

[22] Capinski, Marek, and Kopp, Peter E., *Measure, Integral and Probability*, 2nd ed., Springer-Verlag, London, 2007. ISBN: 978-1852337810

[23] Carathéodory, C., *Vorlesungen über Reelle Funktionen*, 3rd ed., Chelsea, New York, 1968. ISBN: 0-821-83653-6

[24] Carleson, Lennart, On convergence and growth of partial sums of Fourier series, *Acta Mathematica*, 116:135à157, 1966.

[25] Carslaw, Horatio Scott, *Introduction to the Theory of Fourier's Series and Integrals*, MacMillan and Co., London, 1921. ISBN: 978-0559202872

[26] Carter, M., and van Brunt, B., *The Lebesgue-Stieltjes Integral*, Springer-Verlag, London, 2000. ISBN: 978-0387950129

[27] Cauchy, Augustin, *Cours d'Analyse de l'Ecole Royale Polytechnique: Première Partie—Analyse Algébrique*, L'Imprimerie Royale, Paris, 1821.

[28] Collins, Peter, *Differential and Integral Equations*, Oxford University Press, New York, 2006. ISBN: 978-0199297894

[29] Conway, John B., *Functions of One Complex Variable*, 2nd ed., Springer-Verlag, New York, 1978. ISBN: 0-387-90328-3

[30] Courant, Richard, and Hilbert, David, *Methods of Mathematical Physics, Vol. I*, Wiley-VCH, Weinheim, Germany, 1989. ISBN: 978-0471504474

[31] Cowen, Carl C., Linear fractional composition operators on H^2, *J. Integral Equations Operator Theory*, 11:151–160, 1988.

[32] Cowen, Carl C., and MacCluer, Barbara D., *Composition Operators on Spaces of Analytic Functions*, CRC Press, New York, 1995. ISBN: 0-8493-8492-3

[33] Cowen, Carl C., and Gallardo-Guttiérez, Eva, The adjoint of a composition operator, preprint January 31, 2005. www.math.iupui.edu/~ccowen/Downloads/42CGCompOpsAdj.pdf.

[34] Daniell, P. J., A general form of the integral, *Annals of Math.*, 19:279–294, 1917.

[35] Dirac, Paul A. M., *Lectures on Quantum Mechanics*, Dover, New York, 2001. ISBN: 978-0486417134

[36] Lejeune-Dirichlet, P. G., Sur la convergence des séries trigonométriques qui servent à représenter une fonction arbitraire entre des limites données, *J. Reine Agnew. Math.*, 4:157–159, 1829.

[37] Dunham, William, *The Calculus Gallery*, Princeton University Press, Princeton, N.J., 2004. ISBN: 978-0691095653

[38] Dym, H., and McKean, H. P., *Fourier Series and Integrals*, Academic Press, New York, 1972. ISBN: 0-12-226450-9

[39] Edmunds, David E., Kokilashvili, Vakhtang, and Meskhi, Alexander, *Bounded and Compact Integral Operators*, Springer Netherlands, Dordrecht, 2009. ISBN: 978-9048160181

[40] Felscher, Walter, Bolzano, Cauchy, epsilon, delta, *The American Mathematical Monthly*, 107(9):844à862, 2000.

[41] Feynman, Richard, Hibbs, Albert R., and Styer, Daniel F., *Quantum Mechanics and Path Integrals: Emended Edition*, Dover, New York, 2010. ISBN: 978-0486477220

[42] Fisher, Stephen D., *Complex Variables*, 2nd ed., Dover, New York, 1990.

[43] Fourier, Joseph, *Théorie Analytique de la Chaleur*, Firmin Didot (translated by Alexander Freeman in 1878 and rereleased by Cambridge University Press in 2003), Paris, 1822. ISBN: 978-1-108-00180-9

[44] Furuta, Takayuki, *Invitation to Linear Operators: From Matrices to Bounded Linear Operators on a Hilbert Space*, Taylor & Francis, London, 2001. ISBN: 978-0415267991

[45] Garnett, John B., *Bounded Analytic Functions*, Academic Press, New York, 1981. ISBN: 0-12-276150-2

[46] Gohberg, Israel, and Goldberg, Seymour, *Basic Operator Theory*, Birkhäuser, Boston, 1981. ISBN: 3-7643-3028-7

[47] Goldberg, Richard R., *Methods of Real Analysis*, John Wiley and Sons, New York, 1976. ISBN: 0-471-31065-4

[48] Griffiths, David J., *Introduction to Quantum Mechanics*, Prentice-Hall, Upper Saddle River, N.J., 1995. ISBN: 0-13-124405-1

[49] Guicciardini, Niccolò, *Reading the Principia*, Cambridge University Press, New York, 2003. ISBN: 978-0521544030

[50] Halmos, Paul R., *Introduction to Hilbert Space*, Chelsea, New York, 1957. ISBN: 0-8284-0082-2

[51] Halmos, Paul R., *Measure Theory*, Springer, New York, 1974. ISBN: 978-0387900889

[52] Halmos, Paul R., Shifts on Hilbert spaces, *J. reigne agnew. Math.*, 208:102–112, 1961.

[53] Halmos, Paul R., What does the spectral theorem say?, *The American Mathematical Monthly*, 70.3:241–247, March, 1963.

[54] Hardy, G. H., and Littlewood, J. E., Notes on the theory of series (XVII): some new convergence criteria for Fourier series, *J. London Math. Soc.*, s1-7:252–256, 1932.

[55] Hawkins, Thomas, *Lebesgue's Theory of Integration*, American Mathematical Society, Providence, R.I., 2001. ISBN: 978-0821829639

[56] Hermann, Jacob, *Phoronomia, sive de viribus et motibus corporum solidorum et fluidorum libri duo*, Rod. & Gerh. Wetstenios, Amsterdam, 1716.

[57] Hermite, Charles, Sur un nouveau développement en série de fonctions, *Compt. Rend. Acad. Sci. Paris*, 58:93–100 and 266–273, 1864.

[58] Hilbert, David, and Bernays, Paul, *Grundlagen der Mathematik. I, Die Grundlehren der mathematischen Wissenschaften*, 40, 2nd German ed., Springer-Verlag, Berlin, 1968. ISBN: 978-3-540-04134-4 Online with translation at www.ags.uni-sb.de/čp/p/hilbertbernays/goal.htm.

[59] Hilbert, David, and Bernays, Paul, *Grundlagen der Mathematik. II, Die Grundlehren der mathematischen Wissenschaften*, 50, 2nd German ed., Springer-Verlag, Berlin, 1970. ISBN: 978-3-540-05110-7

[60] Jackson, Dunham, *Fourier Series and Orthogonal Polynomials*, Dover, New York, 2004. ISBN: 978-0486438085

[61] Jahnke, Hans Neils, *A History of Analysis (History of Mathematics, V. 24)*, American Mathematical Society, Providence, R.I., 2003 ISBN: 978-0821826232

[62] James, Ioan, *Remarkable Mathematicians*, Cambridge University Press and the Mathematical Association of America, Cambridge and Washington, D.C., 2002. ISBN: 0-521-52094-0

[63] Johnston, William, A condition for absence of singular spectrum with an application to perturbations of self-adjoint Toeplitz operators, *American Journal of Math.*, 113:243–267, 1990.

[64] Johnston, William, *An Introduction to Statistical Inference*, Mohican Press, Perrysville, Ohio 2001. ISBN: 0-923231-40-4

[65] Johnston, William, The weighted Hermite polynomials form a basis for $L^2(\mathbb{R})$, *The American Mathematical Monthly*, 121(3):249–253, 2014.

[66] Johnston, William, and McAllister, Alex M., *A Transition to Advanced Mathematics*, Oxford University Press, New York, NY, 2009. ISBN: 978-0-19-531076-4

[67] Jordan, Camille, Sur la série de Fourier, *Comptes Rendus* 92:228à230, 1881.

[68] Kac, Mark, Hugo Steinhaus—A reminiscence and a tribute, *The American Mathematical Monthly*, 81(6):572–581, 1974.

[69] Kahane, Jean-Pierre, *Trigonometric Series*: book review, *Bulletin of the AMS*, 41(3):377–390, 2004.

[70] Kanas, Stanislawa and Tatarczak, Anna, Generalized Meixner-Pollaczek polynomials, *Advances in Difference Equations*, 2013:131.

[71] Kato, Tosio, *Perturbation Theory for Linear Operators*, Springer, New York, 1995. ISBN: 978-3540586616

[72] Katznelson, Yitzhak, Sur les ensembles de divergence des sèries trigonométriques, *Studia Mathematica*, 26:301à304, 1966.

[73] Kendall, David G., Kolmogorov as I remember him, *Statistical Science*, 6(3):306–312, 1991.

[74] König, Heinz, *Measure and Integration*, Springer, New York, 1997. ISBN: 978-3540618584

[75] Kirkwood, James R., *An Introduction to Analysis*, 2nd ed., PWS Publishing Company, Boston, 1995. ISBN: 0-534-94422-1

[76] Lagrange, Joseph-Louis, *Mecanique Analytique*, Courcier, Paris, 1811 (reissued by Cambridge University Press, 2009). ISBN: 978-1108001748

[77] Lalescu, Traian, *Introduction à la théorie des équations intégrales*, Nabu Press, Paris, 2010. ISBN: 978-1149419762

[78] Larson, Ron, and Edwards, Bruce H., *Calculus*, 9th ed., Brooks Cole, New York, 2009. ISBN: 978-0547167022

[79] Lax, Peter David, Translation invariant spaces, *Acta Math.*, 101:163–178, 1959.

[80] Lebesgue, Henri, Intégrale, longueur, aire, *Thése*, Université Henri Poincaré Nancy, in *Ann. Mat. Pura Appl.*, 3:231–359, 1902.

[81] Lebesgue, Henri, Sur une généralisation de l'intégrale définie, *Comptes Rendus*, April 29, 1901.

[82] Legendre, Adrien Marie, and Davies, Charles, *Elements of Geometry and Trigonometry from the Works of A.M. Legendre*, A. S. Barnes & Co., New York, 1853. quod.lib.umich.edu/m/moa/AAW9179.0001.001?view=toc.

[83] Luxemburg, W. A. J., Arzelà's Dominated Convergence Theorem for the Riemann integral, *The American Mathematical Monthly*, 11:970–973, 1971.

[84] *Mathematical Genealogy Project*, Internet site: genealogy.math.ndsu.nodak.edu/id.php?id=34254, January 1, 2011.

[85] McShane, E. J., *Unified Integration*, Academic Press, Orlando, Fla., 1983. ISBN: 0-12-486260-8

[86] Naylor, Arch W., and Sell, George R., *Linear Operator Theory in Engineering and Science*, Springer, New York, 2000. ISBN: 978-0387950013

[87] Newton, Isaac, *The Method of Fluxions and Infinite Series : With its Application to the Geometry of Curve-lines*, Henry Woodfall, London, 1736. Online at archive.org/details/methodoffluxions00newt.

[88] Newton, Isaac, I. Bernard Cohen and Anne Whitman (translators), *The Principia: Mathematical Principles of Natural Philosophy*, University of California Press, Berkeley, 1999. ISBN: 0-520-08817-4

[89] Pearson, D.B., Singular continuous measures in scattering theory, *Commun. Math. Phys.*, 60:13–36, 1978.

[90] Polking, John, Boggess, Albert, and Arnold, David, *Differential Equations with Boundary Values*, Pearson Education Press, Upper Saddle River, New Jersey, 2002. ISBN: 0-13-091106-2

[91] Porter, David, and Stirling, David S. G., *Integral Equations*, Cambridge University Press, New York, 1990. ISBN: 978-0521337427

[92] Powers, David L., *Boundary Value Problems*, 3rd ed., Academic Press, New York, 1979. ISBN: 0-12-563760-8

[93] Priestley, H. A., *Introduction to Integration*, Oxford University Press, New York, 1997. ISBN: 0-19-850123-4

[94] Radjavi, Heydar, and Rosenthal, Peter, The invariant subspace problem, *The Mathematical Intelligencer*, 4:(1):33–37, 1982.

[95] Reed, Michael, and Simon, Barry, *Methods of Modern Mathematical Physics, Volumes I-IV*, Academic Press, New York, 1972. ISBN: 978-0125850018

[96] Reid, Constance, *Hilbert*, Springer, New York, 1996. ISBN: 978-0387946740

[97] Rickey, V. Frederick, Isaac Newton: man, myth, and mathematics, *The College Mathematics Journal*, 18:362–389, November 1987.

[98] Riesz, Frigyes, and Sz.-Nagy, Bela, *Functional Analysis*, Dover, New York, 1990. ISBN: 0-486-66289-6

[99] Rosenblum, Marvin, Self-adjoint Toeplitz operators and associated orthonormal functions, *Proceedings of the AMS*, 13:590–595, August, 1962.

[100] Rosenblum, Marvin, The absolute continuity of Toeplitz's matrices, *Pacific Journal of Math.*, 10:987–996, 1960.

[101] Rosenblum, Marvin, and Rovnyak, James, *Hardy Classes and Operator Theory*, Oxford University Press, New York, 1985. ISBN: 0-486-69536-0

[102] Rosenblum, Marvin, and Rovnyak, James, The factorization problem for nonnegative operator valued functions, *Bulletin of the AMS*, 77:287–318, 1971.

[103] Rosenblum, Marvin, and Rovnyak, James, *Topics in Hardy Classes and Univalent Functions*, Birkhäuser Verlag, Boston, 1994. ISBN: 3-7643-5111-X

[104] Roto, Gian-Carlo, On models for linear operators, *Comm. Pure Appl. Math.*, 13:469–472, 1960.

[105] Royden, H. L., *Real Analysis*, 3rd ed., Prentice Hall, New York, 1988. ISBN: 978-0024041517

[106] Rudin, Walter, *Principles of Functional Analysis*, 3rd ed., McGraw-Hill, New York, 1976. ISBN: 0-070-54235-X

[107] Rudin, Walter, *Functional Analysis*, 2nd ed., McGraw-Hill, New York, 1991. ISBN: 0-070-54236-8

[108] Schappacher, Norbert, and Schoof, René, Beppo Levi and the arithmetic of elliptic curves, July 2010. hal.archives-ouvertes.fr/docs/00/12/97/19/PDF/95010.pdf.

[109] Schmidt, Erhard, Zur Theorie der linearen und nichtlinearen Integralgleichungen. 1. Entwicklung willkriger Funktionen nach Systeme vorgeschriebener, *Math. Ann.*, 63:433–476, 1907.

[110] Seeley, Robert T., *An Introduction to Fourier Series and Integrals*, Dover, New York, 2006. ISBN: 978-0486453071

[111] Simmons, George F., *Differential Equations with Applications and Historic Notes*, 2nd ed., McGraw-Hill, New York, 1991. ISBN: 0-07-057540-1

[112] Stewart, C. A., P. J. Daniell, *J. London Math. Soc.*, 22:75–80, 1947.

[113] Stillwell, John, *Mathematics and its History*, 2nd ed., Springer-Verlag, New York, 2004. ISBN: 978-0387953366

[114] Streater, Raymond F., and Wightman, Arthur S., *PCT, Spin and Statistics, and All That*, Princeton University Press, Princeton, N.J., 2000. ISBN: 978-0691070629

[115] Steinhaus, Hugo, Stephan Banach, *Studia Mathematica, Seria Specjalna, Z. I.*, 1(1):7–15, 1963.

[116] Taylor, Henry Martin, Newton, Sir Isaac, *Encyclopaedia Britannica*, 11th ed., XIX:582–592, Cambridge University Press, New York, 1911.

[117] Tricomi, F. G., *Integral Equations*, Dover, New York, 1985. ISBN: 978-0486648286

[118] Wagner, Jack, Barrow's fundamental theorem, *The College Mathematics Journal*, 32(1):58–59, 2001.

[119] Wang, Fu Traing, A remark on C summability of Fourier series, *J. London Math. Soc.*, s1-22 (1):40–47, 1947.

[120] Weir, Alan J., *General Integration and Measure*, Cambridge University Press, Cambridge, 1974. ISBN: 0-521-20407-7

[121] Weir, Alan J., *Lebesgue Integration and Measure*, Cambridge University Press, Cambridge, 1996. ISBN: 0-521-09751-7

[122] Weyl, H., Singuläre Integralgleichungen mit besonderer Berucksichtigung des Fourierschen Integraltheorems, *Hermann Weyl Gesammelte Abhandlungen*, Band I, Springer-Verlag, New York, 1968.

[123] Wheeden. Richard L., and Zygmund, Antoni, *Measure and Integral: An Introduction to Real Analysis*, Marcel Dekker, New York, 1977.

[124] Zygmund, A., *Trigonometric Series*, 3rd ed., Cambridge University Press, Cambridge, 2003. ISBN: 0-521-89053-5

Index

Almost everywhere (a.e.), 22, 46, 51, 52, 61, 69, 74, 79, 90, 99, 104, 108, 113, 130, 139, 164, 186
Analytic function, 112, 144, 230, 237, 238, 240–242, 251
Aronszajn, Nachman, 256
Arzelà's theorem, 103
Axiom of choice, 128, 247
Axiom of completeness, 40, 68

Banach space, xi, 125, 126, 131, 137, 173, 178, 258
Banach, Stefan, 123, 126, 178, 216
Barrow, Isaac, 61, 63
Basis, 110, 148, 156
   orthonormal, 126, 137, 140, 148, 151, 156–158, 163, 169, 173, 200, 202, 221, 252
Bernays, Paul, 179
Bernoulli, Daniel, 109
Bernoulli, Jacob, 37, 99
Beurling's theorem, 239, 241, 244
Beurling, Arne, 216, 238, 241, 256
Beurling-Lax theorem, 245
Blaschke product, 239
Bohm, David, 181
Bolzano, Bernhard, 37
Borel measure, x, 191, 192, 194, 196, 200, 202, 206, 216, 229, 249
Borel, Émile, 74, 180
Boundary function, 230, 231, 238–240, 247, 250
Bounded
   function, 4, 42, 50, 65, 67, 69–71, 78, 79, 96, 100, 105, 115, 122, 132, 171
   integral, 86, 87, 89
   sequence, 40
Bounded convergence theorem, 100, 103, 106
Bounded total measure, 19, 22, 23, 26, 55, 86, 186
Bounded variation, 112, 116, 117, 120, 211

Branges, Louis de, 216, 243
Bressoud, David, 64
Brown, Scott, 256
Brunt, Bruce van, 64
Burkill, John Charles, 122

Cantor function, 195, 196
Cantor set, 21, 23, 25, 188, 199
Cantor's first diagonalization method, 10
Cantor's second diagonalization method, 17
Cantor, Georg, 6, 8, 10–14
Carathéodory, Constantin, 212
Carleson, Lennart, 122, 123, 157
Cartan, Elie, 74
Carter, Michael, 64
Cauchy sequence, 28, 136, 137, 154
Cauchy, Augustin Louis, 37, 40, 62, 67
Characteristic function, 27
Closed set, 84, 149, 152, 154–156, 232, 233
Cohen, Paul Joseph, 13
Compact operator, 231
Continuity, 3, 5, 23, 26, 37, 42–44, 46, 50, 52, 63, 65, 81, 89, 98, 102, 105, 106, 110, 115, 121, 122, 130, 157, 164, 171
Continuum hypothesis, 12
Convergence
   in L^p, 135, 136, 156, 157, 218
   of a function sequence, 28, 44, 65, 80
   of a function series, 186, 193
   of a sequence, 39, 40, 131
   of a series, 5, 112, 117, 122
Cooley, James, 121
Countable union theorem, 9, 14, 86
Courant, Richard, 179
Cowen, Carl, 255

d'Alembert, Jean le Rond, 109
Daniell, Percy John, 1, 211

Daniell-Riesz approach, x, 1, 18, 49, 50, 52, 64, 183, 192, 196, 211, 212
Darboux's theorem, 76, 80
Darboux, Jean Gaston, 74, 123
Derivative, 66, 89, 92, 93, 105, 115, 122, 134, 151, 180, 195, 212, 215, 229, 241
    partial, 105
Dirichlet function, 22, 44, 73, 84, 89, 102, 139
Dirichlet kernel, 114, 118
Dirichlet, Peter Gustav Lejeune, 44, 112, 116, 122
Dirichlet-Jordan theorem, 112, 117, 120
du Bois-Reymond, Paul David Gustav, 122, 123
Dunham, William, 64

Eigenvalue, 173, 195, 204, 218, 226–228, 232, 235, 237, 249, 253
Eigenvector, 227, 228, 233, 235, 249, 253
Erdős, Paul, 212
Euclidean space, x, 125, 140, 142, 143, 145, 148, 178, 180
Euler's formula, 119
Euler, Leonhard, 109, 111, 179

Fatou's Lemma, 100, 104
Fatou's lemma, 103, 104, 106
Fejér, Lipót, 122–124, 212
Fourier series, 67, 108–118
Fourier, Jean Baptiste Joseph, 4
Fubini's Theorem, 98, 99, 172
Fubini's theorem, 176, 177
Function space
    H^2, 144, 221, 230
    L^0, 49, 50, 69
    L^1, 55, 87, 122, 126, 127, 129, 177, 178, 183
    L^2, 126, 133, 137, 154, 156, 165, 178, 193, 198, 215, 216
    L^p, 122
    $W^{1,2}$, 229
    ℓ^2, 144, 159, 163
Fundamental theorem of calculus, 3, 62, 65, 67, 85, 89, 91–93, 115

Gödel, Kurt Friedrich, 13
Gallardo-Guttiérez, Eva, 255
Gamma function, 98
Gauss, Carl Friedrich, 66, 121, 256
Gibbs' phenomenon, 118
Gibbs, Josiah Willard, 118
Goursat, Édouard, 74
Gram, Jorgen Pedersen, 180
Gram-Schmidt process, 149–157
Greatest lower bound, 68

Hölder's Inequality, 133
Hölder's inequality, 133, 134, 140, 164
Halmos, Paul, 216, 243
Hardy space, 144, 230, 237, 238
Hardy, Godfrey Harold, 122, 144, 216
Harmonic function, 151
Harmonic oscillator, 165, 166, 168, 173, 180
Harmonic series, 96, 118
Hawkins, Thomas, 64
Hermann, Jacob, 63
Hermite differential equation, 166–168
Hermite, Charles, 161, 166, 168, 180
Hilbert space, 143
Hilbert space isomorphism, 157–159, 200, 251
Hilbert, David, 12, 179
Hunt, Richard, 122

Infimum, 68, 72–77, 82, 83, 104, 121, 217
Infinity
    continuum hypothesis, 12
    countable, 7
    countable unions, 9
    uncountability of $\mathbb{R}$, 11
    uncountable, 7
Inner product, 140–143
Inner-outer factorization, 239
Integral
    Lebesgue, 5, 6, 14, 22, 23, 26–28, 43, 46, 49–61, 63–66, 70, 71, 73, 81, 85, 89, 90, 96, 99, 100, 103, 104, 112, 125, 127, 131, 132, 136, 173, 183, 184, 189, 191–193, 215
    Riemann, 4, 22, 44, 57, 64, 65, 67, 68, 71–74, 76–85, 88, 89, 91, 96, 103, 177
Integration by parts, 90, 94, 95, 98, 111, 165, 169, 174
Invariant subspaces, 232, 257
Isometry, 221, 254

James, Ioan, 179
Jordan Form Theorem, 227
Jordan, Camille, 112, 116, 123, 183, 227

Katznelson, Yitzhak, 122
Kepler's Laws, 3, 61, 63
Kepler's laws, 3, 4
Kolmogorov, Andrei, 5, 122

Lagrange, Joseph Louis, 109, 110
Laguerre, Edmond, 180
Lalescu, Traian, 123
Least upper bound, 68

Lebesgue dominated convergence theorem, 100, 103, 106, 108
Lebesgue, Henri, 1, 4, 5, 18, 64, 79, 278
Legendre, Adrien-Marie, 122, 180
Leibniz, Gottfried, 3, 64
Levi, Beppo, 86, 87, 100, 122
Levi-Cevità, Tullio, 122
Limit, 37, 41
   of a function, 41–44
   of a function sequence, 44–46, 54
   of a sequence, 38
Linear transformation, 201
Littlewood, John Edensor, 122
Lower bound, 68

Maclaurin series, 144, 166, 167, 175, 230, 268
Matrix, 201, 215, 220, 221, 224–228
Mean value theorem, 92, 105, 117, 242
Measurable
   function, 2, 127, 129–131, 136, 185, 186, 191, 193, 200, 204
   set, 127, 128, 177, 178, 184, 185, 187, 190, 203
Measure, 18, 19, 22, 26, 57, 64, 74, 127, 128, 145, 173, 178, 183–185, 187, 188
   absolutely continuous, 193, 194, 202, 207–209, 250, 253
   point mass, 188, 201
   singular continuous, 195, 199
Minkowski's inequality, 134, 135
Minkowski, Hermann, 134
Monotone convergence theorem, 86, 87

Naylor, Arch, 64
Neumann, John von, 212, 216, 256
Newton, Isaac, 3, 4, 61, 63
Newton-Raphson method, 180
Nondecreasing convergence theorem for L^0, 85, 108
Norm, 125, 126, 131–133, 137, 138, 149, 193, 218
   operator, 217, 226, 235, 243, 248
Norm limit, 138, 139, 149

Ohm, Georg, 122
Operator, 216–251
   bounded linear, 216, 217, 222, 231, 238, 243
   compact, 237
   composition, 255
   Hilbert-Schmidt, 218
   integral, 217
   rank one, 217
   Toeplitz, 220, 250
   Volterra, 232
   Wiener-Hopf, 219, 221
Orthonormal basis, 148, 170
Outer function, 239

Pólya, George, 212
Parseval's identity, 159
Partial isometry, 244
Picard, Émile, 74, 124
Poisson, Simeon, 122
Polynomial
   Chebyshev, 252
   Hermite, 161, 162, 166–173, 175, 176, 194, 205
   Laguerre, 150, 180, 203, 221
   Legendre, 150, 162, 180, 202, 252
Priestley, Hilary, 64
Probability
   experiment, 205
   measure, 206, 209
   sample space, 205
Projection Theorem, 152
Projection theorem, 154–156, 160, 163

Quantum mechanics, 165–173

Radon-Nikodym derivative, 193
Random variable, 206, 209
Real numbers, 11, 28
Reid, Constance, 179
Rickey, Frederick, 63
Riemann sums, 67, 72, 77
Riemann, Bernhard, 1, 4, 65–67
Riemann-Lebesgue Lemma, 113
Riemann-Lebesgue theorem, 79
Riesz, Frigyes, 1, 64, 85, 87, 89, 127, 129, 136, 192, 199, 211, 216
Riesz-Fischer theorem, 136, 137
Rodrigues formula, 151, 161–163, 168–170, 175
Rodrigues, Olinde, 151
Rosenblum, Marvin, 221, 243, 244, 249
Rota, Gian-Carlo, 243
Rota-de Branges-Rovnyak Universal Model, 244
Rovnyak, James, 216, 243, 244
Rudin, Walter, 216

Schauder, Juliusz, 179
Schmidt, Erhard, 180, 218, 232
Schrödinger equation, 165, 166, 168, 173
Schrödinger, Erwin, 165
Schwartz, Laurent, 212
Schwarz Inequality, 140, 172
Sell, George, 64

Shift operator, 238
Sigma-algebra, 192
Sobolev, Sergei, 213
Spectral theorem, 249
Standard construction, 52, 70
Steinhaus, Hugo, 123, 179
Step function, 23, 26, 28, 46, 52, 69
Stieltjes, Thomas, 74
Supremum, 68, 69, 72–74, 76–78, 80,
    82, 83, 85, 116, 120, 217, 218, 226,
    239, 247

Tarski, Alfred, 178, 179
Toeplitz, Otto, 257
Total variation, 116
Triangle inequality, 61, 133, 136, 137
Trigonometric series, 67, 109, 110, 119, 122, 124
Tukey, John, 121

Uncountable, 11
Upper bound, 68

Vitali function, 129
Vitali set, 127
Vitali, Guiseppe, 127
Volterra, Vito, 25, 122

Wallis, John, 6
Wang, Fu Traing, 122
Weierstrass, Karl, 37, 40, 122
Weir, Alan, 1, 2, 54, 64, 121, 129, 212
Weyl, Hermann, 179, 216, 253
Wheeden, Richard, 170
Wien, Wilhelm, 180

Zaremba, Stanislaw, 74
Zygmund, Antoni, 124, 170